37th Annual Conference and Expo of the Association of Avian Veterinarians (AAV 2016)

Held at ExoticsCon 2016

Portland, Oregon, USA
27 August – 1 September 2016

Editor:

Charlotte Kirk Baer

ISBN: 978-1-5108-3879-6

Printed from e-media with permission by:

Curran Associates, Inc.
57 Morehouse Lane
Red Hook, NY 12571

Some format issues inherent in the e-media version may also appear in this print version.

Printed by Curran Associates, Inc. (2017)

For permission requests, please contact Association of Avian Veterinarians at the address below.

Association of Avian Veterinarians
PO Box 9
Teaneck, NJ 07666
USA

Phone: 720-458-4111
Fax: 720-398-3496

office@aav.org

Additional copies of this publication are available from:

Curran Associates, Inc.
57 Morehouse Lane
Red Hook, NY 12571 USA
Phone: 845-758-0400
Fax: 845-758-2633
Email: curran@proceedings.com
Web: www.proceedings.com

BUILDING EXOTICS EXCELLENCE: ONE CITY, ONE CONFERENCE

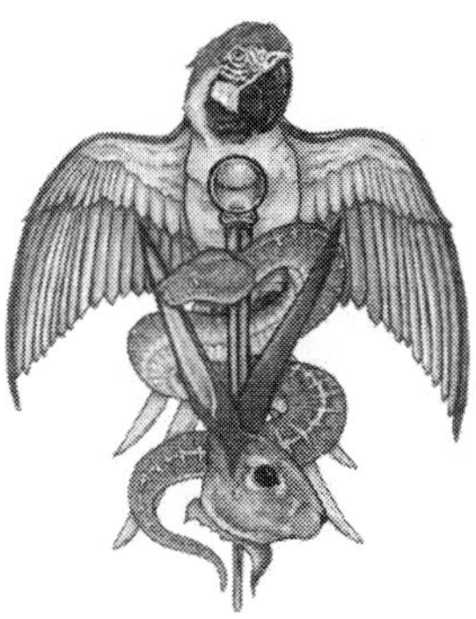

2016 PROCEEDINGS EXOTICSCON

AUGUST 27- SEPTEMBER 1, 2016 (EXHIBITS 28–31)
PORTLAND, OREGON | HILTON PORTLAND & EXECUTIVE TOWER

Association of Avian Veterinarians

AAV VETERINARY REVIEWERS

Natalie Antinoff, DVM, DABVP (Avian)
Heather Barron, DVM, DABVP (Avian)
Hughes Beaufrere, Dr. Med Vet, PhD, DACZM, DABVP (Avian), DECZM (Avian)
Greg Burkett, DVM, DABVP (Avian)
Sue Chen, DVM, DABVP (Avian)
Robert Dalhausen, DVM, MS, PhD (candidate)
Laurel Degernes, DVM, MPH, DABVP (Avian)
Scott Echols, DVM, DABVP (Avian)
Thomas Edling, DVM, MSpVM
Jesse Fallon, DVM, MS
Scott Ford, DVM, DABVP (Avian)
Jennifer Graham, DVM, DAVCP (Avian and Exotic Companion Mammal), DACZM
Cheryl Greenacre, DVM, DABVP (Avian and Exotic Companion Mammal)
Robert Groskin, DVM
David Sanchez-Migallon Guzman, LV, MS, DECZM (Avian and Small Mammal), DACZM
Michelle Hawkins, VMD, DABVP (Avian)
Sharman Hoppes, DVM, DABVP (Avian)
Angela Lennox, DVM, DABVP (Avian and Exotic Companion Mammal)
Elizabeth Mackey, DVM
Christoph Mans, Dr Med Vet, DACZM
Steven Metz, DVM
Anna Osofsky, DVM, DABVP (Avian)
Juli Ponder, DVM, MPA
Lauren Powers, DVM, DABVP (Avian), DABVP (Exotic Companion Mammal)
Jeff Rhody, DVM, DABVP (Exotic Companion Mammal)
Marcy Souza, DVM, MPH, DABVP (Avian), DACVPM
Brian Speer, DVM, DABVP (Avian), DECZM (Avian)
Anneliese Strunk, DVM, DABVP (Avian)
Thomas Tully, Jr, MS, DVM, DABVP (Avian), DECZM
Paul Welch, DVM
Kenneth Welle, DVM, DABVP (Avian)

TECHNICAL EDITOR

Charlotte Kirk Baer, MS

Introduction

We invite you to enjoy the benefits provided by these AAV Proceedings of ExoticsCon 2016, a combined conference of the Association of Avian Veterinarians (AAV), Association of Exotic Mammal Veterinarians (AEMV), and the Association of Reptilian and Amphibian Veterinarians (ARAV). The 2016 combined conference provides access to leading specialty veterinary researchers and practitioners world-wide sharing their experience and knowledge in person and through these proceedings, which are furnished to members of the three associations.

AAV recognizes and thanks the following committees and staff for their time and hard work in assuring a quality program and publication:

ExoticsCon Trilateral Conference Committee

- Cheryl Greenacre, DVM, DABVP (Avian), DABVP (Exotic Companion Mammal) (AAV Conference Chair)
- Chris Griffin, DVM, DABVP (Avian) (AEMV)
- Dan Johnson, DVM, DABVP (Exotic Companion Mammal) (AEMV Conference Chair)
- Kenneth Welle, DVM, DABVP (Avian) (AAV)
- Nicole Johnson, DVM (ARAV)
- Susan Horton, DVM (ARAV Conference Chair)

AAV Veterinary Manuscript Reviewers

Natalie Antinoff, DVM, DABVP (Avian); Heather Barron, DVM, DABVP (Avian); Hughes Beaufrere, Dr. Med Vet, PhD, DACZM, DABVP (Avian), DECZM (Avian); Greg Burkett, DVM, DABVP (Avian); Sue Chen, DVM, DABVP (Avian); Robert Dalhausen, DVM, MS, PhD (candidate); Laurel Degernes, DVM, MPH, DABVP (Avian); Scott Echols, DVM, DABVP (Avian); Thomas Edling, DVM, MSpVM; Jesse Fallon, DVM, MS; Scott Ford, DVM, DABVP (Avian); Jennifer Graham, DVM, DAVCP (Avian and Exotic Companion Mammal), DACZM; Cheryl Greenacre, DVM, DABVP (Avian and Exotic Companion Mammal); Robert Groskin, DVM; David Sanchez-Migallon Guzman, LV, MS, DECZM (Avian and Small Mammal), DACZM; Michelle Hawkins, VMD, DABVP (Avian); Sharman Hoppes, DVM, DABVP (Avian); Angela Lennox, DVM, DABVP (Avian and Exotic Companion Mammal); Elizabeth Mackey, DVM; Christoph Mans, Dr Med Vet, DACZM; Steven Metz, DVM; Anna Osofsky, DVM, DABVP (Avian); Juli Ponder, DVM, MPA; Lauren Powers, DVM, DABVP (Avian), DABVP (Exotic Companion Mammal); Jeff Rhody, DVM, DABVP (Exotic Companion Mammal); Marcy Souza, DVM, MPH, DABVP (Avian), DACVPM; Brian Speer, DVM, DABVP (Avian), DECZM (Avian); Anneliese Strunk, DVM, DABVP (Avian); Thomas Tully, Jr, MS, DVM, DABVP (Avian), DECZM; Paul Welch, DVM; Kenneth Welle, DVM, DABVP (Avian)

Technical Editor

Charlotte Kirk Baer, MS

Conference Office Staff — North American Veterinary Community

Peter Scott, MBA, APR, Chief Operations Officer; Meghan R. Golden, MBA, Vice President of Industry Services & International Relations; Joe Sorrentino, MBA, CMP, Vice President of Operations; Summer Graziano, Senior Marketing Director; Sheri Matheny, Customer Service Manager

Proceedings Layout and Design — BrightLizard, Inc.

Elizabeth Smith

Disclaimer

The material appearing in this publication comes exclusively from the authors and contributors identified in each manuscript. The techniques and procedures discussed reflect the personal knowledge and experience of the authors and contributors, and demonstrate their views of the methods that may be used for these medical procedures. The procedures demonstrated do not incorporate all known techniques, are not exclusive, and other techniques and technology may also be available. Any questions or requests for additional information concerning any of the manuscripts should be addressed directly to the authors.

AAV does not research, review, or otherwise verify any of the information contained in this publication. Opinions expressed in this publication are those of the authors and contributors and not necessarily those of AAV. AAV is not responsible for errors or for opinions expressed in this publication.

AAV expressly disclaims any warranties or guarantees, expressed or implied, and shall not be liable for damages of any kind in connection with the material, information, techniques, or procedures set forth in this publication.

Manuscripts contained in these conference proceedings vary in length and format based on author intentions for possible future publication. Authors are given the option to publish their complete manuscript herein as the final, official publication of the manuscript. Authors who wish to retain the right to publish a complete manuscript in another professional publication at a later date have limited their submission herein to an abstract only.

For permission to reproduce or order copies of this publication, contact the AAV office, www.aav.org, PO Box 9, Teaneck, New Jersey 07666, USA, e-mail pubs@aav.org.

For information on ExoticsCon educational conferences, visit http://www.exoticscon.org. For information on membership and other association programs visit www.aav.org.

Table of Contents

Saturday Sessions

Section 1: Technician Sessions

Section 2: Practice Management Sessions

Association of Avian Veterinarians

Section 3: AAV Plenary Sessions

Section 4: AAV Hot Topics

Section 5: AAV Wildlife and Conservation

Section 6: AAV Master Classes Track 1

Section 7: AAV Master Classes Track 2

Section 8: AAV Master Classes Track 3

Section 9: AAV Master Classes Track 4

Section 10: AAV Research

Section 11: AAV Clinical

Section 12: AAV Case Reports

Section 13: AAV Practical Information

Section 14: AAV Posters

Hybrid Sessions

Section 32: Saturday Hybrid Master Classes

Section 33: Wednesday Hybrid Track

Section 34: Hybrid Posters

Section 1

Technician Sessions

A Chicken Just Walked In Through the Front Door, Now What?

Janet L. Jones, LVMT

Session #080

Affiliation: From the Veterinary Medical Center, University of Tennessee, Knoxville, TN 37996, USA.

Abstract: Chickens have become a popular backyard pet. Many cities have now allowed people to keep chickens in their backyard. It has become more common for the veterinary practice to have a chicken present for illness or emergency. A chicken is not triaged any differently than any other domestic animal. It is important to understand what is normal, including behavior, before being able to identify what is abnormal. The history is the most important starting point. The physical exam starts with visual observation, then with gentle restraint, start with the head and work your way to the back end. Treatments will vary depending on if the bird is for egg production, meat production, show, or as a pet only.

When that client walks into your clinic with a sick chicken, that sick bird should be triaged just like any other domestic animal. Owners are often very attached to their pet chickens so emotions can run high with some clients. The most important part of the examination is obtaining a thorough history. The next step is to do a complete physical examination.[1] Most chickens will tolerate minimal physical restraint. By starting with the head and working down to the tail, you should be able to examine all the important body parts of the bird.

The most important part of the physical examination is to obtain a complete history. Questions to be asked include the signalment of the bird and determination of its use. It is important to know how many are in the flock, how long they have been in the flock together, where they were obtained, and if any vaccinations were given. Housing questions include size and composition, space of the coop and range, substrate and perch size, are they brought in at night? Questions to ask about nutrition include type, brand, amount, where purchased, how old is the food, any treats or supplements, and any medicated foods? If the patient is a laying hen, questions to be asked should include how many boxes and sizes, and the hens laying history? When addressing the presenting complaint, ask how long it has been occurring, any other birds are affected, and if the condition has progressed?

Before restraining the bird, a visual examination should be performed from a distance. Note the mentation of the bird. Chickens will try to act normal even when they are very ill. They put on this show so as not to get picked on by other birds in the flock or get picked off by a predator. A healthy bird will be curious and alert of their surroundings. A sick bird will hold its head down, eyes partially closed, and not standing upright. Observe in the cage the droppings for consistency, blood, or abnormal size or smell. There are three components to the droppings; the clear urine portion, the dark colored stool portion and the white urate portion. Chicken restraint starts with controlling their wings. Gently hold their wings to the body in a natural folded position. Do not place heavy restraint across the keel as chicken do not have a diaphragm. Usually the least amount of restraint possible is best for the birds comfort. Place a towel down on the exam table to prevent foot slippage. Examine from head to tail and leave the oral exam last as most chickens do not like having an oral exam and will get stressed during this part of the examination.

Examine the head for symmetry of the beak, eyes, sinuses and nostrils while looking for discharge, crusts, scratches, scabs, swellings, and discolorations. The beak should come to a point. The iris should both be the same color. Examine the ears by gently moving away the feathers and look for any discharge, blood, or parasites. The comb should be firm and red. A capillary refill time can be performed on the comb with a refill time of 2 seconds or less. The comb should not be pale or flopped over to one side, not unless it is normal for the bird. Hydration status can be obtained from the basilic vein be pressing on the vein and noticing refill time. A hydrated bird will refill instantly, in a dehydrated bird you will be able to see the vein refill and if the bird is in shock or over 10% dehydrated there is a one second refill time. Palpate the crop; it should feel soft and fluctuant. Birds lack organized lymph nodes so there should be no subcutaneous masses. The keel should be straight and you should be able to feel the pectoral muscles on either side.

Auscult the heart and listen to heart sounds by placing the stethoscope on either side of the keel. Birds have a four chambered heart and their rate tends to be faster than most mammals of about 150-250 beats per minute. You should not hear any lung sounds in birds and their respiratory rate is usually 15-30 breaths per minute. Temperature is typically not taken in birds. The feathers should lay down smooth. Lift them up and check for ecto parasites. All joints should be examined and palpated for range of motion, stiffness, crackles, and clicks. The coelomic cavity should be palpated and should be soft and doughy. It should not be extended or fluid filled. The vent should be clean and dry. The bird should be lifted up to expose the plantar surface of the feet looking for scabs, callouses, swellings, dark areas, cuts, or keratin overgrowth.

The last part of the exam is the oral exam. Gently open the bird's mouth to evaluate for white plaques, masses, abnormal smell, and evaluation of the choanal slit and papillae. Accurately weigh the bird on a gram scale to calculate dosages.

When you come up with your treatment plan, keep in mind that there are certain drugs that are illegal to use in chickens. You can access all of these drugs from Food Animal Residue Avoidance Databank (FARAD) website at www.farad.org.[2] They have a databank listing the withdrawal times for drugs used in food-producing animals in the United States.

References

1. Greenacre CB, Morishita TY. *Backyard Poultry Medicine and Surgery: A Guide for the Veterinary Practitioner*. Ames, IA: John Wiley & Sons; 2015.

2. Food Animal Residue Avoidance Databank (FARAD). Available at: http://www.farad.org. Accessed May 24, 2016.

Chicken Breeds: Who Lays The Best Eggs?

Janet L. Jones, LVMT

Session #081

Affiliation: From the Veterinary Medical Center, University of Tennessee, Knoxville, TN 37996, USA.

Abstract: Chicken breeds can be categorized as egg, meat, dual purpose, and ornamental breeds. All of today's many breeds of chickens came from a single origin, the red jungle fowl. The most productive egg layers are the hybrids. Some hybrid pullets start laying when they are 17 weeks of age, but most breeds take more than 26 weeks before they mature and start laying. Egg coloration ranges from white, brown, to greenish or bluish eggs. Other considerations when choosing the right chicken is temperament of the hen and temperature tolerance.

Chickens have coexisted with people for centuries providing both meat and eggs. Today's chickens are now also used as a companion animal. All of today's breeds of chickens came from the red jungle fowl of Southeast Asia. Today's breeds of chickens can be divided into egg breeds, meat breeds, dual purpose breeds, and ornamental breeds. A survey conducted in 2010 found that 47% of chicken owners have them for eggs; whereas, less than 10% sell them for meat.

The most productive egg layers are hybrids, including the Hy-Line Brown, California White, Golden Comet, Cherry Egger, and the Indian River. The Brown sex Links lay zmostly extra-large eggs. Some hybrid pullets can start laying when they are only 17 weeks old. Other breeds take more than 26 weeks to mature and start laying eggs. Some of the breeds that lay the darkest brown eggs are the Marans, Barnevelders and Welsummers. The Ameraucanas and Aracanus lay greenish or bluish eggs. In most cases, chickens with red ears lay brown eggs and hens with white ears lay white eggs. Chickens that are allowed to range on pasture produce the most nutritious and flavorful eggs.

Some specific traits of the more common egg breeds are the Ameraucanas are the Easter egg layers. They lay about 3 medium green to blue colored eggs per week. They have a calm and gentle disposition and are a very hardy bird. The Hamburg lays about 4 small to medium white eggs per week and tend to be a bit more flighty and love to free roam. The Lakenvelder lays about 3 medium white eggs per week and the hens are non-broody but can be good flyers. The leghorn is the best of the layers as they lay about 5 white extra-large eggs per week. They are a very docile breed that handle confinement well. The Maran produces 3 large dark brown eggs per week and the hens have a calm temperament and adapt to confinement well. The Sicilian Buttercups lay about 2 small white eggs per week and the hens are very active birds.

Other considerations when choosing the right chickens for a backyard flock is temperature tolerance. The breeds that tolerate cold weather the best are Chanteders, Buckeyes, Brahmas and Javas. Where heat is more of a problem, those hens that tolerate hot environments are Jungle Fowl, Malays, Sumatras, Javas and Cubalayas. The breeds that tend to be the most broody are the Silkies, Aseels, Modern Games, Old English Games and Cochins. The tendency to be broody with chickens has been bred out of many breeds because hens stop laying eggs for a period of time when they are broody even if they are not hatching eggs.

There are many different factors to consider when keeping chickens for egg consumption and there are many different breeds as well.[1-2] By taking into account the environment, size of enclosure, and if chickens are wanted as pets as well as eggs production; you now have the information available to help make the decision as to what is the best chicken for desired traits.

References

1. Greenacre CB, Morishita TY. *Backyard Poultry Medicine and Surgery: A Guide for the Veterinary Practitioner*. Ames, IA: John Wiley & Sons; 2015.

2. Griepentrog, T. Best chicken breeds for backyard flocks: our latest survey results can help you choose the best chickens for eggs, meat, temperament and more. *Mother Earth News.* April-May 2010: 30.

Chicken Diseases: What the Backyard Chicken Owner Needs To Know

Janet L. Jones, LVMT

Session #083

Affiliation: From the Veterinary Medical Center, University of Tennessee, Knoxville, TN 37996, USA.

Abstract: When chickens present at the clinic they are often showing general signs of disease. A compete and through history are the most important first step in diagnosis of the patient. Respiratory diseases are the most common diseases of poultry with non-specific respiratory disease being extremely common. Other very common presentations are trauma and husbandry issues. Other conditions range from musculoskeletal, dermatologic, gastrointestinal, reproductive and zoonotic diseases.

Backyard poultry present to the clinic when the patient has not been doing well and often for a number of days. General signs of illness with backyard poultry are usually decreased appetite, lethargic, distance from the rest of the flock, not perching, and sitting with the head down and eyes closed. The most important part of the initial examination is getting an accurate, compete, and through history from the owner.

The most common respiratory disease of chickens is non-specific respiratory disease. This usually occurs in the spring and fall of the year when the weather changes. The patient presents with tracheal rales, snicks, sneezes, and coughing. They continue to be alert, active, and continue to eat and drink. If left untreated, the respiratory signs usually subside in 7-10 days. Mycoplasmosis is also very common in chickens. They present with a cough and with secondary E. coli infection, they may also have decreased activity, look "puffed up", and appear sleepy and reluctant to move. Clinical signs can reoccur after discontinuation of antibiotics. With infectious coryza there is acute death or the development of respiratory signs including sick birds, oculonasal discharge, facial edema, and/or infraorbital sinuses that are swollen. The last common respiratory disease is fowl cholera. These birds present with subcutaneous masses, swollen eyes, ears, or wattles. Depopulation is usually recommended to remove disease carriers.

Trauma is a very common presenting complaint usually as a result of predator attack. It is important during your physical examination that while you are palpating the patient that you part the feathers and check the skin carefully. Chickens have very thin skin and small lesions and puncture wounds can be very easily overlooked. Fly strike is common in debilitated birds with open wounds or fecal accumulation near the vent. Those birds that are severely ill need to be stabilized and correction of shock before extensive wound management under sedation or anesthesia is to be performed. For husbandry presentations, usually pododermatitis due to moisture in the bedding is a common problem as well as those birds that are kept on concrete for long periods of time. High levels of ammonia and dust can also cause non-specific respiratory disease.

For musculoskeletal diseases, when husbandry has been ruled out for pododermatitis, septic joint can be a common presentation for lameness in a chicken. This is usually the result of bacteria in a joint due to trauma, diarrhea, or respiratory disease that allowed bacteria access to the joint. These birds present with lameness either in one or multiple joints, and an enlarged, warm joint. Prognosis can be fair to poor and long term therapy can result in arthrodesis and limited range of motion. Merek's disease affects only chickens and these birds present with either unilateral or paralysis with one leg stretched forward and the other leg pointing back. These birds are typically between 10-20 weeks of age. The cause of this disease is a herpesvirus and once a bird is infected they shed the virus indefinitely. Hens with reproductive diseases have a distended coelom due to fluid or soft tissue/

organ enlargement. They can present with lameness resulting from weakness, illness, and simply as a result of an enlarged coelomic cavity that prevents normal ambulation. Valgus deformities result from lateral deviation of the intertarsal joint, tibiotarsus, the hip, or a combination of these. Treatment can involve slowing down the growth of the bird, or in the case of splay leg by bandaging the leg in its normal position. Slipped tendon is luxation of the gastrocnemius tendon forcing the leg into a vulgus position with a swollen hock. Surgery can be performed in young birds if caught early enough.

For dermatologic diseases in chickens, avian pox presents as round "pox-like" lesions of the integument of the mostly unfeathered region of the head or neck. Lesions can also be seen on the feet or vent. The patient may also have decreased weight gain and egg production. Affected birds should be isolated. Lice signs include hyperemia and irritation of the skin with small scabs and clots. Feathers can look moth-eaten and nits are laid in clumps at the base of the feathers along the ventrum. Mites are much smaller than lice, and feed on blood, feathers, and skin. Appearance is variable. Treatment for both lice and mites is usually ivermectin.

The most common gastrointestinal disease of hens is either a pendulous crop which the cause is unknown or the patient presents with an enlarged and fluid filled crop. Crop impaction can occur when chickens are exposed to new environmental substrates, such as grass, newspaper, sawdust shaving/wood chips, and feathers. Surgical removal is usually necessary. Coccidiosis is a common intestinal parasite of chickens and they present with pale combs and wattles from blood loss in the intestines, ruffled feathers, depression, blood in droppings, and shivering. Mortality rate is high in young birds. Husbandry can play a role in control of this parasite along with anticoccidial drugs.

For zoonotic diseases, *Salmonella* needs to be addressed with owners of backyard chickens. Most of the estimated 1.4 million annual cases of human salmonellosis in the United States result from ingestion of contaminated food including eggs. It is transmitted via the fecal-oral route. *Salmonella* naturally lives in the intestine of poultry and is considered normal flora in healthy, live poultry.

This list of diseases is the most common that are typically seen in practice.[1,2] There is a wide array of both common and uncommon diseases of chickens. There are a number of resources in both book format and online resources to help guide you through the diagnosis pathway. The University of Florida has a viewable table that can be helpful in determining diseases based on clinical signs.

References

1. Greenacre CB, Morishita TY. *Backyard Poultry Medicine and Surgery: A Guide for the Veterinary Practitioner*. Ames, IA: John Wiley & Sons; 2015.

2. University of Florida, IFAS Extension. *Common Poultry Diseases*. Available at: http://edis.ifas.ufl.edu. Accessed May 24, 2016.

Chicken Husbandry: What Owners Need To Know About Housing and Predators

Janet L. Jones, LVMT

Session #084

Affiliation: From the Veterinary Medical Center, University of Tennessee, Knoxville, TN 37996, USA.

Abstract: There are many considerations when backyard chickens are kept as pets. Owners must be aware of the time, effort, and knowledge required taking care of their chickens. Not only must the coop provide adequate space and ventilation, it must also be predator proof as just about any predator will take poultry if they get a chance. On the list of predators are dogs, they tend to kill birds as a result of play with the chicken and are typically not maimed as in a predator attack.

Basic management of chickens includes housing, chick purchasing, brooding, daily care, and predator options.[1-2] Good husbandry and housing is as vital for the wellbeing of backyard chickens as it is for other livestock. When choosing a breed of chicken, the outside environment must be taken into consideration. Most chickens prefer to have free range during the day and secure housing to roost at night.

Housing design will affect the chickens care, comfort, welfare, and well-being. Housing should provide shelter from the wind and rain, protection against predators when the birds are shut in, a dry and safe area to sleep, and a secluded area where hens can lay eggs. The coop can range from a simple design of a small plastic pod to an elaborate house. The ideal structure would have a cement floor and be insulated with washable walls. Depending of the size of the hens, usually a 12" × 12" laying box is sufficient for most hens. Any openings in the chicken coop should be covered with hard wire cloth and the opening door should have a screen door. This may not always be possible with an outdoor run and the risks must be weighed.

Ventilation is important for moisture removal, excess heat removal, exchange of gases produced by the litter, and providing fresh air. The winter months can be the most difficult to maintain ventilation and moisture control. Usually with air inlets along the roof line and an opening where air can come in will achieve this. The temperature of the coop should be between 50° and 70°F. Birds can get frost bite with their heads and feet being the most common to get frostbite. Birds that get heat stress will pant, spread their wings, increase water consumption, and decrease food consumption. Most chickens can grow sufficiently with normal daylight and do not need any special lighting. Birds that are egg producers should have a day length of 15 hours and 8 hours of dark. Egg production will naturally increase in the spring and decrease in the fall and winter months. The ideal bedding material should be absorbent, loose, and fairly inexpensive. The most common used is pine wood shavings. Hard wood shavings are not recommended due to potential presence of molds and fungus which can result in respiratory infections.

When purchasing chicks, obtaining them from a reputable hatchery is recommended along with paying the extra few dollars and having them vaccinated for Merek's disease. The brooder pen should be set up 48-72 hours prior to chick arrival. The heat bulb should be centrally located with feed and water alternating around the heat source. Start with a 95°F temperature for the first week and decrease by 5°F each week until outdoor temperature is met. The chicks should be evenly distributed throughout the brooder. They should be fed a commercially made medicated starter diet. When they are in their grow-out stage at about 6 weeks of age, they can be given limited outdoor access as they will be predominantly feathered and can tolerate the environment better. The birds will learn to enter the coop at night as sunset approaches. At adulthood, usually 16-24 weeks of age, the hen will start laying. No additional management changes need to be made. Birds establish a social hierarchy and disrupting this can cause aggressive fighting between birds. Hens may stop laying eggs while going through a molt which can last up to 6 weeks. They should be fed a commercially made laying formula.

As a general rule, the space requirements for most birds are 377 square inches per adult bird. If chickens are too densely populated, they can become cannibalistic and peck at each other if they do not have enough feed, water, nest boxes, incorrect lighting, abrasions or tears as a result of injury or mating, dietary deficiencies, prolapse, or meanness of a breed. Usually increasing the space and additional enrichment may help reduce these behaviors. A broody hen will stop laying eggs, sit on a nest, and become more aggressive as she protects her eggs. It is recommended to remove eggs on a daily basis. Egg eating can develop from overcrowding, uneven nest space, nutritional deficiency, too bright of light, or disposing of broken or cracked eggs in the coop.

For predator control, ensure that all holes or other access into the chicken coop is secured. Raccoons, opossums, mink, skunks, foxes, coyotes, or weasels may find ways to enter the chicken coop and pull chickens through wire and holes to consume them. Wire fences should be buried 8 inches deep to ensure that predators are unable to dig under the fence. Bushes and shrubs can provide access to aerial attacks by raptors. Dogs can be responsible for what appears to be a predator attack but is the result of trying to play with the chickens.

References

1. Greenacre CB, Morishita TY. *Backyard Poultry Medicine and Surgery: A Guide for the Veterinary Practitioner*. Ames, IA: John Wiley & Sons; 2015.

2. Wallace-Houghton, J., Lister, S. Backyard poultry husbandry and general management. *In Practice*. 2012;34:136-145.

Chicken Anesthesia: Are Chickens Different Than Other Birds?

Janet L. Jones, LVMT

Session #082

Affiliation: From the Veterinary Medical Center, University of Tennessee, Knoxville, TN 37996, USA.

Abstract: There may be many circumstances as to why a chicken may need to be under anesthesia. Some of the most common reasons may be for obtaining diagnostic samples, or for medical and surgical treatments. The same basic principles of anesthetic management apply to the chicken patient as well as any other avian patients with small differences in restraint of the poultry patient. Use of monitoring equipment and anesthetic depth are important components to the anesthetic procedure.

Monitoring the depth of anesthesia for your poultry patient can be one of the most challenging tasks during the procedure. There are many different monitors that should be used during the anesthetic procedure as well as using your own observations. The challenge is to keep the patient at a surgical level of anesthesia. By checking the pinch reflex as well as monitoring trends of decreasing or increasing respiratory and cardiac trends, will help keep your poultry patient at that preferred anesthetic plain.

Perioperative considerations include evaluation of all aspects of husbandry including the source and length of ownership, diet and dietary preferences, and previous or current illnesses and therapies. A thorough physical examination is critical in assessing the chickens' physical condition, their cardiopulmonary status, and the severity of illness. Minimum diagnostics should include a packed cell volume (PCV), total solids, and blood glucose. Preferred diagnostics would be a complete white blood cell count (CBC) and a biochemistry. Another preanesthetic consideration is to fast the chicken. Generally a 4-6 hour fast is adequate for crop empting. The crop should be gently palpated prior to an anesthetic event and the head should be elevated during the surgical procedure.

To restrain the chicken for physical examination, generally the least amount of restraint with chickens is the best. By gently placing hands around the chicken's wings and allowing the chicken to stand on an examination table on a towel will eliminate foot slippage and most chickens should stand comfortably. Examination of the oral cavity should be performed last as most chickens do not like having their mouth manually opened. The bird can be picked up by placing the hands around the wings and gently lift off the table to examine the feet. Any premedications should be administered at this time. Once the bird is sedated, wrap a towel around the body to prevent the wings from flapping and using gas anesthesia is preferred. Once the bird is anesthetized, it should be placed on a heated surface where the procedure is to be performed and should not have a righting reflex or flap its wings. The bird should not respond to a firm toe pinch. The respiratory rate and effort must be visualized during the whole anesthetic procedure.[1-3]

Once the bird is laid on the table it is preferred that an intravenous catheter be placed in either the basilic vein or the medial metatarsal vein. An intraosseous catheter can also be placed in either the distal ulna or the proximal tibotarsus. The bird should be intubated using a non-cuffed, appropriately sized endotracheal tube. Using a heated table or additional heat source to place the bird on will help prevent the patient from becoming hypothermic. Once laid in recumbency, the monitoring equipment should be placed on the bird. The most important piece of equipment is the Doppler. The crystal can be placed in the mouth along the palatine artery, on the basilic vessel on the wing, or on the medial metatarsal vessel on the leg. A hand held pulse oximeter should be placed on either the toe or the beak while the rest of the monitoring equipment is being placed on the bird. The use of a multifunction monitor is very beneficial as monitoring ECG, respiratory rate with end–tidal CO2, and pulse oximetry are recommended. Fluids should be administered during the anesthetic period and emergency drugs should be readily available in case of an emergency.

During the anesthetic procedure the respiratory rate and effort must be visually monitored at all times. Birds will usually go into respiratory distress before they go into cardiac distress. If a decrease in respiratory or cardiac rate is appreciated, the anesthetic gas should be adjusted, usually in 0.25% increments. Any avian patient can get cold during an anesthetic event and will need to have the anesthesia turned down during the procedure. Listening to the Doppler is one of the most valuable monitoring methods as any change in cardiac rate or rhythm will be immediately heard. Blood pressure can be monitored during the procedure by placing the appropriate sized cuff on the distal humorous or dorsal metatarsal artery.

Once the anesthetic procedure is completed, the gas should be turned off and the patient left on oxygen. Chickens will feel more comfortable upon recovery if they are placed in sternal and once they start to shake their head, the endotracheal tube should be removed. The endotracheal tube should be examined for any signs of blood. Once the chicken is able to stand and ambulate well on its own, the patient can be place in its cage with food and water. Post-operative analgesics must be considered and appropriate for the surgical procedure performed.

References

1. Gunkel C, Lafortune M. Current techniques in avian anesthesia. *Semin Avian Exot Pet Med.* 2005;14: 263-276.

2. Nevarez JG. Monitoring during avian and exotic pet anesthesia. *Semin Avian Exot Pet Med.* 2005;14: 277-283.

3. Raftery A. Avian anaesthesia. *In Practice.* 2013;35:272-278.

Complete Blood Counts and Normal Hematologic Values in Young Bobwhite Quail

Ian Kanda, RVT,
Jessica Robertson, DVM,
James Meinkoth, DVM, MS, PhD, Dipl ACVP,
Joao Brandao, LMV, MS

Session #056

Affiliation: From the Department of Veterinary Clinical Sciences (Kanda, Robertson, Brandao) and the Department of Veterinary Pathobiology (Meinkoth), Center for Veterinary Health Sciences, Oklahoma State University, 2065 W Farm Road, Stillwater, OK 74078, USA.

The Northern bobwhite quail *(Colinus virginianus)* is a game bird listed as near threatened by the International Union for Conservation of Nature and it is experiencing decline throughout much of its range.[1-5] Species and age-specific hematologic data for bobwhite quail is limited.[6] The purpose of this study is to report hematologic values in healthy young Northern bobwhite quail, using different white blood cell count methodologies.[7,8] In 18 49-day-old captive raised quail, complete blood cell counts were performed as part of a health assessment prior to release into the wild. Data were reviewed and analyzed for the purpose of this study. Values for packed cell volumes (0.36 [0.25-0.45] L/L), total solids (36 [23-41] g/L), mean corpuscular volume (240 [166-327] fL), red and white blood cell counts (1.56 [0.93-2.04] x 10^{12}/L, and 13.8 [3.05-22.0] x 10^9/L, respectively) were obtained. A dilution with Natt-Herrick's stain in a manual hemocytometer was used to obtain the red blood cell count. Cell differentials were recorded for each animal. With considerable variability, bobwhite quail of this age were found to be primarily lymphocytic (57 [23.8-81.8] %). White blood cell counts were obtained from blood smears, estimating on both 400x (16.2 [9.2-23.4] x 10^9/L) and 1000x (10.8 [4.4-18.4] x 10^9/L) magnification. Counts were also made using a manual hemocytometer with blood diluted in phloxine B stain (13.8 [3.05-22.0] x 10^9/L). Findings suggest that 1000x magnification used for white blood cell counts underestimates results and 400x magnification overestimates results when compared to the current gold standard utilizing Phloxine B stain.

Acknowledgments: The costs associated with this project were supported by the Kristie Plunkett Wildlife Fund.

References

1. International Union for Conservation of Nature. *Colinus virginianus. The IUCN Red List of Threatened Species*. Available at: http://www.iucnredlist.org/details/22728956/0. Accessed July 3, 2016.

2. Brennan LA. How can we reverse the northern bobwhite population decline? *Wildl Soc Bull.* 1991:544-555.

3. Lusk JJ, Guthery FS, DeMaso SJ. Northern bobwhite (*Colinus virginianus*) abundance in relation to yearly weather and long-term climate patterns. *Ecol Modell.* 2001;146:3-15.

4. Williams CK, Guthery FS, Applegate RD, et al. The northern bobwhite decline: scaling our management for the twenty-first century. *Wildl Soc Bull.* 2004;32:861-869.

5. Williams CK, Lutz RS, Applegate RD, et al. Habitat use and survival of northern bobwhite (*Colinus virginianus*) in cropland and rangeland ecosystems during the hunting season. *Can J Zool.* 2000;78:1562-1566.

6. Ernst R, Coleman T, Kulenkamp A, et al. The packed cell volume and differential leucocyte count of bob-white quail (*Colinus virginianus*). *Poult Sci.* 1971;50:389-392.

7. Campbell TW, Ellis CK. *Avian and Exotic Animal Hematology and Cytology*. 3rd ed. Ames, IA: Blackwell Publishing;2007.

8. Dein FJ, Wilson A, Fischer D, et al. Avian leucocyte counting using the hemocytometer. *J Zoo Wildl Med.* 1994:432-437.

Section 2

Practice Management Sessions

Creating Great Content for Social Media

Laurie Hess, DVM, Dipl ABVP (Avian)

Session #344

Affiliation: From Veterinary Center for Birds & Exotics, Bedford Hill, NY 10507, USA.

First, we have to come to an understanding. Everything we do in the digital space is done to advance your practice. Whether your goals are to grow financially, to treat more pets, to increase adoptions, or to create a more informed community of pet owners, we must proceed with a purpose. That purpose is to have your story heard. This talk will help you understand the factors that you can control (and some you can't) when you consider sharing your message.

Keyword Roots

For years and years, you have heard talk of keywords. In the early days of the Internet, websites could simply drop keywords into a designated section of the website's code appropriately named keywords. This practice helped the web spiders that crawled and inspected websites so that they could be categorized for searches by knowing which sites contained what information. This practice of simply listing keywords was short lived, in that once websites all began listing the same keywords, search engines needed to rely on different indicators. For example, if every veterinary practice listed veterinarian in their keywords, the search engine would return about 30,000 results for practices. How would these be sorted?

Although the game of how keywords are used has changed, they are still at the root of any good search engine optimization (SEO) campaign. I know - you thought this talk was going to be about social media. As I mentioned earlier, every social media effort requires a focus on SEO, for it is about sharing your story with as many current and potential customers as possible. If no one is listening, how valuable is your message?

Know Your Keywords

It is very important that thorough research be conducted to determine what words are being used to find practices like yours. This should be done by a professional firm that has the ability to scour the Internet for information about terms and phrases that are being used successfully in both your campaigns and those of your competitors and to determine the actual phrases used in searches that are leading pet owners in your area to practices like yours. A reliable firm should be able to tell you how much traffic is being generated for specific websites. With this information in hand, you should be able to create a list of desirable keywords. Here are some examples of keyword categories that may work for your practice:

- Practice name (including shortened varieties)
- Best veterinarian (insert town, county, or other geographic identifiers)
- New puppy/kitten/bird/bunny, or other care (insert town, county, or other geographic identifiers)
- Veterinary (insert town, county, or other geographic identifiers)

Do you see a theme developing? Yep...it's all about local search. Google and the other search engines compete to return the best results to the web user's search query. Better results for Google means that a site has greater credibility; greater credibility leads to more frequent usage, which ultimately leads to greater ad sales opportunities for Google. By the way, I refer to Google so frequently and use that term generically to represent all search engines, because most searches (some studies say over 85%) are done using the almighty Google. If Google's goal is to return great information, and they can tell that the searcher is located in a particular geographic region, doesn't it make sense that they would want to respond with businesses that are accessible to the user? You betcha. So, why not make it easy on Google? Tell Google where you are located.

Key Phrases

Research has shown that most searches involve more than one word. Think about your own behavior. If you were looking online for a company to install a privacy fence around your yard, would you search fence? If you did, you would wind up with many undesirable results. Instead, if you are like most, you would be more detailed in your request. Something like: privacy fence companies in Chicago or best privacy fence company in Chicago would work better. Consider this fact when developing your strategy.

Keyword and Key Phrase List

Once you have worked with a professional firm to develop the list of valuable keywords, you will have a system to guide your practice's online content creation efforts. Let's keep this list handy, but put it off to the side while we look at developing a strategy.

Posting Calendar

One of the biggest problems that practices have is knowing what to post when. In many cases, posting is a reaction to the realization that the practice Facebook, blog, or Twitter accounts have been inactive. Inevitably, the person in charge of posting on the practice's social media sites sits in front of the computer to create activity. This type of content creation often results in very low interaction.

Posts should be written weeks in advance and scheduled on a calendar, so that the daily activity involves copying very well developed content from a spreadsheet or word document onto the practice's social media sites. This activity prevents underdeveloped posts but allows for the timely posting of great activities that happen in the practice on a daily basis.

Right about now, a hand goes up in the audience with a question regarding how often a practice should post. My answer is: It depends. I bet you were looking for a more detailed response. The truth is that there is no specific recipe for posting frequency, but one rule that should always apply is: if you don't have anything of value to share, don't say anything at all. Practices should start by sharing 3 times a week. Check your interaction, and make adjustments as needed. If you increase posting frequency, and the level of interaction with your followers does not decrease, you are in good shape. With that said, let's be sure to keep this in perspective. By having someone follow or like your social media site, you have been given the honor of having access to the stream of information that sites share with your follower on a daily (hourly) basis. In their stream or newsfeed, you are listed with many other brands, organizations, and causes. Finding the balance between being topical and a nuisance can be difficult. Ease into the process, and keep an eye on how many people unlike your sites on a monthly basis.

Relevance

As mentioned above, please remember that the person receiving your message voluntarily followed your practice with the understanding that you would be providing information about your practice and caring for pets. Deviating from the expected subject matter can be offensive and will cause followers to unlike your sites.

Determining what to write on any specific day requires us to revisit the posting calendar. Your practice's social media posting calendar should be developed by examining the following:

- Practice Marketing Plan (considering monthly promotions or educational focus)
- Pet focused days/months that are promoted on a national basis (mentions in news media will support your posts)
- National non-pet holidays/events/milestones (create a pet spin on the event) Local events or concerns that affect pet owners
- Local personalities who can contribute to topics

With the above guides in place, you can develop annual, monthly, and weekly themes. These themes should be listed in a calendar form allowing for posts to be written and developed over time. Time will allow your practice to gather pictures, to recall real stories to support the message, to identify clients who can offer testimonial support, and even to alert the media that you will be discussing those matters (we'll discuss the media in a later lecture...but this is all connected).

Where Do I Get Content?

Great question. The best scenario is for you to create it yourself, but while this sounds great, the reality of day to day practice life might cause social media content creation to slide to the back burner. Some practices are large enough to hire dedicated marketing staff members who can handle this task, but those situations are unreasonable for most.

Many practices seek professional assistance in creating content. Make certain that the firm creating the content is dedicated to the veterinary space and is creating CUSTOM content for you. Use of generic or scraped content can be very dangerous, often leading to negative results. The pitfalls of using this content are that Google considers this practice to be duplicate content.

Because Google is always looking to credit the original author of the content, it gets confused and actually penalizes websites and social media channels for using content published on other sites, because it looks at the practice as scraping content or stealing content that is not owned by the site. If you would like to learn more, simply Google the term duplicate content, and you'll see what I mean. Besides being penalized (which is a huge deal), why would you want to look like or sound like every other veterinary practice? Your message should be unique. A professional content provider should consult with you on a monthly basis, learn about your practice's needs, and provide custom content that aligns with your SEO plans.

Combining this activity with your practice interaction on the sites provides a very stable environment. Before you think that you cannot afford such a service, this service can be found for around $200-300 a month. What would you pay a staff member to perform such a task?

Keyword Stuffing

So, you are armed with your keyword and key phrase list. The goal is to get as many of those words/phrases into every paragraph as possible, right? Nope. You need to develop a balance considering frequency of usage, positioning of the text, and the relationship these factors have on the search engine algorithm. Professional assistance may be helpful here. Reputable firms will have the ability to analyze content and determine the optimal blend. The rule of thumb is to always write for the person you are trying to affect. Do so with the search engines in mind, but let the user experience lead the way. How much respect would you have for a practice that wrote a blog that was stuffed with keywords and phrases?

Key Takeaways

- Understand the relationship between content used in social media and in search engines.
- Know which keywords and key phrases are being used to affect practices.
- Develop a posting calendar and come up with content well in advance.
- Write for the user experience; Google will follow good content.
- Avoid duplicate content. Be as custom as possible.

Building and Growing Your Exotic Animal Practice

Dan H. Johnson, DVM, Dipl ABVP (Exotic Companion Mammal)

Session #345

Affiliation: From Avian and Exotic Animal Care, 8711 Fidelity Blvd, Raleigh, NC 27617, USA.

Exotic animals represent about 20% of the companion animal market for veterinary services.[1] Pet owners readily seek out veterinary care for birds, small exotic mammals, reptiles and amphibians. The greatest obstacles to growing your exotic animal practice are: (1) helping potential clients find you and (2) convincing them that your training and expertise matter.

Marketing

Properly naming your practice plays an important role in building public confidence. To do so, a stand-alone exotic animal hospital will usually incorporate the terms avian, bird, or exotic into the practice name. Similarly, a companion animal hospital that offers competent avian and exotic animal care should distinguish that service with a name that effectively communicates the level of expertise offered (e.g., Westover Animal Hospital and Exotic Animal Clinic or Pet Bird and Exotic Animal Service of Westover Animal Hospital). Having a separation of the exotic practice from your dog and cat practice is less threatening to your referral base, is more appealing to your exotic-animal clientele, and allows you to position yourself as the avian and exotic animal expert in your market.

Let your current clients know that you see exotic animals. Your practice logo should communicate that birds, reptiles, and small exotic mammals are seen at your hospital. The reception area should contain artwork and products that reflect your interest in exotic animal medicine. Place exotic animal brochures and client education materials in the exam room and on the countertops. There should me reminders that you treat exotic animals throughout the office. Membership plaques from the Association of Avian Veterinarians (AAV), Association of Reptilian and Amphibian Veterinarians (ARAV), and Association of Exotic Mammal Veterinarians (AEMV) should be mandatory.

Your exam room computer can passively communicate to clients which species you treat and what services you offer. Place a number of your best, most interesting photographs together in a single folder, and use the computer's screensaver to play a slideshow of those photographs. This also works on a large, flat-screen television placed in your waiting room, and most current models will play your slideshow from a flash drive (no external computer required). You need nice cards and brochures, but a refrigerator magnet never, ever gets discarded. An attractive refrigerator magnet will live on for many years!

Setting Appointments

Triage for exotic animals requires skill. In general, birds and exotic animals are much sicker than dogs and cats are before they begin to show signs of disease. Compounding this, many exotic pets do not get the daily close attention that traditional pets do. Often, exotic animal owners have had to call 2-3 veterinary practices before they find one that will treat their pet. As a result, a high percentage of avian and exotic cases present as emergencies.

Staff should be trained to properly triage exotic animal cases and schedule these pets accordingly. Most sick animal calls should be seen the same day as work-in emergencies. A large percentage of surgical and medical cases will not materialize if treatment appointments are delayed until the next workday. Exotic animals are inconvenient to treat, and if time will not allow an emergency same-day appointment, a referral should be offered.

Exotic Animal Hospital Setup

Exam rooms should be small, with a low ceiling, and lights that can be dimmed. It is best if there are no windows, or high places such as planters for birds to escape to. Exam rooms need to be ferret-proofed to prevent small animals from getting beneath and behind your base cabinets. Check the space under the door to make sure a small rodent or ferret couldn't possibly sneak through. Given the opportunity, they will. Birds, rabbits, and small exotic mammals are very sensitive to noise. In the hospital wards, limit noise due to barking and other things to reduce patient stress. Create a visual barrier between prey animals and predatory species including dogs and cats. Reduce drafts from air conditioning vents. Choose the warmest part of the building for birds and exotic animals. The hospital will also need a sink or tub large enough for waterfowl and iguanas to swim in. The isolation ward should be dark, warm, and away from all noise. Under ideal circumstances, there should be an incubator with a nebulizer, oxygen, heat, and humidity control.

Sharing What You Know

Many avian and exotic practitioners are reluctant to share knowledge with local colleagues. This is understandable since it took time, money and effort to gain all of the exotic animal knowledge and expertise that you have acquired. After all, "Why should I be the CE for some vet who calls when he has a rabbit spay on the table and realizes he has a question?" But this author believes that taking calls from area practitioners is one of the best ways to show neighboring veterinarians how much more you know than they do. Look at every phone call from a veterinarian as a chance to earn a future referral. Answering a drug dose question for a colleague is likely to earn you a medicine or surgery referral in the future. If you are not willing to come to the phone, how else is the neighboring vet community going to learn how much you know? If you think that sharing a little info with a colleague will take away from your business, then you vastly underestimate the barrier to competition than we enjoy as exotic animal practitioners. Sharing a dose or technique with a colleague will help you both.

Showcase Your Exotic Animal Practice

If you are trying to grow your exotic animal caseload, then invite everyone back to see what you do. Ask your clients behind the scenes to check out the facility and see you working. Say "yes" to school tours.

Pre-veterinary and veterinary students are your future referral base. Host as many as you can! Roll out the red carpet for the veterinarians and staff at other veterinary hospitals. Allow interested veterinarians to come over and get a taste for what it is you do.

Getting the Word Out

Letting the public know that you see birds and exotic animals is half the battle. Of course you need a good website and an active Facebook page. But, should you advertise? The author has chosen to only advertise in highly targeted areas such as the state veterinary medical association quarterly newsletter because it goes directly to other area veterinarians. No magazines or newspapers, and no coupon packs because the general public is not targeted enough. A few ways that the author markets his practice directly to his target audience are:

1. Being active in, an exhibitor at, and a sponsor of area exotic animal events. Examples from the author's local geographic area: reptile and exotic animal show, aquarium society meetings, koi club meeting, backyard chicken enthusiasts annual home show, bird club meetings, ferret club and rescue events, rabbit meetup and rescue events, museum events (reptile and amphibian day, bug festival), annual turtle festival at local park;

2. Speaking at schools, libraries and summer camps (especially your client's kids!);

3. Speaking and writing in veterinary meetings/publications (getting your name on anything that goes to the general vet referral base is excellent!);

4. Staying active with the veterinary school; and

5. Teach exotic animal medicine at the area technician school, hosting technician interns, and helping technician candidates who attend school online.

References

1. American Pet Products Association. 2015-2016 APPA National Pet Owners Survey Statistics: Pet Ownership & Annual Expenses. Available at: http://www.americanpetproducts.org. Accessed July 6, 2016.

Setting Up a Private Practice Residency

Angela M. Lennox, DVM, Dipl ABVP (Avian), Dipl ABVP (Exotic Companion Mammal), Dipl ECZM (Small Mammal)

Session #346

Affiliation: From Avian and Exotic Animal Clinic, 9330 Waldemar Road, Indianapolis, IN 46268, USA.

Definition of Residency in Medicine

In the medical field, a residency is defined as a "period of advanced medical training and education that normally follows graduation from medical school and licensing to practice medicine and that consists of supervised practice of a specialty in a hospital and in its outpatient department and instruction from specialists on the hospital staff." The concept of the residency program was inaugurated in the medical field 127 years ago. The American Veterinary Medical Association (AVMA) received applications for the very first two veterinary specialties in 1950; these were the American College of Veterinary Pathologists and the American Board of Veterinary Public Health. AVMA created the Advisory Board on Veterinary Specialties in 1959, which became the American Board of Veterinary Specialists (ABVS) in 1992. ABVS is the umbrella organization for US veterinary specialties, and is composed of one member from each AVMA-recognized specialty plus other advisors. ABVS regulates current veterinary specialties, and evaluates applications for newly created specialties.[1]

Pathways for Private Practice Residency in the United States

The American Board of Veterinary Practitioners (ABVP) was established in 1978 as a means for veterinarians in private practice to obtain specialist status without the burden of completing a concentrated residency. Requirements match the natural progression of private practice, emphasizing years of clinical practice along with publication and testing requirements.

ABVP offers a residency pathway as well as private practitioner pathway. Specialties of interest to exotic practitioners include Avian, Exotic Companion Mammal and Reptile and Amphibian. There are currently 18 ABVP-approved residency sites in Avian Medicine (7/18 at Universities) and 7 in Exotic Companion Mammal (3/7 at Universities). Not every site may offer a residency each year. A Residency Program Guide provides instructions and guidelines for establishment of a residency program.[2]

The American College of Zoological Medicine established a specialty program in 1983, which was approved by ABVS in 1988. A residency program was initiated in 1992. A residency emphasizing Zoological Companion Animals is available. To the best of the author's knowledge there is no private practice ACZM Companion Animal Residency.

Intern and Residency Matching Programs

The Veterinary Intern and Residency Matching Program (VIRMP) is sponsored by the American Association of Veterinary Clinicians and was established as a matching program for candidates seeking intern and residency opportunities and both University and private practice programs. VIRMP follows similar principles to those adopted by the Physicians National Intern and Resident Matching Program. According to VIRMP:

"The applicant completes and application packet and submits the application via this website. The VIRMP uses a central computer facility for matching prospective interns and residents with programs. Applicants create a rank order list of programs to which they want to apply. Institutions and private practices create a rank order list of applicants who they are willing to accept for their program. Both rank order lists are confidential. The VIRMP utilized sophisticated computer software to match applicants and programs in a way that insures each receives the highest available option on their rank order lists."

Internship programs are not regulated by any entity; however, AVMA has developed the Veterinary Internship Guidelines to "establish reasonable expectations for internship providers as well as participants." Despite the guidelines, there is no governing board; therefore, the structure, goals and mentorship are entirely self-regulated by the program offering the internship. For this reason, quality varies significantly leading to the warning: buyer beware. The assumption is that the intern will receive less compensation in exchange for more intense, focused mentorship; however, some internships have earned a reputation for deficient or absent mentorship and appear to be a way to obtain cheap labor.

Residencies are regulated by the sponsoring organizations. The VIRMP lists residencies in the categories of Ambulatory, Anesthesia, Avian Medicine, Behavior, Cardiology, Clinic Nutrition, Dentistry, Dermatology, Exotic/Wildlife/Zoo, Poultry Medicine, Radiation Oncology/Therapy, Small and Large Animal Surgery, Theriogenology, and more.[3]

Obligations of Private Practice Residencies

Residency training programs in private practice present distinct advantages and disadvantages over university-based programs. Advantages include the benefits of a higher caseload, in some cases significantly higher. Disadvantages are the lack of exposure to the resources of a university, including access to multiple specialists in a single location.

Prior to application for residency, the prospective mentor must consider the following:

1. Will the mentor be able to provide significant contact hours with the resident?
2. Will the mentor be available by cell phone or electronic communication nearly 24 hours per day, 7 days per week when not directly mentoring?
3. How will the mentor provide planned, direct teaching moments?
4. How will the resident gain exposure to specialists outside the mentor's specialty?
5. How will the mentor objectively evaluate the resident during the course of the residency?

Modifications to the routine of private practice may be significant when mentoring a resident. These may include adjusting schedule to ensure residents are alone in clinics as seldom as possible, the establishment of planned staff training, case rounds and journal clubs. All of these may represent a significant paradigm shift and change in priorities for the mentor.

References

1. American Veterinary Medical Association. American Board of Veterinary Specialties. Available at: www.avma.org/ProfessionalDevelopment/Education/Specialties/Pages/default.aspx. Accessed July 6, 2016.

2. American Board of Veterinary Practitioners. Residency Program Guide. Available at: http://www.abvp.com/why-residency. Accessed July 6, 2016.

3. American Association of Veterinary Clinics. Veterinary Internship and Residency Matching Program. Available at: www.virmp.org. Accessed July 6, 2016.

Updates on Veterinary Drug Compounding

Lance M. Roasa, DVM, MS, JD

Session #347

Affiliation: From Superior Veterinary Care, Lincoln, NE 68504, USA.

Abstract: Veterinary practitioners use compounded medications extensively, but the legal status of the drug is often poorly understood. The Food and Drug Administration's (FDA) regulations don't allow compounded from bulk ingredients, which is necessary for many medications treating commonly presented conditions. The FDA has released draft guidance, which help outline new thinking on the status of compounding legally. A best-practices approach will help exotics practitioners stay up to date on FDA publications, know the legal status of the drugs they use in practice, and reduce liability and adverse events.

What is a Drug?

Although veterinarians use the term drug in common language, technically it is a legally defined term with a very broad definition. The legal definition of drug is found in the Food Drug and Cosmetic Act of 1938 (FD&C Act) as "articles intended for use in the diagnosis, cure, mitigation, treatment, or prevention of disease in man or other animals."[1] Further broadening the definition Congress also names drugs as "articles (other than food) intended to affect the structure or any function of the body of man or other animals."[2] This expansive definition encompasses virtually every substance that a veterinarian gives or prescribes, other than food or water and brings them under the jurisdiction of the Food, Drug and Cosmetic Act with regulation and enforcement of the Food and Drug Administration (FDA).

After defining the term drug, the FD&C Act mandates that all animal drugs that are new to the market undergo the New Animal Drug Application (NADA).[3] The NADA process is a long and potentially expensive process that tests the new animal drug for safety, effectiveness, the manufacturing process, and stability among other characteristics. Once a drug is tested and is approved through the process and also is protected by a patent, it is called a legend (or pioneer) drug. A pioneer drug must be labeled with the statement (or legend): "Caution: Federal law restricts this drug to use by or on the order of a licensed veterinarian."[4] When a veterinarian reaches for a drug that bears this label, and they follow the label requirements, they can rely on the fact that the contents have been tested, approved and manufactured in accordance with the FDA's regulations.

What is a Generic Drug?

The FD&C Act contains further specific requirements for generic drugs. After a drug has cleared the NADA process and is not protected by a patent, another manufacture can submit for the Abbreviated New Animal Drug Application (ANADA) process.[5] The ANADA bypasses some of the safety and efficacy testing that a New Drug must undergo and concentrates on the manufacturing processes of the generic manufacturers.[6] To carry the term generic drug, the drug must undergo and pass the ANADA process. Thus, a compounded drug that is not ANADA approved is not a generic drug.

The term generic name also is different from generic drug. A drug's generic name is its established name, which is also usually the active ingredient.[7] The word generic can easily be confused with a drug that has been sponsored through the ANADA process. There are surprisingly few true generic drugs. The public can access drugs that are approved as pioneer drugs or generic drugs the FDA's database.[8]

What is Extralabel Use?

According to the FD&C Act every drug that does not bear the legend default is deemed unsafe and is adulterated or illegal to administer and or prescribe.[9] Every time a drug is used not according to label requirements on the legend drug, it is said to be extralabel use. Unfortunately, veterinary practice is not that black and white. In most veterinary practices the patients don't always meet the label's requirements for size, species and condition. Very few drugs for exotic species have undergone the NADA process, because the return on investment for the drug sponsor (manufacturer) is minimal. In an attempt to solve these grey areas of practice, the Animal Medicinal Use Clarification Act (AMDUCA) was enacted in 1994. AMDUCA allows veterinarians, with a valid veterinary client patient relationship (VCPR) to use drugs in an extra label manner if the FDA regulations are followed.[10] Those regulations define extralabel use as:

> "Actual use or intended use of a drug in an animal in a manner that is not in accordance with the approved labeling. This includes, but is not limited to, use in species not listed in the labeling, use for indications (disease and other conditions) not listed in the labeling, use at dosage levels, frequencies, or routes of administration other than those stated in the labeling, and deviation from labeled withdrawal time based on these different uses."
>
> Title 21 Code of Federal Regulations, part 530.3(a)

This regulation gives veterinarians wide leniency to use drugs in extralabel manner in their practice with as long as human safety in the form of withdrawal times is not compromised. This does not allow the drug to be adulterated or mixed in a manner that is not on the label.

Where Does Compounding Come In?

Even with wide ability to use drugs in an extralabel manner, there are still species and conditions that cannot be treated without additional measures. This may mean mixing the drug in a more dilute or stronger concentration, adding flavoring or an additional substance, or changing the formulation of the medication (i.e., grinding a pill into a power then liquefying). The FDA regulation allows this, under strict guidance. The animal compounding regulations give these requirements, among others, for compounding under AMDUCA:

- There is no approved animal or human drug that can be used for the condition diagnosed under labeled or extralabel uses,
- a human drug cannot be used when an animal drug is available,
- the compounding is done by a licensed veterinarian or pharmacist,
- adequate procedures are followed to ensure safety and effectiveness, and
- it is in compliance with applicable state laws.

Most importantly, the regulations clearly state, "Nothing in this part shall be construed as permitting compounding from bulk drugs."[11] A bulk drug is a means the starting point is an active pharmaceutical product (API) and not product that has gone through the NADA, ANADA process or the human equivalent. In other words, a bulk drug is the pure compound, without the inert ingredients for stability, absorption, taste or other reasons in pills, liquids and other consumer ready drugs.

Currently, according to the FD&C Act and FDA regulation, compounding can be performed in very controlled circumstances, and not starting with a bulk drug. Still practitioners are faced with scenarios that require a bulk starting point. Many times a legend or generic drug contains too many inert ingredients or is not at a concentration that is conducive to compounding for very small or very large patients. Some drugs that veterinarians reach for have no legend or generic preparations at all. The prohibition on bulk starting points seems to fly in the face of what is happening in everyday practice and pharmacies all over the country. Compounding pharmacies seem to offer every medication imaginable, in every concentration and administration route, and many are made from bulk-sourced drugs. Further confusing the topic of compounding is differing laws and regulations in every state and selective enforcement of the rules by the FDA and state agencies that oversee pharmacies and veterinary practices. State laws vary on the stocking, dispensing and resale of compounded medications.

What about State Laws and Regulations?

State laws vary greatly on the issue of compounded formulations, and a practicing veterinarian should check with local sources before relying on this data. The state pharmacy boards are typically concerned with the licensing and regulation of pharmacists, and most states have specific requirements regarding veterinary drug compounding. The American Veterinary Medical Association (AVMA) has compiled these resources. An issue with an extremely wide range of variance at the state law level is whether a veterinarian can dispense or resale compounded drugs. Two states, New York and New Mexico, do not allow compounded drugs to be used, either in office or as dispensed.[12] As of January 2016, 21 states had specific provisions allowing in office use, but prohibiting dispensing or reselling.[13] Nine states specifically allow administration in office-use and for dispensing to clients.[14] Twelve states specifically allow administration of a compounded medication but are not clear on dispensing or reselling.[15] The remaining states have unclear laws.[16] On a national level, this has contributed to confusion as to when and how a veterinarian can administer and dispense compounded drugs. A practitioner must look to local resources for the laws and regulations for compounded medications.

What about MUMS?

The Minor Use/Minor Species Act of 2004 (MUMS) was intended to make more medications available to veterinarians treating minor species and uncommon diseases of the major species.[17] Minor species are defined all animals other than humans, horses, dogs, cats, cattle, pigs, turkeys, and chickens. MUMS modifies the FD&C Act in three categories; conditional approval, designation and indexing.

Conditional approval

With a Conditional Approval under the MUMS act a sponsor can make the drug available before full efficacy testing, but after proving safety. Conditional approval speeds the drug to market, as long as there is a reasonable expectation of effectiveness. The drug can stay on conditional approval up to 5 years, while effectiveness data is being collected.

Designation status

Designation provides drug sponsors incentives for approval with grants for safety and effectiveness testing and provides 7 years of exclusive marketing once approval is received. Currently 137 drugs are on the designation list.

Indexing

An Index drug is for drugs that are use in species that are too rare or too varied to be subjected to well-controlled studies. The FDA maintains a list, the Index of Legally Marketed Unapproved New Animal Drugs for Minor Species (the Index), of drugs that can be used without full approval. Currently the Index lists 11 drugs for unapproved usage.

Legally, Where is Compounding Heading?

In May of 2015, the FDA release draft guidance for industry (GFI #230) addressing compounding from bulk drugs.[18] Although the document is not finalized and it is not regulation, it does outline the FDA's current thinking and when they will take action regarding compounded drugs. GFI #230 contains 11 steps to follow when prescribing drugs from compounding pharmacies.

1. The compounding is to be done at a state licensed pharmacy.

2. A valid prescription identifies an individual animal, and the compounding is done after the receipt (unless specific guidelines are met).

3. The drug is not intended for use in food-producing animals, and the prescription or accompanying paperwork contains the statement: "This patient is not a food-producing animal." Chickens, swine, sheep and goats are always considered food-producing regardless of whether the individual is a family pet. This means that this guidance cannot apply to pot-bellied pigs or pet-chickens or goats.

4. To compound from a bulk substance, the veterinarian has to expect a clinical difference from a compounded substance from an FDA-approved starting point. Also, the prescription or accompanying paperwork has to specify why that clinical difference exists.

5. If an FDA-approved starting point exists, then the pharmacist has to make a determination the desired compounded drug cannot be made from the FDA-approved drug.

6. The prescription or accompanying paperwork must state the animal's species and the statement:

   "There are no FDA-approved animal or human drugs that can be used as labeled or in an extra-label manner under section 512(a)(4) or (5) and 21 CFR part 530 to appropriately treat the disease, symptom, or condition for which this drug is being prescribed."

7. The bulk drug source has to be manufactured by a registered establishment, and a valid certificate of analysis has to accompany the bulk drug.

8. The drug has to be compounded following USP guidance, and sterile drugs have to be prepared in USP ISO Class 5 standards.

9. The drug is not sold or transferred by any entity other than the compounding pharmacy. A veterinarian can administer the drug, but not resale or transfer the drug.

10. Within 15 days of becoming aware of an adverse event, the pharmacy has to report to the FDA.

11. The label contains the species and name of the patient and the name of the owner or caretaker of the animal patient.

What are Best Practices for Using Compounded Drugs?

It is the author's opinion that following these FDA-outlined steps when reaching for a compounded medication will not draw the scrutiny of the FDA. Further, additional steps should be taken to ensure a best-practices approach to compounded medications to ensure maximal patient safety and liability protection.

1. Know the legal status of the drug you are administering or prescribing.
   - Is the drug an FDA-approved animal drug? If so, are you following label instructions?
   - Is the drug a generic drug that has undergone the ANADA process?
   - Is the drug a compounded drug, but under an AMDUCA exemption?
   - Is the drug compound from a bulk drug source but you deem it is necessary?

2. Only use extralabel and compounded drugs when medically necessary.

3. Document the legal status of the drug in the medical record.

4. Consult with the client about the legal status of the drug, its safety, efficacy, and potential for adverse events.
   - Obtain the client's consent to use of extralabel and compounded drugs, as well as their understanding of safety and adverse events.
   - Document the client's consent and understanding in the medical record.

5. Ensure follow-up regarding the patient's response to extralabel and compounded drugs.
   - Document the response, any adverse events, and/or owner compliance.
   - Report adverse events to the pharmacy.

6. Know your compounding pharmacist(s).
   - Ensure the compounding pharmacy you are using is Pharmacy Compounding Accreditation Board (PCAB) accredited.[19]
   - Talk to your compounding pharmacists, ask questions (e.g., how the prescription is made, bulk or FDA-approved product?), and check the certificates of analysis.
   - Understand their protocols for sterility and shipping in hot and cold weather.
   - Confirm that they carry liability insurance that will cover adverse events.

7. Know and follow the beyond-use-date (BUD); know how the BUD was derived.

8. The use of compounded formulations in animals to be used for food is not allowed.

9. Know your state laws regarding stocking, reselling, dispensing, administering, and/or labeling compounded medications.

10. Stay informed of policy changes, and watch for FDA regulation changes and safety/efficacy data.

References

1. Title 21 US CODE § 321(g)(1)(B).

2. Title 21 US CODE § 321(g)(1)(C).

3. Title 21 US CODE § 360b.

4. Title 21 Code of Federal Regulations, part 201.105(b)(1).

5. Title 21 US CODE §§ 360b(b), (c).

6. Title 21 US CODE § 360b(c)(2)(A).

7. See Generic Animal Drugs: Approved or Unapproved? Available at: http://www.fda.gov/AnimalVeterinary/GuidanceComplianceEnforcement/ComplianceEnforcement/UnapprovedAnimalDrugs/ucm249392.htm. Accessed May 9, 2016.

8. Animal Drugs @ FDA Last Accessed May 9 at 2016 Available at: http://www.accessdata.fda.gov/scripts/animaldrugsatfda/index.cfm?gb=1. Accessed July 6, 2016.

9. Title 21 US CODE § 360b(a)(1).

10. Title 21 US CODE § 360b(a)(4).

11. Title 21 Code of Federal Regulations, part 530.13.

12. American Veterinary Med Assoc. State Summary Report. Aug 2015. Available at: https://www.avma.org/Advocacy/StateAndLocal/Pages/compoundinglaws.aspx. Accessed January 28, 2016.

13. State laws and regulations that allow veterinary offices to administer compounded products but specifically prohibit them from dispensing or reselling products compounded by a pharmacy: Alabama, Alaska, Arizona, Arkansas, Connecticut, Georgia, Iowa, Kentucky, Louisiana, Maryland, Missouri, North Carolina, North Dakota, Ohio, Oklahoma, Rhode Island, Utah, Vermont, Washington, West Virginia, Wyoming. Available at: https://www.avma.org/Advocacy/StateAndLocal/Pages/compoundinglaws.aspx. Accessed January 22, 2016.

14. State laws and regulations that allow veterinary offices to administer compounded products and dispense the products to clients but may be subject to conditions or limitations in some cases: California, Florida, Maine, Michigan, Minnesota, Nebraska, Tennessee, Texas, Virginia. Available at: https://www.avma.org/Advocacy/StateAndLocal/Pages/compoundinglaws.aspx. Accessed January 22, 2016.

15. State laws and regulations that allow veterinary offices to administer compounded products but do not specifically address or the law is not clear regarding dispensing products compounded by a pharmacy: Colorado, Delaware, Idaho, Illinois, Massachusetts, Mississippi, Montana, Nevada, New Hampshire, New Jersey, Oregon, South Carolina. Available at: https://www.avma.org/Advocacy/StateAndLocal/Pages/compoundinglaws.aspx. Accessed January 22, 2016.

16. States where with no laws and regulations addressing compounding: Hawaii, Indiana, Kansas, Pennsylvania, South Dakota, Wisconsin. Available at: https://www.avma.org/Advocacy/StateAndLocal/Pages/compoundinglaws.aspx. Accessed January 22, 2016.

17. Title 21 US CODE §§ 360ccc.

18. Draft Guidance for Industry #230, Compounding Animal Drugs from Bulk Substances, May 2016, Available at: http://www.accessdata.fda.gov/scripts/animaldrugsatfda/index.cfm?gb=1. Accessed July 6, 2016.

19. Pharmacy Compounding Accreditation Board, Find a PCAB Accredited Pharmacy Available at: http://www.pcab.org/accredited-locations.html. Accessed May 9, 2016.

Section 3

AAV Plenary Sessions

Kenneth Welle, DVM, Dipl ABVP (Avian)
AAV President
Moderator

Validation of a Novel Radiograph Density Determination Process: A Multicenter Study

M. Scott Echols, DVM, Dipl ABVP (Avian), Boston Clark Terry, Samer Merchant, Ed Hsu, PhD, Doug Folland, DVM, Dipl ABVP (Avian), Brian Speer, DVM, Dipl ABVP (Avian), Dipl ECZM (Avian), Kris Alhgrim, DVM, Jose Biascoechea, DVM, Dipl ABVP (Avian), Greg Burkett, DVM, Dipl ABVP (Avian), Todd Driggers, DVM, Sally Haddock, Hugh Hays, DVM, Susan Kelleher, DVM, Angela M. Lennox, DVM, Dipl ABVP (Avian), Dipl ABVP (Exotic Companion Mammal), Dipl ECZM (Small Mammals), Larry Nemetz, DVM, Mike Putnam, DVM, Russell Welfare, DVM

Session #115

Affiliation: Echols Veterinary Services, Salt Lake City, UT 84054, USA (Echols), Medical Center for Birds, Oakley, CA 94561, USA (Echols, Speer), Department of Bioengineering, University of Utah, Salt Lake City, UT 84112, USA (Terry, Merchant, Hsu), Tracy Aviary, Salt Lake City, UT 84105, USA (Folland), GoldenView Veterinary Hospital, Golden, CO 80401, USA (Alhgrim), Exotic Vet Care, Mount Pleasant, SC 29464, USA (Biascoechea), The Bird Hospital, Avian Veterinary Services, Durham, NC 27077, USA (Burkett), Avian and Exotic Animal Clinic, Mesa, AZ 85204, USA (Driggers), St Marks Veterinary Hospital, New York, NY 10003, USA (Haddock), Summertree Animal Clinic, Dallas, TX 75244, USA (Hays), Broward Avian & Exotic Animal Hospital, Deerfield Beach, FL 33442, USA (Kelleher), Avian and Exotic Animal Clinic of Indianapolis, Indianapolis, IN 46268, USA (Lennox), The BIRD Clinic, Orange, CA 92866, USA (Nemetz), Richmond Veterinary Clinic, Richmond, KY 40475, USA (Putnam), North Powers Animal Hospital, Colorado Springs, CO 80920, USA (Welfare).

Tissue and especially bone density abnormalities are increasingly being recognized as indicators of various diseases. Currently, computerized tomography (CT) is the only non-invasive means to accurately measure tissue densities (via Hounsfield Units or HU). Dual-energy x-ray absorptiometry (aka DXA scan) is the established standard for measuring bone mineral density. Such systems are typically only validated for adult humans, expensive and not readily available. In contrast, radiographs are readily accessible, but unlike HU, their intensities are highly variable depending on instrument hardware and settings used. The intensity variability complicates repeatability and reproducibility of radiograph-based studies, especially ones that involve multi-site participation. Here, a novel methodology to standardize radiograph intensity was developed and tested in a multi-site study. Identical calibration bars and phantoms consisting of discrete densities were constructed and scanned using CT to obtain HU numbers. Radiographs of standardized phantoms and calibration bars were obtained using different hardware and source output settings at different sites, and analyzed using image processing techniques such as background subtraction and linearized to the intensities of the calibration bar. Preliminary results show correction reduced the intensity variability of same regions of the phantom, as measured by its standard deviation, by up to a factor of 4. Moreover, correlation of the radiographic intensities to their HU numbers, as indicated by the coefficient of determination R^2, improved from weak (0.4) to strong (0.8). These findings suggest that the proposed method of calibration may provide a means for reliable radiograph density determination and standardization critical for multi-site studies.

The Development of an Effective Vaccine Against Proventricular Dilatation Disease

Samer S Hameed, DVM, MSc,
Ian Tizard, DVM, PhD, DSc, Dipl ACVM,
Susan Payne, PhD

Session #077

Affiliation: From Schubot Exotic Bird Health Center, Texas A&M University, College of Veterinary Medicine and Biomedical Sciences, College Station, TX 77843, USA.

Abstract: Proventricular dilatation disease (PDD) is a fatal disease that affects mainly psittacine birds (parrots).[1] Gastrointestinal signs and/or neurologic signs can be seen in PDD- affected birds as a result of non-suppurative inflammation of the central, peripheral and autonomic nervous systems.[1-3] In 2008, parrot *Bornavirus* (PaBV) was proposed as a causative agent of PDD.[4,5] The prognosis of PDD is poor once clinical signs have developed and currently no etiologic treatments or vaccines are available.[6]

Two experiments were completed using cockatiels (*Nymphicus hollandicus*) to evaluate a new vaccine against PDD. In the first experiment, inactivated parrot *Bornavirus* vaccine (PaBV-2) was administered intramuscularly (IM) three times at 3-week intervals, followed by one time (IM) recombinant N protein vaccine (PaBV4). In the second experiment the recombinant N protein vaccine was used alone two times (IM) 3 weeks apart. Vaccinated and unvaccinated cockatiels were challenged with virulent PaBV-2. Over 28 weeks post-challenge, cloacal swabs were collected and body weight and clinical signs were recorded. Histopathology, immunohistochemistry (IHC) and quantitative real-time PCR (qRT-PCR) were used to detect the presence of any microscopic lesions, PaBV-antigen and genomes.

The results of this study showed that recombinant N protein vaccine protected the vaccinated challenged group from gross and histopathologic lesions as well as PDD-related mortality in comparison to the unvaccinated challenged control group. However, PaBV-antigen was detected in several organs as well as cloacal swabs for both groups indicating that the vaccine does not prevent infection. This study demonstrates for the first time the efficacy of a new vaccine against PDD.

References

1. Gregory CR, Latimer KS, Niagro FD, Ritchie BW. A review of proventricular dilatation syndrome. *J Assoc Avian Vet.*1994;8(2):69-75.

2. Doneley RJ, Miller RI, Fanning TE. Proventricular dilatation disease: an emerging exotic disease of parrots in Australia. *Aust Vet J.* 2007;85(3):119-123.

3. Keller DL, Honkavuori KS, Briese T, Lipkin WI, Muthuswamy A, Steinberg H, Sladky KK. Proventricular dilatation disease associated with avian *Bornavirus* in a scarlet macaw *(Ara macao). J Vet Diagn Invest.* 2010;22(6):961-965.

4. Kistler AL, Gancz A, Clubb S, Skewes-Cox P, Fischer K, Sorber K, Chiu CY, Lublin A, Mechani S, Farnoushi Y, Greninger A, Wen CC, Karlene SB, Ganem D, DeRisi JL. Recovery of divergent avian bornaviruses from cases of proventricular dilatation disease: identification of a candidate etiologic agent. *Virol J.* 2008;5:88.

5. Honkavuori KS, Shivaprasad HL, Williams BL, Quan PL, Hornig M, Street C, Palacios G, Hutchison SK, Franca M, Egholm M, Briese T, Lipkin WI. Novel borna virus in psittacine birds with proventricular dilatation disease. *Emerg Infect Dis*. 2008;14(12):1883-1886.

6. Staeheli P, Rinder M, Kaspers B. Avian bornavirus associated with fatal disease in psittacine birds. *Virol J*. 2010;84(13):6269-6275.

Pharmacokinetics and Egg Residues After Oral Meloxicam Use in Poultry

Marcy J. Souza, DVM, MPH, Dipl ABVP (Avian),
Molly S. White,
Kristin I. Gordon,
Lillian Gerhardt, LVMT,
Joan Bailey,
Sherry K. Cox, MS, PhD

Session #018

Affiliation: From the Department of Biomedical and Diagnostic Sciences (Souza, White, Gordon, Bailey, Cox) and Department of Small Animal Clinical Sciences (Gerhardt), College of Veterinary Medicine, University of Tennessee, Knoxville, TN 37996, USA.

Backyard poultry are becoming increasingly popular; however, few guidelines exist for appropriate dosing or egg withdrawal time for commonly used medications. Appropriate dosing of meloxicam for domestic chickens and the length of time drug residues persist in eggs is unknown. With this study, we sought to determine the pharmacokinetics of meloxicam following a single oral administration, as well as how long drug residues persist in eggs. Eight white leghorn hens were administered meloxicam orally once at 1 mg/kg. Blood samples were collected just prior to 48 hours after drug administration. Eggs were collected for 3 weeks after drug administration. Plasma, egg whites, and egg yolks were analyzed by high-performance liquid chromatography. The half-life, maximum concentration, and time to maximum concentration in plasma was 2.8 hours, 7.21 µg/ml, and 2 hours, respectively. Drug was detected in egg whites up to 4 days after dosing and in egg yolks up to 8 days after dosing. Based on these results, meloxicam would need to be administered orally twice daily at 1 mg/kg in chickens to maintain plasma concentrations reported to be therapeutic for humans.

Section 4

AAV Hot Topics

James Carpenter, DVM, Dipl ACZM
Moderator

Evaluation of the Effects of 4.7-mg Deslorelin Acetate Implants on Egg Laying in Cockatiels *(Nymphicus hollandicus)*

Noémie M. Summa, Dr. med. vet., IPSAV,
David Sanchez-Migallon Guzman, LV, MS, Dipl ECZM (Avian), Dipl ACZM,
Kirk C. Klasing, MS, PhD,
Emma Wils-Plotz, PhD,
Nerisa E. Riedl,
Philip H. Kass, DVM, PhD, Dipl ACVPM (Epidemiology),
Michelle G. Hawkins, VMD, Dipl ABVP (Avian)

Session #277
AAV 2016 House Officer Manuscript Competition Award Recipient

Affiliation: From the William R. Pritchard The Veterinary Medical Teaching Hospital (Summa), Department of Veterinary Medicine and Epidemiology (Guzman, Hawkins), Department of Animal Science (Klasing, Wils-Plotz, Riedl), Department of Population Health and Reproduction (Kass), University of California, Davis, CA 95616, USA.

GnRH agonists are the drug of choice for medical management of avian reproductive disorders.[1-3] In previous studies, 4.7-mg deslorelin implants prevented egg laying for approximately 70 days in Japanese quails (*Coturnix coturnix japonica*),[4] 5 weeks in pigeons (*Columba livia*),[5] and up to 6 months in chicken (*Gallus domesticus*).[6] No controlled study has been performed in psittacine birds. The goal of this study was to evaluate the efficacy and duration of action of a 4.7-mg deslorelin implant on egg laying in cockatiels *(Nymphicus hollandicus)*. Twenty-six compatible breeding pairs were monitored for egg laying over 90 days on a 16:8 light:dark cycle, followed by a 90-day period in a 12:12 light:dark cycle without nest to prevent egg laying. All females (n = 13) then received a 4.7-mg deslorelin implant (Suprelorin®, Virbac, Carros, France) or a placebo implant (n=13) SC between the scapulae and were returned to a 16:8 light:dark cycle with nest box and housed individually in close contact with a male in an adjacent cage. Egg production and quality were monitored daily over 180 days. Eleven out of 13 control birds laid following implantation. At the time of writing this abstract (82 days after implantation), none of the treated birds have laid an egg. No adverse effects associated with the deslorelin implant have been observed. Further data are pending to evaluate its reversibility in cockatiels.

Acknowledgments: The study was supported by a grant from the Center for Companion Animal Health, School of Veterinary Medicine, University of California, Davis. The authors would also like to thank Peptech, Inc. for their donation of the deslorelin and placebo implants and Kristy Smith, Lauren Whitfield, Fatemeh Alemi for their technical assistance

References

1. Joyner K. Theriogenology In: Ritchie B, Harrison G, Harrison L, eds. *Avian Medicine: Principles and Applications*. Lake Worth (FL): Wingers Publishing Inc.;1994:748-773.

2. Keller KA, Beaufrère H, Brandão J, et al. Long-term management of ovarian neoplasia in two cockatiels (*Nymphicus hollandicus*). *J Avian Med Surg*. 2013;27:44-52.

3. Rosen LB. Avian reproductive disorders. *J Exot Pet Med.* 2012;21:124-131.

4. Petritz OA, Sanchez-Migallon Guzman D, Paul-Murphy J, Fecteau K, Mete A, Kass PH, Hawkins MG. Evaluation of the efficacy and safety of single administration of 4.7-mg deslorelin acetate implants on egg production and plasma sex hormones in Japanese quail (*Coturnix coturnix japonica*). *Am J Vet Res.* 2013;74:316-323.

5. Cowan ML, Martin GB, Monks DJ, et al. Inhibition of the reproductive system by deslorelin in male and female pigeons (*Columba livia*). *J Avian Med Surg.* 2014;28:102-108.

6. Noonan B, Johnson P, Matos Rd. Evaluation of egg-laying suppression effects of the GnRH agonist deslorelin in domestic chickens. *Proc Annu Conf Assoc Avian Vet*. 2012;312.

Evaluation of Dexmedetomidine-Alfaxalone-Butorphanol and Dexmedetomidine-Ketamine Anesthesia-Butorphanol in Domestic Dove *(Streptotelia risoria)*

G Zoller, DVM, IPSAV (Medecine zoologique),
L Chassang, DVM, IPSAV (Medecine zoologique),
P Loos, DVM, IPSAV (Medecine zoologique),
J Botman, DVM,
L Duroyaume, DVM,
M Huynh, DVM, Dipl ECZM (Avian)

Session #283

Affiliation: From Centre Hospitalier Vétérinaire Frégis, 43 Avenue Aristide Briand 94110 Arcueil, France (Zoller, Huynh), Faculté Vétérinaire de Liège, Boulevard de Colonster 20, 4000 Liège, Belgique (Chassang, Loos, Botman, Duroyaume).

Two injectable intramuscular anesthetic protocols were evaluated in a group of 10 privately owned domestic doves (*Streptotelia risoria)* for endoscopic examination. The birds received randomly dexmedetomidine 0.4 mg/kg and alfaxalone 20 mg/kg (Protocol A, *n* = 5) or dexmedetomidine 0.4 mg/kg, ketamine 40 mg/kg and butorphanol 1 mg/kg (Protocol B, *n* = 5).

Heart rate and respiratory rate were collected every 5 minutes by the same observer. Once the bird was sedated, the bird was positioned for a left lateral endoscopic procedure. Whenever the injectable anesthetic protocol was not sufficient to achieve proper positioning, the bird was anesthetized with isoflurane.

Mean time to loss of consciousness was 102.8 ± 48.13 seconds and 79.6 ± 44.36 seconds for protocol A and B, respectively. Mean time to loss of righting reflex was 240.2 ± 134.69 seconds and 162.6 ± 102.26 seconds for protocol A and B, respectively. Two birds could not undergo the endoscopic procedure in each protocol. One bird undergoing protocol A died following a prolonged recovery period. All birds undergoing protocol B experienced a prolonged recovery period (> 1 h). The cardiac rate and respiratory rate presented a high variability within the sample group for both protocols. Glycemia decreased during the procedure with each protocol but remained within normal range for the species.

Due to high variability of effect, dexmedetomidine-ketamine and dexmedetomidine-alfaxalone cannot be recommended as anesthetic protocols to perform minimally invasive procedures based on our findings.

Pharmacokinetics of a Sustained-Release Formulation of Meloxicam in Hispaniolan Amazon Parrots *(Amazona ventralis)*

David Sanchez-Migallon Guzman, LV, MS, Dipl ECZM (Avian),
Dipl ECZM (Small Mammal), Dipl ACZM,
Michael Court, BVSc, PhD, Dipl ACVA,
Noémie Summa, Dr. med. vet., IPSAV,
Joanne Paul-Murphy, DVM, Dipl ACZM, Dipl ACAW

Session #301

Affiliation: From the Department of Veterinary Medicine and Epidemiology (Guzman, Paul-Murphy), and the Veterinary Medical Teaching Hospital (Summa), School of Veterinary Medicine, University of California, Davis, CA 95616, USA, Department of Veterinary Clinical Sciences, College of Veterinary Medicine, Washington State University, Pullman, WA 99164, USA (Court).

Meloxicam is a COX-2 selective nonsteroidal anti-inflammatory drug commonly prescribed for extra-label use in birds.[1] Several studies have investigated the pharmacokinetics of meloxicam in avian species, including chickens,[2] ostriches,[2] ducks,[2] turkeys,[2] pigeons,[2] vultures,[3] ring-necked parakeets (*Psittacula krameri*),[4] Hispaniolan Amazon parrots (*Amazona ventralis*),[5] African grey parrots (*Psittacus erithacus*),[6] red-tailed hawks (*Buteo jamaicensis*),[7] and great-horned owls (*Bubo virginianus*).[7] Pharmacodynamic[8] and safety[9] studies have also been performed previously in Hispaniolan Amazon parrots. The current study evaluated the pharmacokinetic profile of a sustained-release formulation of meloxicam (Meloxicam SR, ZooPharm, Fort Collins, CO 80550 USA) in Hispaniolan Amazon parrots. Parrots (n=12) received 3 mg/kg of meloxicam SR (5 mg/ml) SC in the flank and blood samples were collected at 10 time points over 120 hours. Plasma concentrations were determined by a validated HPLC assay and pharmacokinetic analysis was performed using a noncompartmental approach. Plasma concentrations reached a mean Cmax of 23.4 µg/ml (range: 14.7-46.0 µg/ml) at 1.8 hours (range: 0.5-6 hours), with a terminal half-life of 7.4 hours (range: 1.4-40.9 hours). There was noticeable individual variation, such that some parrots (n=4) showed very fast plasma elimination of meloxicam, similar to the high variability reported in a previous meloxicam pharmacokinetic study with the same birds.[5] Based on these results, meloxicam SR could be administered SC every 12 hours for birds that rapidly eliminate meloxicam or every 96 hours for birds that eliminate meloxicam slower, precluding accurate frequency of administration recommendations in this species. Further studies are needed to evaluate the safety and pharmacokinetics of meloxicam SR in birds.

Acknowledgments: This study was supported by the American Association of Zoo Veterinarians research fund, the Center for Companion Animal Health at the University of California Davis and the The Richard M. Schubot Parrot Wellness and Welfare Program.

References

1. Food and Drug Administration. Adverse Drug Experience (ADE) Reports. Available at: http://www.fda.gov/AnimalVeterinary/SafetyHealth/ProductSafetyInformation/ucm055369.htm. Accessed July 6, 2016.

2. Baert K, De Backer P. Comparative pharmacokinetics of three non-steroidal anti-inflammatory drugs in five bird species. *Comp Bioch Physiol C*. 2003;134:25-33.

3. Naidoo V, Wolter K, Cromarty AD, et al. The pharmacokinetics of meloxicam in vultures. *J Vet Pharmacol Ther*. 2008;31:128-134.

4. Wilson HG, Hernandez-Divers S, Budsberg SC, et al. Pharmacokinetics and use of meloxicam in psittacine birds. *Proc Annu Conf Assoc Avian Vet*. 2004;7-9.

5. Molter CM, Court MH, Cole GA, et al. Pharmacokinetics of meloxicam after intravenous, intramuscular, and oral administration of a single dose to Hispaniolan Amazon parrots (*Amazona ventralis*). *Am J Vet Res*. 2013;74:375-380.

6. Montesinos A, Ardiaca M, Juan-Salles C, et al. Effects of meloxicam on hematologic and plasma biochemical analyte values and results of histologic examination of kidney biopsy specimens of African grey parrots (*Psittacus erithacus*). *J Avian Med Surg*. 2015;29:1-8.

7. Lacasse C, Gamble KC, Boothe DM. Pharmacokinetics of a single dose of intravenous and oral meloxicam in red-tailed hawks (*Buteo jamaicensis*) and great horned owls (*Bubo virginianus*). *J Avian Med Surg*. 2013;27:204-210.

8. Cole GA, Paul-Murphy J, Krugner-Higby L, et al. Analgesic effects of intramuscular administration of meloxicam in Hispaniolan parrots (*Amazona ventralis*) with experimentally induced arthritis. *Am J Vet Res*. 2009;70:1471-1476.

9. Dijkstra B, Guzman DS, Gustavsen K, et al. Renal, gastrointestinal, and hemostatic effects of oral administration of meloxicam to Hispaniolan Amazon parrots (*Amazona ventralis*). *Am J Vet Res*. 2015;76:308-317.

Pharmacokinetics of Itraconazole, Voriconazole, and Terbinafine Delivered by an Impregnated Subcutaneous Implant in Japanese Quail *(Coturnix japonica)*

Marcy J. Souza, DVM, MPH, Dipl ABVP (Avian), Dipl ACVPM,
Patrick Redig, DVM, PhD,
Sherry K. Cox, MS, PhD

Session #019

Affiliation: From the Department of Biomedical and Diagnostic Sciences, College of Veterinary Medicine, University of Tennessee, Knoxville, TN 37996, USA (Souza, Cox) and The Raptor Center, College of Veterinary Medicine, University of Minnesota, St. Paul, MN 55108 USA (Redig).

Aspergillosis is a common fungal infection diagnosed in both wild and pet birds. Although effective antifungal medications are available, treatment of aspergillosis can require months of medication administration, which entails handling, and thus stress, one or more times per day. This study examined the delivery of the antifungal drugs itraconazole, voriconazole, and terbinafine to Japanese quail via impregnated implant. Implants were manufactured to contain 0.5 mg (n = 3 for each drug), 3 mg (n = 3 for each drug), 8 mg (n = 3 for each drug), or 24 mg (n = 6 for each drug) of either itraconazole, voriconazole, or terbinafine and were administered subcutaneously over the dorsum and between the scapulae. A total of 7 birds with no implant were used as controls. Blood was collected from birds prior to and at 2, 7, 21, 42, and 56 days after implant placement. Plasma was analyzed by high-performance liquid chromatography for concentrations of itraconazole, voriconazole, or terbinafine, as appropriate. Over the course of the study, target terbinafine concentrations were achieved in some birds at various time points, but concentrations were not consistent within each individual bird or within groups. Itraconazole and voriconazole concentrations were also inconsistent and did not reach target plasma concentrations. Currently, the implant examined in this study cannot be recommended for treatment of aspergillosis in avian species.

The Use of Alfaxalone in Quaker Parrots *(Myiopsitta monachus)*

Michelle C. Whitehead, BSc, DVM,
Sharman M. Hoppes, DVM, Dipl ABVP (Avian),
Jennifer L. Perkins, LVT,
Mauricio A. Lepiz, DVM, MS, Dipl ACVA

Session #030

Affiliation: From the Department of Small Animal Clinical Sciences, College of Veterinary Medicine & Biomedical Sciences, Texas A&M University, College Station, TX 77843-4474, USA.

Pet birds commonly require sedation and/or anesthesia to perform even simple diagnostic or therapeutic procedures such as venipuncture, radiographs, and beak and nail trims. The objective of this study was to evaluate the clinical safety and efficacy of alfaxalone (Alfaxan). Due to observed hyperexcitability with alfaxalone as the sole agent, premedication with midazolam was instituted. Ten adult quaker parrots (*Myiopsitta monachus*) were randomly assigned to three groups: 1) low dose alfaxalone 10 mg/kg (LD), 2) high dose alfaxalone 25 mg/kg (HD), and 3) alfaxalone 10 mg/kg with midazolam (1 mg/kg) premedication (AM), administered intramuscularly. Anesthetic monitoring recorded induction time, sedation quality, duration of action and vital parameters. All protocols achieved sedation however, muscle tremors and hyperexcitation was variable. The LD group had a significantly longer mean ± SD induction time (13.5 ± 4.5 min) as compared to the HD (6.0 ± 1.3 min, $P = 0.006$) and AM (6.5 ± 2.9 min, $P = 0.016$) groups, while recovery time was significantly longer in the HD group (86.2 ± 13.4 min) than the LD group (44.4 ±10.8 min, $P = 0.001$). Midazolam premedication resulted in reduction of both muscle tremors and hyperexcitation associated with alfaxalone administration, but the recovery time was significantly longer (103.5 ± 15.1 min, $P < 0.001$) than the LD group. Alfaxalone as a sole agent resulted in heavy sedation along with muscle tremors and hyperexcitation, which was attenuated by premedication with midazolam. Further investigation is warranted to characterize the role of alfaxalone and premedications in quaker parrots.

Impact of Site on Mesenchymal Stem Cell Yield in Chickens *(Gallus gallus domesticus)*

Jennifer E. Graham, DVM, Dipl ABVP (Avian), Dipl ABVP (Exotic Companion Mammal), Dipl ACZM,
Dawn Meola, BS,
Darleen Lessard, MS,
Andrew M. Hoffman, DVM, DVSc, Dipl ACVIM

Session #035

Affiliation: From the Department of Clinical Sciences, Cummings School of Veterinary Medicine, Tufts University, 200 Westboro Road, North Grafton, MA, 01536 (Graham, Meola, Hoffman); Department of Quantitative Health Sciences, University of Massachusetts Medical School, 368 Plantation Street, Worcester, MA, 01605 (Lessard).

Bone marrow mesenchymal stem-stromal cells (BM-MSCs, heretofore 'MSCs') are used in human and veterinary medicine to treat a variety of conditions including diseases of tendons, ligaments, or cartilage/joint injuries. MSCs can be induced to differentiate into adipocytes, chondrocytes, tenocytes, and osteocytes. Although MSCs have been isolated from chicken bone marrow successfully, described protocols thus far involve sacrificing the animals, removing limbs, and transecting entire bones to obtain marrow. A viable way of obtaining MSCs in avian species that does not involve euthanasia is needed to apply this technology to benefit all avian species, including those that may be threatened or endangered. Bone marrow aspirates were taken from the proximal tibiotarsus and sternum (keel) of five 60-day-old chickens in an attempt to harvest MSCs. Each bird also had cells flushed from the bone marrow cavity of the harvested contralateral tibiotarsus as has been described in previous studies to serve as a baseline 'control' for MSC yield. Results show that colony forming unit (CFU) numbers obtained by tibiotarsal aspiration were significantly greater for passage 1 than from keel aspirate or tibiotarsal flushing. Seeding density also had an impact on CFU numbers with higher CFU yield found with higher plate seeding density at passage 1 and 3.

References

1. Fortier L, Travis A. Stem cells in veterinary medicine. *Stem Cell Res Ther*. 2011;2(9):1-6.

2. Khatri M, O'Brien TD, Sharma JM. Isolation and differentiation of chicken mesenchymal stem cells from bone marrow. *Stem Cells Dev*. 2009;18(10):1485-92.

3. Bai C, Hou L, Ma Y, Chen L, Zhang M, Guan W. Isolation and characterization of mesenchymal stem cells from chicken bone marrow. *Cell Tissue Bank*. 2013;14(3):437-51.

Median Toxic Dose of Intravenous Bupivacaine in Isoflurane-Anesthetized Ross-708 Broiler Chickens

Peter M. Midgley-DiGeronimo, DVM,
Max Rinaldi, DVM,
Anderson da Cunha, DVM, Dipl ACVA,
Javier G. Nevarez, DVM, PhD, Dipl ECZM (Herpetology),
Joao Brandao, LVM,
Rhett Stout, DVM, Dipl ACLAM,
Thomas N. Tully, Jr., DVM, MS, Dipl ABVP (Avian), Dipl ECZM (Avian)

Session #050

Affiliation: From the Department of Veterinary Clinical Sciences, School of Veterinary Medicine, Louisiana State University, Baton Rouge, LA 70803, USA.

The objective of this study was to determine the median toxic dose of intravascular (IV) bupivacaine associated with 50% probability of causing clinically relevant cardiovascular effects in chickens anesthetized with isoflurane. After induction of anesthesia with isoflurane and placement of monitoring equipment, including invasive blood pressure and ECG, the 13 Ross-708 broiler chickens (*Gallus gallus domesticus*), weighing 1.7-2.75 kg, received bupivacaine IV over 2 minutes. Using an up-and-down design the median toxic dose was determined. For the up-and-down design, each animal received a variable dose of bupivacaine based on the response of the previous animal. Dose was decreased when heart rate (HR) and/or mean arterial blood pressure (MAP) changed less than or equal to 30% after IV bupivacaine administration. Dose was increased when HR or MAP changed less than 30% after bupivacaine administration. The TD_{50} was defined as the dose that would cause 30% or greater variation in HR or MAP in 50% of the population studied. Based on the findings of this study, the intravenous TD_{50} of bupivacaine in isoflurane anesthetized chickens was determined to be 1.94 mg/kg.[1] There currently are no published effective bupivacaine doses in birds. These results suggest that 1.94 to 1.96 mg/kg of bupivacaine (IV) are associated with a 50% probability of a clinical significant change in blood pressure in isoflurane-anesthetized chickens. Using cardiovascular safe doses of local anesthetic agents opens a wide range of potential investigations of therapeutic applications for avian species.[2,3]

References

1. Lipnick RL, Cotruvo JA, Hill RN, et al. Comparison of the up-and-down, conventional LD_{50}, and fixed-dose acute toxicity procedures. *Food Chem Toxicol.* 1995;33:223-231.

2. Lierz M, Korbel R. Anesthesia and analgesia in birds. *J Exot Pet Med.* 2012;21:44-58.

3. Machin KL, Livingston A. Plasma bupivacaine levels in mallard ducks (*Anas plathyncos*) following a single subcutaneous dose. *Proc Annu Conf Am Assoc Zoo Vet.* 2001;159-163.

Regional Limb Perfusion in Avian Species

Cameron M. Ratliff, DVM

Session #063

Affiliation: From the Department of Veterinary Clinical Sciences, Lloyd Veterinary Medical Center, Iowa State University, College of Veterinary Medicine, Ames, IA 50011 USA.

Abstract: Regional limb perfusion, previously unreported in avian species, is described in a chicken that presented for progressive swelling and lameness of the right foot for several months duration. Radiographs of the right foot demonstrated soft tissue swelling and periosteal reaction of metatarsal and phalangeal bones. Complete blood count showed an inflammatory leukogram. Empirical systemic antibiotic and anti-inflammatory therapy was started but the bird had not improved at recheck. A tourniquet was placed at the femoral-tibiotarsal joint of the right leg and regional limb perfusion using amikacin and flunixin meglumine was performed after intravenous catheterization of the medial metatarsal vein. The rooster had improved activity level with decrease in lameness and measurable decrease in swelling of the right foot at follow-up 15 days later. Regional limb perfusion using intravenous antibiotics and non-steroidal anti-inflammatories is a viable treatment modality in avian species for suspected distal limb infection and cellulitis. Fluid support should be provided if using nephrotoxic drugs and uric acid monitored.

Introduction

Regional limb perfusion (RLP) is a technique that is useful in treating distal limb infections as it allows for a higher local concentration of medications within deeper tissues. This technique is used routinely in equine medicine,[1-3] but has also been reported in various other species including rabbits,[4,5] elephants,[6] wallaby and kudu.[7] Indications for RLP range from osteomyelitis to cellulitis, septic arthritis, tenosynovitis, and sole abscesses. Amikacin is a commonly used antibiotic in RLPs, with concurrent administration of anti-inflammatories documented as well.[8] RLP has not been reported in an avian species. Aside from lack of familiarity with a technique that is more commonplace in large animal medicine, RLP may be perceived as difficult to perform. As venous access is necessary for RLP, it becomes technically difficult in smaller animals, however, it is feasible in birds as long as the medial metatarsal vein can be accessed for injection. The author has performed this technique in a bird as small as a 150 gram Eastern screech owl and as large as a 4.5 kg American White pelican. The technique has been used for various morbidities involving the pelvic limb, however, provided proper occlusion of vessels proximally, this technique could reasonably be considered for distal wing lesions through catheterization of the basilic vein. This technique is described in detail in a domestic chicken.

Clinical Report

A 2-year-old male Rhode Island red rooster (*Gallus gallus domesticus*) was presented for chronic swelling of the right foot of several months duration. The rooster had been housed with 12 other hens in a backyard coop and was fed a diet of commercial layer ration pellets. On examination, the chicken showed marked lameness of the right foot, which was severely swollen and warm to the touch. No external lesions were seen on any aspect of either feet. The rooster's body condition was adequate judging by keel musculature and body weight was 3.8 kg. No other abnormalities were noted on physical exam.

Radiographs showed soft tissue swelling at the right distal tarsometatarsus and digits, with a smoothly marginated periosteal reaction primarily effecting the phalanges of digit II. Radiographic findings were more consistent with pododermatitis as opposed to osteomyelitis. A complete blood count and serum biochemistries were run, showing leukocytosis (39.6 X 10^3 cells; reference interval, 4.0-10.3 X 10^3 cells) with mature heterophilia and monocytosis, and increased total protein (6.8 g/dl; reference interval, 2.7-4.3 g/dl). All other blood work parameters were within normal limits. A fine needle aspirate was taken from the plantar aspect of the right foot for cytology, which showed marked heterophilic inflammation and no etiologic agents seen. On the basis of an inflammatory leukogram, the rooster was treated empirically with clindamycin (50 mg/kg PO q12h for 14 days; clindamycin HCl 75 mg, Lannett Company, Philadelphia, PA, USA) and meloxicam (2 mg/kg PO q24h for 14 days; meloxicam 15 mg, Zydus Pharmaceuticals, Pennington, NJ, USA).

At the follow-up appointment one month later, there was no improvement to the swelling of the right foot. The owner reported continued decreased activity and ambulation. Regional intravenous perfusion of antibiotics and anti-inflammatories into the right distal pelvic limb was elected by the owner after presented as an alternative treatment option.

Regional limb perfusion

Blood was collected for measurement of uric acid prior to therapy. External temperatures of the feet were taken using a digital infrared thermometer (Reptitemp®, Zoo Med Laboratories, San Luis Obispo, CA, USA) at multiple locations, with the highest value being recorded for both feet (left foot = 38.3°C [101°F], right foot = 41.9°C [107.5°F]). A tourniquet was placed just proximal to the femoral-tibiotarsal joint of the right leg and the right medial metatarsal vein was catheterized using a 23 gauge butterfly catheter (Fig 1). Amikacin (5.3 mg/kg IV once; amikacin 50 mg/ml, Diamondback Drugs, Scottsdale, AZ, USA) was injected intravenously through the butterfly catheter. The catheter was then flushed with 1 ml of heparinized saline before injection with flunixin meglumine (1.3 mg/kg IV once; Prevail 50 mg/ml, MWI Veterinary Supply, Meridian, ID, USA). After injections were completed, a timer was set for 15 minutes to mark when the tourniquet was to be removed. Medial-lateral and dorsal-plantar dimensions of the feet were measured with digital calipers (left foot = 25.1 X 20.98 mm, right foot = 38.67 X 37.73 mm; Neiko® 01407A, Neiko Tools USA, China). One-hundred-twenty milliliters of warm balanced electrolyte solution (Plasma-Lyte A, Abbott Laboratories, North Chicago, IL) was delivered subcutaneously, and an additional 60 ml of warm tap water was gavaged orally into the crop. The tourniquet was left in place for 15 minutes total before removal. No further treatments or environmental changes were initiated and the rooster was to be reevaluated in 2 weeks.

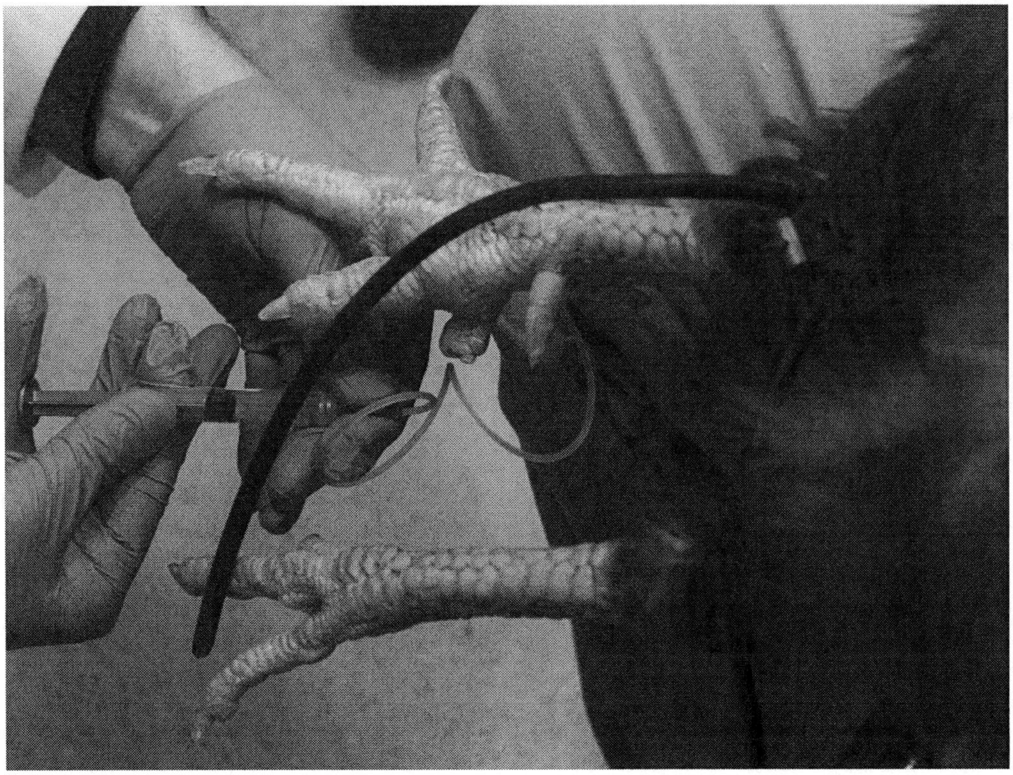

Figure 1. A Rhode Island red rooster with a severely swollen right foot having tourniquet placed and medial metatarsal vein catheterized for regional limb perfusion.

The rooster was seen for reevaluation 15 days later. The owner reported dramatic improvement in the rooster's mobility and noticeable decrease in swelling. The patient had gained 170 g in body weight from 15 days prior. Blood was collected for CBC and serum biochemistry from the right medial metatarsal vein. Total white blood cell count had decreased from 39.6 X 10^3 cells pre-treatment to 34.1 X 10^3 cells, and total protein decreased from 6.8 to 6.0 g/dl after treatment. Surface temperature of both feet (left foot = 35.6°C [96°F], right foot = 38.6°C [101.5°F]) were decreased compared to previous values. Dimensions of the feet were measured (left foot = 22.41 X 17.44 mm, right foot = 30.2 X 28.16 mm) and compared to previous values for the same location.

Discussion

Historically, the presence of a renal portal system in avian species has led to cautious use of potentially nephrotoxic drugs such as aminoglycosides and non-steroidal anti-inflammatories, particularly precluding administration of such drugs in the caudal half of the body where drugs are subject to first-pass through the kidneys.[9] However, there is ongoing debate as to the significance of the presence of a renal portal system, especially in euhydrated animals.[10,11] Fluid support at daily maintenance volume was provided to ensure adequate hydration status and increase blood volume to minimize the impact of the renal portal system and the use of potentially nephrotoxic drugs.

Both amikacin and flunixin meglumine are considered nephrotoxic in avian species, but some species are more sensitive to adverse renal effects. Although less toxic than gentamicin owing to its short half-life, amikacin like other aminoglycosides can still accumulate in the kidneys and cause renal damage.[12] Even when administered together, the nephrotoxic effect of amikacin and flunixin meglumine may be mitigated by instituting fluid therapy and decreasing the dosing frequency. It is important to note that amikacin and flunixin meglumine administration should be separated by complete flushing of the catheter fluid line, as the two drugs precipitate when combined.

Uric acid was measured as an indicator of renal function. Initial uric acid measured immediately before RLP was 4.5 mg/dl (reference interval, 3-11 mg/dl) and was measured again at 15-day recheck after single RLP treatment at 2.0 mg/dl. Despite serum uric acid levels decreasing post-RLP, severe renal compromise must be present to significantly elevate uric acid levels. Normal serum uric acid does not exclude histopathologic changes in the kidneys.[10] Future studies aimed at establishing specific dosages for RPL should pair serial uric acid measurements with renal histopathology in lieu of a more sensitive biomarker for avian renal function. Theoretically, RPL should have less of an impact on the kidneys than systemic administration because it allows for a lower dose of aminoglycosides and non-steroidal anti-inflammatories while still achieving a higher tissue concentration at the target site. Future pharmacokinetic and safety data need to be accumulated to derive proper dosages for RLP, whether in chickens or other avian species.

Despite finding no etiologic agent on cytology, empirical treatment with an oral antibiotic and non-steroidal anti-inflammatory was initiated, and no clinical improvement was seen. Because of the possibility of an infection affecting deep tissues and synovial structures, RPL was chosen to achieve high drug concentrations in these poorly perfused areas.[13] Sampling of the deeper tissues of the right foot for bacterial culture and susceptibility was not performed prior to treatment with RPL for fear of damaging synovial structures or introducing bacteria into a currently non-septic joint. Furthermore, cultures are often negative and negative results would not eliminate the possibility of infection.[14] *Staphylococcus* spp. bacteria are commonly implicated in cases of arthritis and tenosynovitis in poultry, however, this is difficult to sample and culture reliably. *Mycoplasma synoviae*, another organism that can cause infection and swelling of distal limbs in avian species, is difficult to culture without the use of special growth media.

In this study, RLP was performed to treat lameness and distal limb swelling of an undiagnosed etiology. After treatment, decreases were seen in total WBC count, total protein, surface temperature of the limbs, and swelling of the affected foot. RLP may potentially emerge as a viable option in treating distal limb morbidities, including difficult to treat cases of deep pododermatitis or osteomyelitis. When used in tandem with fluid support, RLP should be considered an option for unresolving distal limb infections in any avian species where venous access is available. Further controlled studies with larger sample size including different avian species are warranted.

Acknowledgements: The author would like to thank Harrison's Bird Foods, Inc. for funding this study and Dr. Bianca Zaffarano for clinical guidance and mentorship.

References

1. Rubio-Martinez LM, Cruz AM. Antimicrobial regional limb perfusion in horses. *J Am Vet Med Assoc*. 2006;228:706-712.

2. Rubio-Martinez LM, Elmas CR, Black B. Clinical use of antimicrobial regional limb perfusion in horses: 174 cases (1999-2009). *J Am Vet Med Assoc*. 2012;241:1650-1658.

3. Murphey ED, Santschi EM, Papich MG. Regional intravenous perfusion of the distal limb of horses with amikacin sulfate. *J Vet Pharmacol Ther*. 1999;22:68-71.

4. Finsterbusch A, Argaman M, Sacks T. Bone and joint perfusion with antibiotics in the treatment of experimental staphylococcal infection in rabbits. *J Bone Joint Surg*. 1970;54A:1424-1432.

5. Finsterbusch A, Weinburg H. Venous perfusion of the limb with antibiotics for osteomyelitis and other chronic infections. *J Bone Joint Surg*.1972;54A:1227-1234.

6. Ollivet-Courtois F, Lecu A, Yates RA, et al. Treatment of a sole abscess in an Asian elephant (*Elephas maximus*) using regional digital intravenous perfusion. *J Zoo Wildl Med*. 2003;34(3):292-295.

7. Fiorello CV, Beagley J, Citino SB. Antibiotic intravenous regional perfusion for successful resolution of distal limb infections: two cases. *J Zoo Wildl Med*. 2008;39(3):438-444.

8. Cimetti LJ, Merriam JG, D'Oench SN. How to perform intravenous regional limb perfusion using amikacin and DMSO. *Proc Annu Conv Am Assoc Equine Pract*. 2004;14-29.

9. Smith FM, West NH, Jones DR. Renal portal system. In: Whittow GC, ed. *Sturkie's Avian Physiology*. 5th ed. San Diego, CA: Academic Press;2000:175-176.

10. Burgos-Rodriguez, AG. Avian renal system: clinical implications. *Vet Clin North Am Exot Anim Pract*. 2010;13:393-411.

11. Frazier DL, Jones MP, Orosz SE. Pharmacokinetic considerations of the renal system in birds: part I. Anatomic and physiologic principles of allometric scaling. *J Avian Med Surg*. 1995:9(2):92-103.

12. Frazier DL, Jones MP, Orosz SE. Pharmacokinetic considerations of the renal system in birds: part II. Review of drugs excreted by renal pathways. *J Avian Med Surg*. 1995:9(2):104-121.

13. Lugo J, Gaughan EM. Septic arthritis, tenosynovitis, and infections of hoof structures. *Vet Clin North Am Equine Pract*. 2006;22:363-388.

14. Montgomery RD, Long IR, Milton JL, et al. Comparison of aerobic culturette, synovial membrane biopsy, and blood culture medium in detection of canine bacterial arthritis. *Vet Surg*. 1989;18(4):300-303.

Bone Density Determination Using µCT and a Novel Digital Radiograph System in Parrots

M Scott Echols, DVM, Dipl ABVP (Avian),
Boston Clark Terry,
Samer Merchant,
Ed Hsu, PhD,
Doug Folland, DVM, Dipl ABVP (Avian),
Crystal Wilcox, LVT,
Rebecca Bishop,
Alex Echenberg,
Fred Kromm,
Kate Lyngle-Cowand,
Michiko Wallace, CVT

Session #116

Affiliation: From Echols Veterinary Services, Salt Lake City, UT 84054, USA (Echols), Medical Center for Birds, Oakley, CA 94561, USA (Echols), Department of Bioengineering, University of Utah, Salt Lake City, UT 84112 USA (Terry, Merchant, Hsu), Parrish Creek Veterinary Clinic, Centerville, UT 84014, USA (Folland, Wilcox), Tracy Aviary, Salt Lake City, UT 84105 USA (Bishop, Echenberg, Kromm, Lyngle-Cowand, Wallace).

Bone density has been recognized as an indicator of physical activity, nutritional status and various diseases in humans and animals. Currently, computerized tomography (CT) and dual-energy x-ray absorptiometry (aka DXA scan), are the only non-invasive means for measuring bone mineral density. Such systems are typically only validated for adult humans, expensive and not readily available. Recently a novel method of standardizing tissue density scores from digital radiographs was introduced. To demonstrate its utility and develop an appropriate protocol for studying bone density, the methodology was applied to characterize 2 parrot species, sun conures (*Aratinga solstitialis,* n=11) and quaker parrots (*Myiopsitta monachus,* n=10), housed indoors and outdoors and living within Salt Lake, Davis and Weber counties of Utah. Each bird was physically and biochemically determined to be healthy prior to study. After calibration, intensities were quantified in fourteen matching regions of interest (ROIs). Variability of the intensity for each ROI in both species was quantified by its coefficient of variation (CV), also known as the relative standard deviation. Among the ROIs, the CV ranged from 2.3% to 12.1% for the sun conures, and 3.3% to 20.6% for the quaker parrots. The ROI-averaged intensities between the two species had a relatively high correlation coefficient, 0.89, whereas the correlation for the CV was weak, 0.08. Taken together, these findings suggest that, despite some variability that may reflect factors other than anatomy and site selection, there is intra- and inter-species consistency of the behavior of radiographic intensity used as a measure of bone density.

Comparison of Pentobarbital and Thoracic (cardiac) Compression to Euthanize Anesthetized Sparrows *(Passer domesticus)* and Starlings *(Sturnus vulgaris)*

Joanne Paul-Murphy, DVM, Dipl ACZM, Dipl ACAW,
Andrew Engilis, Jr.,
Peter Pascoe, BVSc, Dipl ACVA,
D. Colette Williams, PhD,
Kate Gustavsen, PhD, DVM,
Tracy L.Drazenovich, DVM,
M. Kevin Keel, DVM, PhD, Dipl ACVP,
Tamsen Polley, MS,
Irene E. Engilis

Session #140

Affiliation: From the Departments of Veterinary Medicine and Epidemiology (Paul-Murphy, Gustavsen, Drazenovich, Polley), Surgical and Radiological Sciences, (Pascoe) and Pathology, Microbiology, and Immunology (Keel), and The William R. Pritchard Veterinary Medical Teaching Hospital, School of Veterinary Medicine (Williams) and Department of Wildlife, Fish and Conservation Biology (Engilis, Engilis), University of California-Davis, Davis, CA 95616, USA.

Evaluation of euthanasia methods for birds is limited, and the available peer-reviewed reports primarily address euthanasia and slaughter methods for commercial poultry.[1] In ornithological field research, euthanasia of birds is done for procurement of tissue or museum and teaching collections or when a bird becomes injured during collection, such that it has a low probability of survival.[2] Euthanasia of wild birds has traditionally used thoracic compression which places digital pressure directly over the heart, therefore rapid cardiac compression may be the more accurate terminology. The current study compared thoracic compression (TC) and intraosseous pentobarbital (IP) on time to circulatory arrest and isoelectric electroencephalogram (EEG) in anesthetized birds.

Wild-caught sparrows (n = 12) and starlings (n = 12) were randomly divided to receive TC (6 sparrows and 6 starlings) or IP (6 sparrows and 6 starlings). Birds were anesthetized and a Doppler placed over the carotid artery. Electrodes were used to record the EEG. Time to cessation of carotid pulse and irreversible isoelectric EEG were measured as indicators of death. Necropsies were performed on all birds. The times to isoelectric EEG were not significantly different for the two methods, in sparrows and starlings. The times to cessation of pulse were shorter for TC compared to IP for both species. Grossly visible coelomic, pericardial, or perihepatic hemorrhage was evident in several of the birds euthanatized by TC.

In conclusion, both methods produced rapid EEG electrical silence and cessation of circulation, suggesting that TC may be an efficient method to euthanize small birds.

Acknowledgments: The authors thank Barry R. Tharp, MD for assistance with EEG interpretation.

References

1. Greenacre CB, Euthanasia In: Speer B, ed. *Current Therapy in Avian Medicine and Surgery*. St. Louis, MO: Elsevier Health Sciences;2015:714-718.

2. Fair JM, Paul E, Jones J, eds. *Guidelines to the Use of Wild Birds in Research*. Washington, DC: Ornithological Council; 2010.

Corneal Thickness in Raptors Measured Using Optical Coherence Tomography

Rüdiger Korbel, Prof., Dr. med. vet., Dr. Habil., Dipl ECZM (Avian), and
Regina Sigg, Dr. med. vet.

Session #223

Affiliation: From the Clinic for Birds, Reptiles, Amphibians and Ornamental Fish, Ludwig-Maximilians-University of Munich, Sonnenstrasse 18, D – 85764 Oberschleissheim, Germany.

Introduction

Knowledge of avian corneal thickness is indispensable for corneal surgery (i.e., perforating lesions,[1] cataract surgery, suturing techniques[1]) and disease-related conditions (i.e., keratoconus, „pop eye syndrome,"[2,3,4] conditional- or light-induced glaucoma[5,6]). Data on corneal thickness are lacking. To compare central (CCT) and peripheral avian corneal thickness (PCT) in diurnal and nocturnal raptors and poultry species (chickens, turkeys) we established normal values for ocular surgery and evaluation of light-related conditions, in comparison with pathologic conditions. Optical coherence tomography (OCT) is a highly sensitive interferometric, image-based technique that measures light reflected from ocular tissue layers. It is used in human ophthalmology and is one of three basic approaches in addition to diffraction and diffuse optical tomography since the early 1980s. Existing information on using OCT in birds is limited to raptors[1] and chickens.[2] Practically, OCT allows in vivo examination of the anterior and posterior eye segments with most detailed 2D and 3D sections through the various ocular layers such as retina and choroid at almost histologic resolution (Fig 1). Using a spectral-domain high-definition optical OCT unit (Spectralis OCT Plus, Heidelberg Engineering Inc., Germany) we investigated the eyes of various turkey breeds as a part of the routine ophthalmologic procedure with evaluation of standard reference values for the anterior eye segment including corneal thickness, lens and filtration angle parameters and for the posterior eye segment including the foveae and the different retinal layer parameters. Reference values were established in healthy eyes to serve as future reference values.

General advances in OCT include contact-free and non-invasive operation. The possibility for continuous and repeated investigation of the same individual over time, and therefore for progressive studies of development of clinical disorders, is of major value as shown in earlier studies[3-7] and avian bornavirus-related ocular conditions. The unit used is equipped with an "active eye tracking" function allowing investigation of specific fundus areas even after head or eye movements. OCT examinations may generally be performed without sedation; however, general anesthesia might be necessary in single individuals prone to stress. Evaluation of the images shows that OCT in the avian eye includes the advantages of high-depth and transversal resolution and a depth resolution that is decoupled from transverse resolution as well as high probing depth in scattering media. For specific use in poultry ophthalmology with patients easily stressed and delicate structures, the contact-free and noninvasive operation are of major value as is the opportunity to create various function-dependent image-contrasting methods.

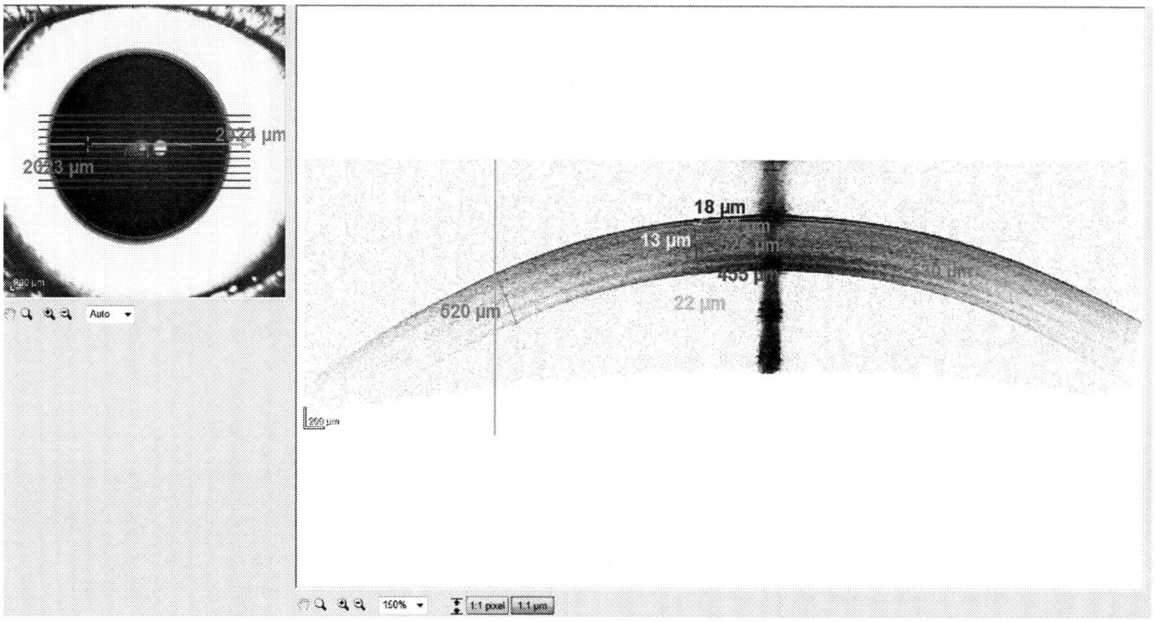

Figure 1. Optical coherence tomography image demonstrating the various layers oft the healthy cornea (OD) of a common buzzard (*Buteo buteo*, Gmel. 1789).

Material and Methods

CCT and PCT were retrospectively measured using fast domain fourier OCT (Spectralis OCT Plus, Heidelberg Engin. Inc.) in diurnal, nocturnal raptor and poultry species. The influence of age, sex and absolute magnitude of corneal thickness also were analyzed. Analysis was performed by multivariate linear and regression estimation equations plots.

Results

OCT showed significant differences in CCT and PCT and interspecific differences with lower values in nocturnal versus diurnal raptors.

Conclusions

OCT, CCT and PCT measurements are reproducible The influence of age on PCT and disease-borne conditions requires further study. Standard reference values measured will serve as a basis for future clinical approaches, disease related and evaluation of light induced corneal conditions.

Various avian-specific capacities, like flying, are closely related to fully functional vision underlining the importance of avian ophthalmology. In addition, ocular lesions in birds are an expression of systemic disorders more than in mammals and therefore represent an important diagnostic criterion.

References

1. Ruggeri M, Major JC Jr, McKeown C, et al. Retinal structure of birds of prey revealed by ultra-high resolution spectral-domain optical coherence tomography. *Invest Ophthalmol Vis Sci.* 2010;51:5789-5795.

2. Moayed AA, Hariri S, Song ES, et al. In vivo volumetric imaging of chicken retina with ultrahigh-resolution spectral domain optical coherence tomography. *Biomed Opt Express.* 2011;2:1268-1274.

3. Korbel R, Barwig S, Rinder M. Vergleichende Untersuchungen zur Ophthalmoskopie, digitalen Scanning Ophthalmoskopie (SDO) und optischen Kohärenztomographie (OCT) unter besonderer Berücksichtigung Bornavirus-assoziierter ophthalmologischer Befunde am Vogelauge. *Proc 17th DVG-Meeting Vogelkrankh, München.* 2012;43-46.

4. Korbel R., Schulze C, Velasco Gallego ME. Optical coherence tomography in the avian eye. *Proc Annu Conf Assoc Avian Vet.* 2013.

5. Korbel, R. Velasco Gallego ML, Schulze C, Rinder M. Avian bornavirus infections in parrots: characterization of ophthalmological lesions using optical coherence tomography (OCT). *Proc Int Conf Avian Herpetol Exot Mammal Med.* 2013;357-358.

6. Velasco Gallego, M. E. Imaging of physiological retinal structures in various raptor species using optical coherence tomography (OCT). Vet. Diss. 2015.

7. Korbel R. Investigations in the turkey eye using optical cohaerence tomography (OCT) as a basis to evaluate light induced alterations in the turkey eye. Proc. 7th Hafez Int Symp Turkey Dis. 2013.

Section 5

AAV Wildlife and Conservation

Glenn Olsen, DVM, PhD
Moderator

A Clinical Update on Chlamydial Infections in Raptors

Brittany A. Seibert,
Branson W. Ritchie, DVM, MS, PhD, Dipl ABVP (Avian), Dipl ECZM (Avian),
Kevin Keel, DVM, MS, PhD,
Michelle G. Hawkins, VMD, Dipl ABVP (Avian)

Session #164

Affiliation: From Masters Student (Seibert), Departments of Medicine & Epidemiology (Hawkins) and Pathology, Microbiology and Immunology (Keel), School of Veterinary Medicine, University of California Davis, Davis, CA 95616, USA, Infectious Disease Laboratory (Ritchie), University of Georgia, Athens, GA 30602, USA.

Chlamydia psittaci is a reportable, zoonotic bacterium with health implications for birds and humans. Due to its zoonotic potential, veterinary staff should be updated on the prevalence and clinical presentations of *Chlamydia* spp. infections in raptors. A retrospective study evaluating raptors testing positive for Chlamydial infections was conducted through the UC Davis Veterinary Medical Teaching Hospital medical records from 2004 –2016. A confirmed or probable case was based on the National Association of Public Health Veterinarians criteria for *C. psittaci*.[1] Specific *C. psittaci* diagnostic testing included real-time PCR (RT-PCR) from a combination mucosal swab, elementary body agglutination testing, indirect fluorescent antibody testing, immunofluorescent antibody staining for *Chlamydia* spp. antigen, and isolation of the organism. Signalment (sex, age, seasonality and species) and clinicopathologic results from positive cases were analyzed. A Chi-squared test was used to statistically evaluate the data, and significance was defined as $p \leq 0.05$. Forty eight red-tailed hawks (RTHA; *Buteo jamaicensis*) and 1 Swainson's hawk (*B. swainsoni*) were identified as confirmed or probable cases. A 22.9% prevalence was identified from the total number of raptors tested for *C. psittaci* and 1.9% prevalence was identified from the total number of all raptors admitted to the VMTH. There was a significant difference in prevalence between age ($p<0.005$), seasonality ($p<0.001$) and species ($p<0.001$). Tissues most commonly identified as positive were lung, liver, air sac, heart and kidney. Genetic sequencing of tissue samples from two RTHA with conflicting IFA/EBA results suggests that these infections may be caused by a novel Chlamydial-like organism.

References

1. Smith KA, Campbell CT, Murphy J, Stobierski MG, Tengelsen LA. 2011. Compendium of measures to control *Chlamydophila psittaci* infection among humans (psittacosis) and pet birds (avian chlamydiosis), 2010 National Association of State Public Health Veterinarians (NASPHV). *J Exotic Pet Med.* 20:32-45.

Prevalence and Associated Hematologic Changes of *Cyathostoma microspiculum* in Double-Crested Cormorants *(Phalacrocorax auritus)*

Stephen Royka, BS,
Molly Lien, DVM,
Missy Fox, CVT,
Heather Barron, DVM, Dipl ABVP (Avian)

Session #219

Affiliation: CROW (Clinic for the Rehabilitation of Wildlife), 3883 Sanibel Captiva Road, Sanibel, FL 33957, USA.

The prevalence and hematologic parameters associated with infection of tracheal worms in wild seabirds in North America have not yet been described. *Cyathostoma* species mature in the lungs and air sacs before moving into the trachea, causing respiratory disease and sometimes death. Double-crested cormorants (*Phalacrocorax auritus*) admitted to CROW (Clinic for the Rehabilitation of Wildlife) in Sanibel, Florida have been found to be occasionally parasitized by *Cyathostoma* species through endoscopy. In the current study, the prevalence of *Cyathostoma* in all admitted cormorants was examined, once worms were speciated via DNA analysis. Upon admission, hemograms were performed to determine packed cell volume, total protein, WBC count and a differential. Fecal examinations were also performed followed by sedated tracheoscopy to determine worm burden. Pearson correlation coefficients were used to evaluate the relationships between tracheal worm burdens (with or without underlying brevetoxicosis) and hematologic parameters. A total of 56 cormorants were examined; 48 diagnosed with presumed brevetoxicosis and 10 found to be infected with *Cyathostoma*. Prevalence of *Cyathostoma microspiculum* in immature cormorants was found to be 11.2%; similar to the 13.7% prevalence in immature great cormorants from north-eastern Poland. Tracheal worm infection correlated with lower total protein values (2.3 g/dl vs. 3.6 g/dl; $P < 0.05$); which may be a useful indicator to either perform tracheoscopy or initiate prophylactic treatment in resource limited facilities. Infection was not found to correlate to any CBC changes, nor was fecal examination found to be a reliable indicator for diagnosis. This is the first study quantifying tracheal worm burden in double-crested cormorants in North America, as well as correlating any potential clinical or hematologic signs.

Evaluation of a Food Web Toxicity After an Invasive Species Eradication Project

Julia B. Ponder, DVM, MPH

Session #309

Affiliation: From The Raptor Center, College of Veterinary Medicine, University of Minnesota, 1920 Fitch Avenue, St. Paul, MN 55108, USA.

Invasive rodents are a leading cause of extinctions on islands worldwide.[1] Eradication of invasive rodents using second-generation anticoagulant rodenticides is an increasingly used conservation strategy to protect biodiversity on islands.[2] These non-species-specific toxicants pose potential risk to native and endemic species. A priori risk assessments are used to predict non-target risk and exposure, and to inform strategies to mitigate potential impacts. While assessments are a reflection of best available knowledge, uncertainty exists in predicting all possible outcomes and can lead to unexpected negative impacts.[3]

After a recently completed eradication program, an unanticipated persistent presence of rodenticide in the food web was documented and associated with impact on an endemic species.[4] Mortality from rodenticide toxicosis was demonstrated in two species of raptors, with one carcass found fresh 773 days post-baiting. Monitoring of an endemic lizard population demonstrated broad population-level exposure to brodifacoum with liver residues persisting for up to 850 days after aerial application of brodifacoum bait to the island. No population level mortality of the lizards was observed.

References

1. Towns DR, Atkinson IAE, Daugherty CH. Have the harmful effects of introduced rats on islands been exaggerated? *Biol Invasions.* 2006;8:863-891.

2. Howald G, Donlan CJ, Galvan JP, et al. Invasive rodent eradication on islands. *Conserv Biol.* 2007;21: 1258–1268.

3. Hoare JM, Hare KM. The impact of brodifacoum on non-target wildlife: gaps in knowledge. *N Z J Ecol.* 2006;157-167.

4. Nicholls H. The 18-km^2 Rat Trap. *Nature.* 2013;497:306-308.

Surgical Repair of Keel Lesions and Lacerations in Aquatic Birds

Rebecca S. Duerr, DVM, MPVM, PhD

Session #205

Affiliation: From International Bird Rescue, Los Angeles and San Francisco Bay Centers, PO Box 2171, Long Beach, CA 90801, USA.

Abstract: Loons, grebes, diving ducks, and other aquatic bird species commonly present with keel lesions or wounds. Keel lesions may develop while birds remain stranded on hard environmental surfaces or may develop during care if birds are not able to be housed in water. Surgical repair is often necessary but the technique used is crucial to rapid re-attainment of waterproof plumage. A general skin closure technique using tightly-spaced, snug, horizontal mattress sutures works well to allow rapid waterproofing restoration in any body area, and may be used as the final closure of keel lesion repair. Keel lesions vary in severity and depth of affected tissues. Debridement of the keel itself is necessary for healing of lesions that expose the keel or that have necrotic skin adhered to the keel. Advancement grafts are needed to close degloving (scalping) injuries to the scalp, and skin autografts work well to allow reacquisition of intact integument and waterproof plumage despite areas of missing skin on limbs. With the surgical techniques described, these injuries can be effectively repaired to have minimal effect on survival to release, even during mass stranding events.

Introduction

Certain aquatic bird species are prone to the development of keel lesions during captive care. Species anatomically unable to stand well when out of water such as loons, grebes, and diving ducks are most vulnerable, and emaciated birds that lack normal pectoral muscle mass are at especially high risk. Lesions may also develop at the keel due to traumatic injuries, chemical burns due to plumage contamination with caustic materials, or as ischemic injuries due to abnormal weight-bearing when housed out of water.[1,2] In the past, development of these lesions was often an indicator for euthanasia, especially during mass stranding events such as oil spills; but with early identification of individuals with keel lesions it is possible to integrate surgical treatment into an overall treatment plan, whether for individuals or groups of birds.

Lacerations and punctures in aquatic birds may be due to predator attack (e.g., dogs, raccoons, etc.), fishing hook or line injuries, boat or propeller strike, impact trauma, or other causes. Many of these species migrate at night and sometimes in foul weather, and are known to suffer impact trauma from collisions with objects such as cars/roads/buildings, or with conspecifics.[3,4] These injuries almost always adversely affect the bird's waterproofing, which is a life-threatening problem of its own, especially for ocean-going species such as loons, grebes, bay and marine ducks, alcids and tubenoses. Even extremely small wounds may ooze enough plasma to cause the bird to be wet at the skin even if the surface of contour feathers appears dry. Loss of waterproofing increases energetic demands due to impaired thermoregulation or changes in the bird's buoyancy, and/or may cause the bird to focus on preening to the exclusion of all other activities such as foraging. Many of these birds strand in an emaciated plane of nutrition or quickly lose body mass to critically low levels prior to capture[5]. The purpose of this paper is to familiarize veterinarians with techniques that are successfully and routinely used to manage keel injuries and lacerations in aquatic wild birds.

Pre-Operative Considerations

At International Bird Rescue we categorize keel lesions as grade 1 (mild: simple superficial linear intact necrotic skin, not adhered to underlying bone), grade 2 (moderate: thickened, scab-like full-thickness linear skin necrosis, not adhered to underlying bone but some may ooze plasma), grade 3 (severe: linear necrotic skin adhered to underlying bone, or an open ulcerated area of skin, may be dry or ooze plasma), and grade 4 (open lesion where necrotic skin and/or muscle acutely ripped open). Grade 1 and some grade 2 lesions may slough off as linear scabs and not cause clinical problems. More complicated grade 2 and grade 3 lesions are highly likely to split open and require surgery. The goal with these hazardous lesions is to identify them early in care, then allow the bird to fatten up and become waterproof while planning surgery to occur prior to the lesion acutely splitting open to become grade 4. This time frame usually allows a few days to a week, contingent on circumstances. Grade 4 lesions require immediate repair. In these lesions, the necrotic skin often remains adhered to the keel and the pectoral muscle may rip parallel to the keel or become avulsed from the keel.

In these species, keel and laceration repairs must take the need for waterproof plumage into account. Most veterinary clinics are not able to manage waterproofing issues in diving birds; hence a close working relationship with a permitted wildlife rehabilitator with appropriate aquatic bird housing and experience is needed. Birds that are emaciated are poor candidates for surgery and do not heal well; correct the bird's nutritional condition prior to surgery or consider euthanasia. Prior to surgery, birds should be well-hydrated and well-nourished, and as waterproof as possible.

Surgical Procedures

Pre-operative skin disinfectants such as povidone iodine or chlorhexidine solutions should be used conservatively to minimize plumage contamination, rinse with 70% ethanol. Do not use surgical scrub or other products with detergents on the bird. Beware chilling the patient with disinfectants or flushing fluids. Pluck downy feathers and a small number of contour feathers to create a 3 mm margin of bare skin along the lesion margins. Minimize plucking to what is needed to allow for the expected closure; sufficient feathers must remain for the bird to become waterproof without waiting for additional feather growth. Consider plucking broken contour feathers along the edges of the injury, as they are often a detriment to waterproofing. Removing an area of bald skin may be the best option in some cases, as a bird with a larger but more easily waterproofed surgical site has a better prognosis than a bird with a smaller incision adjacent to a bald or difficult to waterproof area. Managing a grebe or loon in dry housing for long enough to allow feather growth is extremely difficult in the best of circumstances.

The author's suture of choice is 4-0 Monocryl™ on a taper needle and 3M Micropore™ paper tape is used to hold feathers away from the surgical site. Avoid tapes or adhesive surgical drapes that will pull feathers when removed or that will leave a sticky residue. The author prefers ophthalmic instruments such as Castroviejo forceps for handling skin for these repairs, and blunt-tipped, curved tissue scissors such as Metzenbaums for trimming skin margins and bluntly dissecting along fascial planes.

Closed keel repair

Grade 2 keel lesions with intact skin that do not start to separate at the live/dead interface after site preparation can be repaired without opening the skin. Place closely-spaced, snug, horizontal mattress pattern sutures that evert the necrotic skin into a protuberant lip of tissue; this will later slough off as a scab. Space sutures 2-3 mm apart and 2-3 mm between halves of the mattress stitch, 1-2 mm into healthy skin from the necrotic skin border. Pull all sutures tight.

Open keel repair, simple

Excise linear necrotic skin by sharp dissection. Separate skin and pectoral muscle layers by blunt dissection at least several cm from the wound edge along the fascial plane that separates the skin/fat from the muscle. Do not separate subcutaneous fat from the skin or separate layers of muscle from each other. Chronic lesions with fibrin and granulation tissue present may obscure the separation of the tissue planes; if so, the fascial plane may be easiest to find after narrowly freshening the lesion edge. Avoid trimming the skin too close to the plucked feather edge or feathers will become problematic during closure. With practice, the optimal amount of feather clearance will become apparent. If the live/dead interface is irregularly shaped or not distinct, trim skin as needed to obtain a healthy skin margin and undermine farther from the closure to provide adequate tension reduction.

Once the skin has been prepared and the skin undermined away from the muscle, if any subcutaneous fat or fascia is present that can be sutured as a healthy layer between the keel and the skin, suture it across the keel with simple continuous pattern. Any of the subcutis that does not look healthy should be removed. Close the skin as follows in General Skin Closure.

General skin closure

To close the skin, space everting horizontal mattress sutures 2-3 mm apart and 2-3 mm between halves of each mattress stitch, 1-2 mm into healthy skin from the skin border. Knots should be ~5 mm apart. Pull all sutures tight--do not leave any looseness for swelling, as the skin of these species does not swell like mammal skin. Loose sutures will remain loose and allow weeping of plasma into the plumage and the area will not be able to become waterproof. Do not include any subcutaneous fat in the everted lip of the closure as it may weep lipids into the plumage, also adversely impacting the bird's waterproofing. In birds with thick subcutaneous fat, run the needle 2 mm from the skin edge to the visceral side of the cut edge so that subcutaneous fat is not everted in the lip, just the cut skin edge. To avoid mismatches and puckers when fitting the two sides together, it is helpful to start in the middle and suture ellipse-shaped surgical wounds in the middle, then in quarters, then close each quadrant. This can also help control bleeding by applying general tension to the tissue. Chronic lesions may have sometimes large blood vessels in the visceral skin and fat that may require ligation or cautery. It is not necessary to close subcutaneous spaces in additional layers of closure, these species do not tend to form seromas. Extend the sutures at each end at least half a stitch past the end of the suture line, do not leave even a tiny area of unsutured lip at the ends. Inspect the closure and add additional horizontal mattress sutures if needed to close any gaps. Other suture patterns such simple continuous, simple interrupted, or cruciate typically result in dehiscence or wound weeping that compromises waterproofing.

Open keel repair, complicated

The keel itself is involved in these lesions. Use this method if the skin was adhered to the bone, even if the keel edge appears grossly healthy. There is a very high risk of dehiscence if this is not done, as the ventral edge is often devitalized. After undermining the skin/fat from the pectoral muscle, use rongeurs to trim the ventral edge of the keel. If surgical instruments such as rongeurs are not available, debride the bone edge aggressively by scraping with a scalpel blade until ample capillary bleeding occurs. If any of the bone edge looks lytic or disorganized, trim it deeper. Palpate the new keel edge and remove any sharp areas, leaving a smooth edge. Trimming the bone margin releases the ventral pectoral fascia from its attachment at the keel. Close this fascia across the keel with a simple continuous pattern, pulling the pectoral muscle into becoming a soft tissue pad between keel and skin. Close the skin as with the previously described open closure. Collecting culture samples or placing antibiotic impregnated beads or polymers prior to closure may be helpful in osteomyelitis cases. Osteomyelitis of the keel does not preclude a successful outcome but is certainly an additional factor when considering euthanasia for a severe lesion.

In grade 4 lesions, necrotic skin may remain adhered to the keel while the pectoral muscle rips open parallel to the keel during flapping (e.g. during capture from a pool). These lesions are often highly contaminated but heal well if repaired promptly. Thorough lavage of the lesion after surgical debridement is prudent. Gently appose tissues into as normal position as possible prior to closure. Torn pectoral muscle is often friable and may not hold sutures without tearing, but the associated fascia may hold sutures well, even if torn.

Lacerations

In many diving birds with dense plumage, especially grebes, wounds on the body (as opposed to limbs) are often difficult to find and may be much larger than the initial appearance. Necrotic skin often continues to support normal-looking contour feathers, and one must follow the wound margins before being able to know the true size of the injury. Be aware normal loon and grebe skin is often greyish or bluish in color. Many surprisingly large injuries can be effectively repaired, but the surgeon must be able to close the repair without creating a large gap in feathers. Skin lacerations should be closed using the same technique as the general skin closure described above. Chronic weeping wounds may require excision of granulating skin to enable the bird to become waterproof quickly. Leaving a granulating wound to epithelialize on its own should only be done in rare circumstances. Second intention healing in these species is generally not advisable; primary or delayed primary closure holds the best outcome. Areas of skin where the bird is wet-to-skin may cause the bird to exacerbate the problem by overpreening the feathers at the edge of the affected area to the point of breaking contour feathers, thus creating a patch of broken feathers.

If wounds expose bone such as spinous processes, skull, or synsacrum, debride bone prior to closure by scraping with a #15 or other appropriate scalpel blade until capillary bleeding is seen. Undermine the skin on all sides of the lesion, and freshen the skin edges prior to closing as described in the general skin closure. Chronically exposed bone may be eroded and show a raised lip of periosteal response beneath a granulated skin margin. The granulated skin should be separated from the bone, undermined, edges freshened, and closed. Scalp repairs heal well with simple continuous or cruciate closures if available skin does not allow an everted suture line without excess tension. Advancing skin cranially from the neck works well to repair scalping injuries. There is some risk of graft necrosis with excessive skin tension, but advancement grafts onto the scalp typically heal very well. In ducks, second intention granulation and subsequent scarring of scalping injuries often leads to distortion of the eyelids as the tissue around the exposed skull contracts; hence, allowing scalp defects to heal by second intention in duck species is not advised lest the bird's eyelid apertures become displaced from the globes. Scalping injuries are rare in grebes and loons.

Neck lacerations (or degloving) heal very well when skin is undermined cranially and caudally from the defect, edges freshened, and sutured to bring the cranial and caudal edges together. Neck wounds commonly affect 50-100% of the circumference of the neck, thus closing these wounds longitudinally is rarely an option. Before closing neck wounds, always make sure the esophagus has not been perforated, as it may require repair as well. Highly contaminated neck wounds with copious necrotic material and debris often heal surprisingly well. Prognosis is good if the bird is able to move its neck normally, even if the bird had a fish hook embedded in vertebrae, or extensive muscle damage, or maggots. Debride surgically and close as described in the general skin closure section. If the original wound was not able to be thoroughly debrided (e.g., excess bleeding, unclear live/dead borders, anesthetic complications leading to an abbreviated procedure), these species will often form a granuloma that may be removed in a second procedure a few weeks later when the tissue is healthier.

Most leg or wing lacerations may be closed routinely as described in general skin closure above. Lacerations with deeper necrotic tissue or debris must be thoroughly debrided prior to closure. Performing surgical repair while the wing is held in extension aids in avoiding dehiscence due to excessive tension when the bird later flaps.

Autografts

Areas of missing skin on the patagium, elsewhere on the wing, or on the scaly skin of the legs may benefit from placement of a graft. Grafts collected from elsewhere on an individual's body may be used to repair a defect[6], but in loons, grebes, or diving ducks grafts are best collected above the waterline at dorsolateral flank or caudodorsal neck. Grafts may or may not survive transplant but if the graft dies it still forms an excellent bandage while allowing the bird to have intact integument. Dead grafts allow granulation underneath, and typically exfoliate like a scab. The author has had best results with using skin that has subcutaneous fat attached when placed at the defect. Porcine submucosa is another option for graft material[7] but may be prohibitively expensive for use on wildlife, while autografts only cost surgical time.

Thoroughly debride the site needing the graft, and ascertain the shape and size of the graft that is needed. Surgically prepare the site. Collect the graft from the dorsolateral flank or caudodorsal neck above the bird's waterline, set the graft aside in sterile saline while closing the collection site. If the graft is large enough, secure it to the wound edges with horizontal mattress sutures. 5-0 Monocryl on a small taper needle works well for the author, and magnification may be helpful as the tissue collected for the graft may have seemed large enough prior to collection but may seem similar to a ball of mucoid tissue paper afterwards. Suture the graft at each end of the long dimension first, then place sutures to secure it at each side, thus securing it at four (or more) points. A continuous circumferential horizontal mattress pattern may then be used to secure the rest of the graft edges. If the graft is likely going to be too snug for horizontal mattress attachment, use simple interrupted sutures to anchor it in place and a simple continuous pattern circumferentially. Appropriate moisture to support graft survival can be obtained by a simple covering of Tegaderm. Hydrogel acts as a plumage contaminant and self-adhesive hydrocolloid dressings may cause failure of the graft when removed. Never use oily ointments on these species. Grafts are generally not necessary on the torso of these species. Grafts on legs that will be held in the water do not generally need any covering. Surgical glue can be helpful to anchor the edges of grafts.

Post-Operative Care

Birds with substantial feather contamination may require washing in the days following surgery to remove dried plasma and blood if the bird has difficulty waterproofing the surgical site on its own. Avoid dry-docking birds post-operatively more than overnight. House these species in net-bottom caging when out of water, preferably. Dry housing often leads to new problems with other body areas such as development of toe, hock, or carpal lesions. Post-surgical recovery time for these species is typically 24-48 hours for restoration of waterproofing, with suture removal in 10-14 days. Despite use of absorbable sutures, it is prudent to remove sutures, as it allows inspection of all areas of each lesion. A small proportion of birds may show a deep plug of debris at a suture that may indicate need for a second surgery.

References

1. Mazet JAK, Newman SH, Gilardi KVK, et al. Advances in oiled bird emergency medicine and management. *J Avian Med Surg*. 2002;16(2):146–149.

2. Russell M, Holcomb J, Berkner A. 30-years of oiled wildlife response statistics: effects of oil on wildlife. *Proc Conf Int Effects Oil Wildl*. 2003;1-18.

3. Jehl Jr J. Conspecific collisions can precipitate mortality in migrating eared grebes. *Wilson Bull.* 1998;110(3):409-411.

4. Jehl Jr J.. Observations on the fall migration of eared grebes, based on evidence from a mass downing in Utah. *Condor*. 1998;95:470-473.

5. Duerr RS, Klasing KC. Tissue component and organ mass changes associated with declines in body mass in three seabird species received for rehabilitation in California. *Marine Ornithol*. 2015;43(1): 11-18.

6. Rupley AE. *Manual of Avian Practice.* Philadelphia, PA: WB Saunders Company;1997.

7. Hernandez-Divers SJ, Hernandez-Divers SM. Xenogeneic grafts using porcine small intestinal submucosa in the repair of skin defects in 4 birds. *J Avian Med Surg*. 2003;17(4):224-234.

A Protocol to Establish the Health Status of Young White Storks *(Ciconia ciconia)*

Jan Hooimeijer, DVM, CPBC

Session #231

Affiliation: From Consultancy Practice for Birds, Galgenkampsweg 4 7942 HD Meppel, The Netherlands.

Abstract: In the Netherlands the reintroduction project for the white stork was established in 1969 because the species became extinct, mainly as the result of developments in modern agriculture. Looking at the number of breeding pairs in the Netherlands after 37 years, this conservation project was a success. During the past 10 years, indications that the health status of the white stork population is declining were based on breeding results and general condition of young white storks at the age they receive legbands. Data on breeding results in a local population of white storks in the Netherlands are compared with data from two white stork populations in Poland. The number of young white storks that fledge in the Netherlands is declining. Based on the quality of the feathers, it is possible to ascertain the general condition of young white storks. Reasons for the decline, as well as a protocol to establish the health status of young white storks in the field and the status of the breeding population as a whole, will be discussed.

Introduction

Since the reintroduction program for white storks in the Netherlands started in 1969, the number of breeding pairs in the Netherlands has grown from zero to 850-900. Looking at these numbers, the reintroduction program has been a great success.[1] During the past 10 years indicators suggest reason for concern when considering breeding results and general condition of young white storks at the age of 5-6 weeks when they receive a legband.

The main habitat of white storks is wet grasslands that are shared with other meadow birds like the Eurasian oystercatcher (*Haematopus ostralegus*), Northern lapwing (*Vanellus vanellus*), black-tailed godwit (*Limosa limosa*), Eurasian curlew (*Numenius arquata*), and common redshank (*Tringa totanus*). Data from other meadow birds indicate that present population declines are caused by a decrease in reproduction, not in adult survival, and that reproductive output is presently too low to compensate for adult mortality. The main reason is lack of available insects to feed the youngsters as a result of modern agricultural management.[2] When determining the status of a bird population it is important to monitor breeding results, including the number of youngsters that fledge.

Breeding Results

Data were collected on breeding results from a local population of white storks in the Northeast part of the Netherlands. Data are compared with the data from two breeding populations in Poland. The reason for using the data of these population in Poland is that Poland has had a stable population of white storks for many years without the need for conservation efforts.[3,4]

Feather Quality

Physiologic measures, including hematologic and immunologic variables often are used in avian medicine to determine the health status of an individual.[5,6] Phenotypic traits that are the result of conditions during the development can have long-term fitness consequences. The reliability of physiologic measures to determine an individual's potential is more limited.

In American kestrels, fault bars and broken feathers occur more often in females than in males. Feather problems observed in this study are not only associated with lower body condition but also lower return rates in years following initial capture.[7]

Any stress, including starvation, poor nutrition, environmental changes, dehydration, infection, chilling, overheating, restraint and emotional anxiety in the home of a pet bird can cause development of fault bars in birds.[8,9] Fault bars form on growing feathers and are the result of abnormally formed or missing barbules. They vary in severity such that some vanes show light lines while others have conspicuous holes.[10]

Biologists have given considerable attention to attributes of individuals that could be used as predictors of fitness. In particular, reliable measures of an individual's quality as manifested by physical condition, reproductive performance and attractiveness to mates have been sought.[11]

In Upland buzzards a correlation has been found in the number of fault bars related to the availability of food. When there is an abundance of available food the number of fault bars is lower, making ptilochronology an effective method to evaluate the relative nutritional condition of nestlings.[12]

A survey revealed that female American kestrels with many fault bars had lower survival probabilities. Males and females with many fault bars are less likely to breed.[13] The idea is that flying birds have a natural selection mechanism to reduce fault bar loads on feathers with a high strength requirement during flight. Data on the growth of feathers of nestlings white storks (*Ciconia ciconia*) revealed a consistent allocation of more, and more intense, fault bars in the innermost over the outermost wing feathers, especially the scapulars, which are the feathers with the highest number of fault bars.[14] The same has been shown in the aerial forager barn swallow *Hirundo rustica* where fault bars were more prevalent in tertials than in primaries or secondaries. This finding shows that birds have an evolved adaptive mechanism for reducing fault bar loads on the feathers with the highest function during flight.[15]

Over the past 32 years, working as avian veterinarian in daily practice, it has been always fascinating to see the feathers change in color and sheen within 4-7 days after changing the diet of a bird that is showing typical signs of malnutrition. At the same time, feathers that were showing fault bars were replaced during the moult by feathers without fault bars. Evaluating the quality of a feather is a valuable tool in practice to look at past and current fitness. The fault bars provide information about the bird's condition during the development of the feathers. Evaluating the color and sheen provides information about the bird's current condition.

Discussion

Breeding results from the period 1987-2014 show a decline in hatchability, a decline in number of youngsters that are still alive at the age of 5-6 weeks, and a decline in the number of youngsters that fledge per nest (Table 1). Comparison with the breeding results in two white stork populations in Poland show that during the past years the number of youngsters that fledge per nest is twice the number of fledged youngsters in the Dutch population (Table 2). Considering the decline in number of fledged white storks there is reason for concern regarding the long-term for the same reasons that there are concerns about the status of other meadow birds. Evaluating the quality of feathers can be considered important to collect data that provides information about the bird's general condition. A new protocol is developed, combining the evaluation of fault bars with the evaluation of the color and sheen of the feathers, enabling evaluation of the current status of the white stork population in the Netherlands.

Table 1. Breeding results of the white stork population in the Lokkerij Foundation region.

Item	1987-1990	1991-1994	1995-1998	1999-2002	2003-2006	2007-2010	2011-2014
Pairs of birds	51	141	287	449	629	834	432
Eggs	214	490	975	1612	2349	3086	1616
Eggs per nest	4.1	3.4	3.3	3.5	3.7	3.7	3.7
Youngsters hatched	162	373	756	1172	1566	1986	982
Percentage	79.75	75	77.75	72.50	66	64.25	60.76
Legbanded birds	138	305	581	948	1202	1528	636
Percentage	64.48	62.24	59.58	58.80	58.80	49.51	39.35
Fledged birds	126	303	554	923	1160	1485	607
Percentage	58.87	61.83	56.82	57.25	49.38	48.12	37.56
Fledged birds per nest	2.4	2.1	1.93	2.05	1.84	1.78	1.4

Table 2. Average number of fledged young white storks in two Polish populations.

Year	Fledged birds per nest
2010	2.9
2011	2.7
2012	2.6
2013	2.8

References

1. Hooimeijer J. Twenty five years of white stork (Ciconia ciconia) reintroduction in the Netherlands. Proc Annu Conf Assoc Avian Vet. 2006.

2. Roodbergen M, Werf B, Hotker H. Revealing the contributions of reproduction and survival to the Europe-wide decline in meadow birds; review and meta-analysis. J Ornithol. 2012;153(1):53-74.

3. Koopman FH, Jaarverslag. Stichting De Lokkerij. 2014.

4. Groen C, Geerts E. Profielwerkstuk Veerkrachtig Onderzoek. CSG Dingstede *Meppel.* 2016.

5. Samour J. Diagnostic value of haematology. In: Harrison GJ, Lightfoot T. *Clinical Avian Medicine.* Palm Beach, FL: Spix Publishing, Inc.;2006.

6. Kendal H. Diagnostic value of biochemistry. In: Harrison GJ, Lightfoot T. *Clinical Avian Medicine.* Palm Beach, FL: Spix Publishing, Inc.;2006.

7. Dawson RD, Bortolotti GR, Murza GL. Sex-dependent frequency and consequences of natural handicaps in American kestrels. *J Avian Biol.* 2001;32:351-357.

8. Doolen M. Case presentation. *Compend Contin Educ Pract Vet.* 1999;1:312.

9. Doneley B, Harrison GJ, Lightfoot T. Physical examination In: Harrison GJ, Lightfoot T. *Clinical Avian Medicine.* Palm Beach, FL: Spix Publishing, Inc.;2006.

10. Murphy ME, Miller BT, King JR. A structural comparison of fault bars with feathers defects known to be nutritionally induced. *Can J Zool.* 1989;67:1311-1317.

11. Buchanan KL. Stress and the evolution of condition dependent signals. *Trends Ecol Evol,* 2000;15:156-160.

12. Sundev G, Yosef R, Birazana O. Brandt's vole density affects nutritional condition of Upland buzzards (*Buteo hemilasius*) on the Mongolian grassland steppe. *Ornis Fennica.* 2009;86:131-139.

13. Bortlotti GR,.Dawson RD, Murza GL. Stress during feather development predicts fitness potential. *J Anim Ecol.* 2002;71:333-342.

14. Jovani RM, Blas J. Adaptive allocation of stress-induced deformities on bird feathers. *J Evol Biol.* 2004;17:294-301.

15. Serrano D, Jovani R. Adaptive fault bar distribution in a long distance migratory, aerial forager passerine? *Biol J Linnean Soc.* 2005;85:455-461.

Propatagium Injuries and Surgical Management in Black Kites *(Milvus migrans govinda)*

Stefan Harsch, DVM

Session #037

Affiliation: From the Mississippi Wildlife Rehabilitation, Inc, 9865 Green River Road, Lake Cormorant, MS 38641 and Jivdaya Charitable Trust, Ahmedabad, India.

Abstract: Twenty black kites with comparable propatagial ligament lacerations were surgically repaired after injuries caused by glass-coated kite string. Seventeen animals healed sufficiently to be released and 3 animals died.

Introduction

The Indian city of Ahmedabad, Gujarat holds an International Kite Festival every January as part of celebrating Uttarayan (the end of winter and the arrival of spring). Festival participants use glass powder-coated string to try to cut opponents' strings and bring down their kites. Unfortunately thousands of birds are killed or injured every year as a result of this event in Ahmedabad, alone.[1] A local non-governmental organization (NGO), the Jivdaya Charitable Trust, admitted 3,147 injured birds during the month of January 2016, among them 441 black kites. The purpose of this case series is to report the success of surgical propatagial ligament repair in the black kite.

Materials and Methods

A total of 441 black kites were admitted in January 2016. Between January 14 and January 20, 2016, 51 birds were screened, and 20 met inclusion criteria in this prospective observational case series. The inclusion criteria were as follows: a freshly lacerated propatagial ligament severed in the center of the wing, with major blood vessels and nerves remaining intact.[2] Upon admission, all birds received an initial injection of 5% body weight of lactated Ringer's solution (LRS) subcutaneously, as well as intra-muscular injections of meloxicam (0.5 mg/kg) and enrofloxacin (15 mg/kg). Wings were plucked under isoflurane anesthesia, and wounds were flushed with sterile saline. Severed propatagial ligament ends were identified and approximated with a single interrupted 4-0 PDS suture. The propatagial skin was reinforced just above and below the ligament repair with a simple horizontal mattress suture, and the remaining dorsal and ventral aspects of the propatagium were approximated in a continuous horizontal mattress pattern. The suture line was covered with silver sulfadiazine ointment and a non-adherent pad. The wing was immobilized in a padded "figure of 8" bandage, with dressing and bandage changes every 3 days (awake), until the wound was sufficiently healed.[3] Dehisced wounds were debrided and re-sutured as described above. Patients were given at least 5 days of meloxicam and 10 days of enrofloxacin post-operatively. All patients were kept in an intensive care unit (in small individual cages) for 1-2 days, given 2% body weight LRS subcutaneously every 12 hours, and were tube fed Emeraid Carnivore (Emeraid Cornell, IL) twice daily. After this period, the birds were transferred to small group aviaries (20 animals) and hand fed cut-up beef twice a day. Upon final bandage removal, birds were transferred to an off-site facility (for conditioning in slightly larger aviaries) with free access to food.

Results

Twenty patients met inclusion criteria. All patients received identical surgical ligament repair with reinforced propatagial membrane sutures. Five patients experienced a wound dehiscence required debridement and re-suture. Of these, two died during dehiscence repair. One patient died on the first postoperative day due to unknown causes. Seventeen patients survived and were released, and while 10 animals achieved full flight, 7 animals experienced limited flight capacity. Of those, 2 patients had limited wing extension, and 3 more patients experienced significant wing droop (Table 1).

Table 1. Outcome of surgical propatagial ligament repair in black kites.

Outcome	Number of birds
Died	3 (including 2 dehiscences)
Released:	17
Full flight	10 (including 1 dehiscence)
Limited flight:	7
Wing droop	3
Limited wing extension	2

Discussion

Injuries to the propatagium with a severed propatagial ligament, have a guarded healing prognosis, since elastic fibers in the ligament do not hold sutures well.[4] Outcome is complicated by the very small diameter of this ligament in this species. Unlike that described in the literature,[5] the only practical way to approximate the ligament was to use a single interrupted suture through ligament and surrounding tissue on each end. In order to deflect tension over the ligament repair, a single horizontal mattress suture was needed to reinforce the repair, with dorsal and ventral aspects of the propatagial membrane closed in a normal fashion.

Post-operative complications included wound infection, dehiscence of sutured ligament/propatagial membrane, and tissue necrosis. Wound infections were cleansed and flushed with saline, and wound dehiscences were debrided and re-sutured. However, this resulted in decreased wing extension, in four cases.

Official NGO-reported results of these 20 patients were 17 birds released, with 3 deaths. These release numbers appear significantly greater than expected, compared to rehab facilities in the US. However, this can be explained by the unique situation and religious beliefs practiced in India. This particular NGO, run by the Jain community, does not practice euthanasia, a highly controversial topic in India. Of the 3 reported deaths, only one patient died of unknown causes the day after surgery. The other two deaths, in patients presenting with significant areas of tissue necrosis, did not survive the repair surgery and may have received more medication than wound have been predicted in the United States. Due to the Jain religious beliefs, no patients were subjected to euthanasia, as might have been anticipated, were they treated in the United States. Sufficient flight capability was difficult to assess prior to release, since adequate flight aviaries were not available. Once injuries were completely healed, birds were released, and only then was it possible to assess proper flight. Of the 17 birds that were released, only 10 fit the release criteria of U.S. standards such as the full recovery from the original injury and the ability to exhibit locomotive skills necessary for that species to survive (i.e., proper wing extension and flight).[6] Objective measurements of joint mobility like goniometry to predict successful release were not performed. The remaining 7 animals were able to fly short distances and to land on surrounding trees and structures, but showed significant wing droop. It is likely that these 7 patients would have been considered non-releasable to their native environment due to inability to hunt and would have been euthanized if presenting in the United States. However, Indian black kites live in large communal roosting places and are opportunistic eaters, even consuming carrion. These characteristics allow the NGO to release animals that would have slim chances of survival in the wild, as they are reliably provided with cattle carcasses from a nearby cow and buffalo sanctuary at the release site, a habitat unlikely to be found in the United States.

Conclusion

Indian black kite propatagial ligament lacerations can be successfully repaired and managed following injury caused by glass-coated kite strings during flight. Due to the small ligament diameter in this species, repair may only be successful with a single interrupted suture, reinforced with a single horizontal mattress suture in the overlying membrane. Wound dehiscence is not uncommon, but can be treated with debridement and re-sutured. Because of the unique environment found in northern India that allows these black kites to survive in communal colonies when they potential would not if released to the wild, and the fact that euthanasia is not an accepted practice, release rates are greater than might be predicted in other environments.

References

1. Sutaria P, Sutaria TV. Patagial tendon injury and its surgical management in peacocks (*Pavo cristatus*). *Intas Polivet*. 2012;13(2):257-258.

2. Brown RE, Baumel JJ, Klemm RD. Anatomy of the propatagium: the great horned owl (*Bubo virginianus*). *J Morphol*. 1994;219(2):205-224.

3. Burke HF, Swaim SF, Amalsadvala T. Review of wound management in raptors. *J Avian Med Surg*. 2002;16(3):180-191.

4. Olsen GH, Redig PT, Orosz SE. Limb dysfunction. In: *Manual of Avian Medicine*. St. Louis, MO: Mosby;2000:493-526.

5. Johnson AL, Hulse DA. Management of muscle and tendon injury or disease. In: Fossum TW, ed. *Small Animal Surgery*. 2nd ed. St. Louis, MO: Mosby;2002:1158-1164.

6. National Wildlife Rehabilitators Association. Minimum standards for release of wildlife following rehabilitation. In: Miller EA, ed. *Minimum Standards of Wildlife Rehabilitation*. 4th ed. St. Cloud, MN: National Wildlife Rehabilitators Association;2012:74-75.

Proventricular Dilatation Disease but Negative for Avian *Bornavirus* in Raptors: A Toucan and a Mousebird

H. L. Shivaprasad, BVSc, PhD, Dipl ACPV

Session #062

Affiliation: California Animal Health and Food Safety Laboratory System - Tulare branch, 18830 Road 112, Tulare, CA 93274 USA, School of Veterinary Medicine, University of California, Davis, CA 93274, USA.

Proventricular dilatation disease (PDD) is a disease common in psittacine birds, but also of various other species of birds including passerine birds (canaries, finches), geese, swans, and others.[1-7] The disease is characterized by regurgitation of food, passing of undigested seeds in the feces, neurologic signs, anorexia, weakness, loss of weight and sometimes sudden death. The pathology of PDD includes dilation of the proventriculus (70 % of the time in psittacine birds) and distention of the duodenum with lymphoplasmacytic inflammation in the central, peripheral and autonomic nervous systems, as well as, adrenalitis and myocarditis.[2] The etiology of PDD is avian *Bornavirus*.[7,8]

Recently the family Bornaviridae has been reclassified and avian bornaviruses have been put in to several groups based on biologic characteristics of the virus and their vertebrate hosts.[9] Avian bornaviruses have been classified as Psitticiform 1 *Bornavirus* (ABV genotypes 1-4 and 7), Passeriform 1 (ABV canary 1, 2, 3) and Passeriform 2 (ABV *Estrildid* finch-EF) *Bornavirus* and Waterbird 1 *Bornavirus* (ABV-CG).[9] A few other viruses identified in psittacine birds including Psitticiform 2 *Bornavirus* (ABV genotype 5) and from other birds, as well as, one in a snake remain unclassified.[9,10]

PDD was diagnosed in a 5-year-old peregrine falcon (*Falco peregrinus*), a golden eagle (*Aquila chrysaetos*), a red-tailed hawk (*Buteo jamaicensis*), a 5-year-old female speckled mousebird (*Coliiformes striatus*) and a 5-year-old male toucan (*Rhamphistidae* spp.) based on characteristic microscopic lesions. The lesions included ganglioneuritis of the myenteric ganglia of the gastrointestinal tract, adrenalitis, encephalitis and myocarditis. Most of these birds had various clinical signs ranging from weakness, ataxia, loss of weight (eagle, hawk and mousebird), sudden death (toucan) to odd behavior in the falcon. None of the birds had dilated proventriculus. Brains from these birds examined by PCR for ABV were negative using conventional primers. Immunohistochemistry (IHC) done on the brain of the falcon, eagle and the toucan also were negative for ABV.

Even though these birds had lesions typical of PDD, they were negative for ABV. This may infer that the conventional primers used for detecting ABV here do not work in these species suggesting that these birds may harbor a novel ABV. The other possibility is that the lesions in these birds are probably caused by an unknown virus and awaits further investigation.

Interestingly an adult captive screech owl (*Megascops* spp.) that died suddenly recently revealed myocarditis involving the Purkinje cells and myofibers in the heart by histopathology. Heart and brain were positive for ABV and negative for West Nile virus by IHC. There was no encephalitis. PCR testing on the brain and heart is pending.

References

1. Daoust P-Y, Julian RJ, Yason CV, Artsob H. Proventricular impaction associated with nonsuppurative encephalomyelitis and ganglioneuritis in two Canada geese. *J Wildl Dis*. 1991;27:513-517.

2. Shivaprasad HL, Barr BC, Woods LW, et al. Spectrum of lesions (pathology) in proventricular dilation syndrome. *Vet Pathol*. 1997;34:500.

3. Weissenbock H, Bakonyi T, Sekuli K, et al. Avian *Bornavirus* in psittacine birds from Europe and Australia with proventricular dilatation disease. Emerg Infect Dis. 2009;15:1453-1459.

4. Staeheli P, Rinder M, Kaspers B. Avian *Bornavirus* associated with fatal disease in psittacine birds. *J Virol*. 2010;84:6269–6275.

5. Hoppes S, Gray PL, Payne S, Shivaprasad HL, Tizard I. The isolation, pathogenesis, diagnosis, transmission, and control of avian *Bornavirus* and proventricular dilatation disease. *Vet Clin North Am Exot Anim Pract*. 2010;13::495-508.

6. Delnatte P, Nagy É, Ojkic D, Leishman D, et al. Avian *Bornavirus* in free-ranging waterfowl: prevalence of antibodies and cloacal shedding of viral RNA. *J Wildl Dis*. 2014:;50:512-523.

7. Honkavuori K, Shivaprasad HL, Williams B, Quon PL, et al. Novel *Borna* virus in psittacine birds with proventricular dilatation disease. Emerg Infect Dis. 2008;14:1883-1886.

8. Kistler A L, Gancz A, Clubb S, Skewes-cox P, et al. Recovery of divergent avian Bornaviruses from cases of proventricular dilatation disease: identification of candidate etiologic agent. *Virol J*, 2008;5:88.

9. Kuhn JH, Dürrwald R, Bào Y, et al. Taxonomic reorganization of the family Bornaviridae. *Arch Virol*, 2015;160:621-632.

10. Zimmermann V, Rinder M, Kaspers B, et al. Impact of antigenic diversity on laboratory diagnosis of avian *Bornavirus* infections in birds. J Vet Diagn Invest. 2014;26:769-777.

Generalized Tetanus in a Gyrfalcon *(Falco rusticolus)* with Pododermatitis

Hugues Beaufrère, Dr. med. vet., PhD, Dipl ACZM, Dipl ABVP (Avian),
Delphine Laniesse, Dr. med. vet.,
Paul Stickings, PhD,
Robert Tierney,
Thea Sesardic, PhD,
Durda Slavic, DVM, MS, PhD,
Nicole Compo, DVM,
Dale A. Smith, DVM, DVSc

Session #066

Affiliation: From the Health Sciences Centre (Beaufrère, Laniesse), the Department of Pathobiology (Smith, Laniesse, Compo), and the Animal Health Laboratory (Slavic), Ontario Veterinary College, University of Guelph, 50 Stone Road E, Guelph, N1G2W1, ON, Canada; and the National Institute for Biological Standards and Control (Stickings, Sesardic, Tierney), Potters Bar, Hertfordshire, EN6 3QG, United Kingdom.

A 2-year-old male gyrfalcon was presented for severe and generalized muscle spasticity and pododermatitis. The falcon had been treated for pododermatitis over the previous 4 months. Muscle rigidity and spasms involved the entire bird, but were more severe on the right leg. The bird was also tachypneic and hyperthermic at 45 °C. While the plantar pododermatitis lesions were healed, there was still a small abscess on the lateral aspect of the right foot. Clinical signs were consistent with tetanus. Several bacteria were isolated from the abscess including *Clostridium tetani*. The isolate was confirmed to be toxigenic by PCR. Attempts to detect tetanus toxin in the bird's plasma were unsuccessful.

The abscess was debrided. The gyrfalcon received equine tetanus antitoxin, intravenous metronidazole, methocarbamol, midazolam, a constant rate infusion of fentanyl, active cooling, and supportive care. Inhalant anesthesia with isoflurane was the only treatment that would lower the body temperature and reduce the clinical signs. The gyrfalcon died a few hours after admission.

The characteristic clinical signs and isolation of toxigenic *Clostridium tetani* from a wound are strong supportive evidence for a diagnosis of tetanus. This case constitutes the first reported natural occurrence of tetanus in an avian species. Further information is needed to determine whether gyrfalcons are more susceptible to tetanus than other avian species and whether pododermatitis lesions may be risk factors.

Trends in Plasma Corticosterone in Red-Tailed Hawks Presenting to a Wildlife Hospital

Nicki Rosenhagen, DVM,
Julia K. Whittington, DVM,
Michael Romero, MS, PhD,
Matt Allender, DVM, MS, PhD, Dipl ACZM

Session #079

Affiliation: From the University of Illinois College of Veterinary Medicine 2001 S. Lincoln Ave Urbana, IL 61802, USA (Rosenhagen, Whittington and Allender) and Cummings School of Veterinary Medicine at Tufts University 200 Westboro Rd, North Grafton, MA 01536, USA (Romero).

In avian species, corticosterone is the main glucocorticoid involved in the regulation of energy, immune reactions, and stress responses.[1] Corticosterone levels vary based on season, sex, light cycle and presence of stressors. [2-4] Acutely, increased corticosterone levels may convey adaptive fitness to an individual; however, chronic increases have been documented to have many negative effects.[5]

The first phase of this study aimed to identify the lag time from the onset of an acute stressor to the reflective change in plasma levels in captive red-tailed hawks (*Buteo jamaicensis*). Animals enrolled in the study were restrained for 30 minutes and blood was collected at 90 seconds, 3 minutes, 5 minutes, 10 minutes, 15 minutes and 30 minutes after capture. Plasma corticosterone levels were then measured using radioimmunoassay. We anticipate a significant increase in corticosterone levels in response to the acute stress of handling at approximately 3 minutes based on reports in other species.[6] Results are pending.

The second phase of this study evaluated circadian trends in hormone release in red-tailed hawks. Animals were housed in an isolation area with a controlled 12-hour light-dark cycle. Blood samples were collected in the morning and evening on two separate occasions. The corticosterone levels were evaluated as previously described. We anticipate detection of a distinct circadian rhythm based on reports in other species.[3,4,7] Results are pending.

We believe that focusing treatment of this species during periods of naturally lower corticosterone levels as reflected by circadian patterns will decrease captivity-associated-stress and improve clinical outcome.

References

1. Wingfield JC, Romero LM. Adrenocortical responses to stress and their modulation in free-living vertebrates. In: McEwen, BS and Goodman, HM, eds. *Handbook of Physiology*. New York, NY: Oxford University Press;2001:211-234.

2. Sockman KW, Schwabl H. Plasma corticosterone in nestling American kestrels: effects of age, handling stress, yolk androgens, and body condition. *Gen Comp Endocrinol*. 2001;122:205-212.

3. Dufty AMJ, Belthoff JR. Corticosterone and the stress response in young western screech-owls: effects of captivity, gender, and activity period. *Physiol Zool*. 1997;70:143-149.

4. Romero LM, Remage-Healey L. Daily and seasonal variation in response to stress in captive starlings (*Sturnus vulgaris*): corticosterone. *Gen Comp Endocrinol.* 2000;119(1):52-59.

5. Awerman JL, Romero LM. Chronic psychological stress alters body weight and blood chemistry in European starlings (*Sturnus vulgaris*). *Comp Biochem Physiol A Mol Integr Physiol.* 2010;156(1):136-142.

6. Romero LM, Reed JM. Collecting baseline corticosterone samples in the field: is under 3 min good enough? *Comp Biochem Physiol A Mol Integr Physiol.* 2005;140(1):73-79.

7. Romero LM, Rich EL. Photoperiodically-induced changes in hypothalamic–pituitary–adrenal axis sensitivity in captive house sparrows (*Passer domesticus*). *Comp Biochem Physiol A Mol Integr Physiol.* 2007;147(2):562-568.

Novel Falcon Adenovirus Strains

Marcie Logsdon, DVM,
Dharani Ajithdoss, BVSc, PhD, Dipl ACVP, Dipl ACVM,
Alisha Massa, DVM,
Jessica McCrea, DVM,
Nickol Finch, DVM,
Dan Bradway

Session #173

Affiliation: From the Washington State University College of Veterinary Medicine, 205 Ott Rd., Pullman WA 99164, USA, Washington Animal Disease Diagnostic Lab Pullman, WA 99164-7034, USA, Washington State University Department of Veterinary Microbiology and Pathology, Pullman WA 99164, USA.

In 2015 WSU investigated an outbreak of mortality among Aplomado falcon (*Falco femoralis*) chicks at two breeding facilities. Chicks from both facilities had possible indirect contact with peregrinoids and with each other. Post-mortem findings were consistent with adenoviral infection, which was later confirmed by polymerase chain reaction (PCR). DNA sequencing revealed the adenovirus strain to be nearly identical to that identified during the outbreak amongst Aplomado falcons at the Peregrine Fund breeding facility in 1996.[1,2]

Additional samples were collected in an attempt to identify the source of the infection. Both facilities fed coturnix quail purchased from a common supplier; pooled tissue samples from representative quail tested negative for adenovirus. Cloacal swabs were collected from multiple falcons on premises and tested for adenovirus by PCR. Nearly all samples tested positive for adenovirus. However, once sequenced none of the samples matched the virulent strain that appeared to have been responsible for the chick mortality. A novel strain was identified in a pair of apparently healthy adult aplomado falcons, and several variants of an additional strain were identified in gyrfalcons (*Falco rusticolus*), peregrine falcons (*Falco peregrinus*), peregrine falcon hybrids, aplomado falcons, and a taita/barbary (*Falco fasciinucha/Falco pelegrinoides*) hybrid. These results indicate that adenovirus prevalence may be higher than previously suspected and that there are numerous, potentially host-specific, strains present amongst falcons. These unique strains have unknown virulence outside of their host species and may be responsible for some of the perceived immunity observed in adult falcons.

References

1. Oaks JL, Schrenzel M, Rideout B, Sandfort C. Isolation and epidemiology of falcon adenovirus. *J Clin Microbiol.* 2005;43:3414-3420.

2. Schrenzel M, Oaks JL, Rotstein D, Maalour G, et al. Characterization of a new species of adenovirus in falcons. *J Clin Microbiol.* 2005;43:3402-3413.

San Francisco Bay Mystery Goo Spill: Medical Management During a Crisis

Rebecca S. Duerr, DVM, MPVM, PhD

Session #207

Affiliation: From International Bird Rescue, Los Angeles and San Francisco Bay Centers, PO Box 2171, Long Beach, CA 90801, USA.

After release of an unidentified, epoxy-like substance into San Francisco Bay during January 2015, International Bird Rescue's San Francisco Bay clinic received 313 live, plumage-contaminated aquatic birds for treatment, including 231 surf scoters, 30 horned grebes, 22 greater scaups, 13 buffleheads, 6 Western sandpipers, 4 dunlin, 4 common goldeneyes, 2 ruddy ducks, and 1 red-throated loon. Birds were kept in net-bottom caging when out of water. Fifty-four birds died or were euthanatized in the first 48 hours, largely due to extremely poor condition. After stabilization for 2-5 days, 251 birds were washed. The product was much harder to remove than oil, and 53 birds required additional washes. Once washed, birds were housed in outdoor pools sorted by species, sex, and handling frequency. A large proportion of birds had medical problems requiring attention: 229 had ischemic damage to plantar toe surfaces, 171 had hock lesions, 70 had carpal trauma, and 63 had keel lesions. Lesions were assessed throughout care with a rapid examination scheme ranking each typical lesion location (keel, elbows, hocks, toes, and carpi) on a 0-3 scale (none, mild, moderate, severe). Whole pools of birds with previously identified problems were assessed every 3-5 days for progression of lesions, and surgeries scheduled as time allowed, >100 procedures were performed. Although the proportion of birds surviving to release declined with increased lesion severity, 54% with hock lesions, 60% with toe lesions, 68% with carpal lesions, and 67% with keel lesions requiring surgery survived to release. Overall, 165 birds (52.7%) were released.

Section 6

AAV Master Classes Track 1

Emerging Diseases in Psittacine Birds

H. L. Shivaprasad, BVSc, PhD, Dipl ACPV

Session #317

Affiliation: From California Animal Health and Food Safety Laboratory System-Tulare branch, 18830 Road 112, Tulare CA 93274, School of Veterinary Medicine, University of California, Davis, CA 95616, USA.

Abstract: Emerging and reemerging diseases especially among infectious diseases in humans and animals including birds are a natural phenomenon. These diseases emerge for a various reasons such as mutations (genetic drift), recombination (genetic shift) and reassortment of genes which may be influenced by selection pressure, environment, management, etc. Some of the pathogens can be regional, occurring in a certain geographic location of the world or within the same country. It is also probable that certain pathogen may be present in a bird population but could not be identified because of lack of technology. A good example of this is discovery of avian *Bornavirus*, the causal agent of proventricular dilatation disease, which was only discovered in 2008 by the use of metagenomics even though the disease has been known since 1990's or earlier. This review will deal with certain diseases such as coxiellosis, other bacterial diseases due to *Yersinia pseudotuberculosis*, *Klebsiella pneumoniae*, *Aeromonas hydrophila*, an unidentified silver positive filamentous bacteria and others, as well as, protozoal diseases caused by *Sarcocystis calchasi* and *Trichomonas gallinae* and viral diseases caused by adenoviruses and a novel psittacine herpesvirus.

Coxiellosis

Coxiella burnetii, is an obligate intracellular gram-negative bacterium that causes Q fever in humans and abortions with placentitis and neonatal deaths in cattle, sheep and goats.[1] It is considered a zoonotic disease and the disease is acquired through inhalation of bacteria from infected placenta, bodily fluids and contaminated dust.[2] Ticks are considered to be important vector and reservoir for transmission of *C. burnetii.*

Coxiella spp. other than *C. burnetii* has been identified in many species of psittacines and a toucan associated with systemic disease. The psittacine birds included rosella parrots, lorikeets, eclectus parrot, African grey parrots, hawk-headed parrot, canary-winged parakeet, red-rumped parakeet, red-fronted parakeet, kakariki, Jardine's parrot, lovebird, Senegal parrot, and a budgerigar.[1-3] Age in these birds have ranged from six months to adults and included both males and females, Clinical signs ranged from weakness, lethargy with occasional neurologic signs to sudden death. Most common postmortem findings were severe splenomegaly and mild to moderate hepatomegaly with pale mottling. Histopathology included severe pleocellular or histiocytic splenitis, hepatitis, myocarditis, pneumonia, nephritis, and less commonly encephalitis, myositis, enteritis, adrenalitis, thyroiditis and neuritis. Intralesional bacteria were identified in the cytoplasm of macrophages and transmission electron microscopy revealed bacteria that were round (0.45 µm) and elongated (1.0 µm). Immunohistochemistry for *C. burnetii* were negative. PCR on the liver and 16S rRNA gene sequences revealed 96% to 97% similarity to *Coxiella,* a novel *Coxiella.*

Other Bacteria

Numerous cases of septicemia due to *Yersinia pseudotuberculosis* was diagnosed in various species of psittacines (10 macaws, 4 cockatiels, 2 cockatoos, 2 rosella parrots, 4 parakeets and 1 each in a conure, lorikeet, eclectus and a pionus parrot) over a 25-year period.[4] Most of the birds were from outdoor aviaries and had either nonspecific clinical signs or found dead. Significant gross lesions were enlarged liver with pale foci of necrosis and disseminated septicemia and *Y. pseudotuberculosis* was isolated from various organs. It is suspected that rodents were the source of bacteria.

Aeromonas hydrophila was isolated from the liver of an adult female African grey parrot and *Klebsiella pneumoniae* was isolated from a parrot that died after a brief period of illness. Livers were enlarged with pale foci of necrosis in both birds. The owner of the African grey parrot also kept snakes and frogs which were most likely the source of *A. hydrophila.* Other bacteria that caused septicemia in various species of psittacine birds include *Avibacterium gallinarum* in a green-winged macaw, *Ornithobacterium rhinotracheale* in a parrot, and *Pasteurella multocida* in a few psittacine birds that were either attacked by dogs or occasionally by cats.

A 30-month-old golden lory died after a prolonged period of illness characterized by ill thrift, ataxia, low calcium, bone problems and high WBC's. Necropsy and histopathology revealed enlarged liver with pale foci of necrosis and inflammation with numerous intralesional long segmented filamentous silver positive bacteria. Attempts to isolate the bacteria both aerobically and anaerobically were unsuccessful. To identify the bacteria DNA was extracted from the liver and 16S rRNA gene was amplified using universal eubacterial primers. Sequence of the PCR amplicon closely matched (99.9%) to *Geobacillus* spp. This bacterial species has not been previously described in birds.

Sarcocystis calchasi

Sarcocystis spp. is a ubiquitous protozoa that requires two hosts, with herbivores as intermediate hosts and carnivores as definitive hosts.[5,6] *Sarcocystis falcatula* uses various bird species such as Psittaciformes, Passeriformes, Columbiformes as their intermediate hosts and American opossum as a definitive host. *S. falcatula* infections are common in various species of psittacine birds but Old World psittacine birds are particularly susceptible resulting in sudden death due to pulmonary edema and congestion. Neurologic signs can also be observed in some birds due to encephalitis. Myocarditis, hepatitis and splenitis can also occur. If the birds survive, matures cysts will be present in the skeletal muscles. Diagnosis can be made by demonstrating characteristic sinuous schizonts in the lungs or zoites in other organs including the brain. Immunohistochemistry (IHC) can be used to confirm *S. falcatula.*

In the past several years, several cases of sarcocystosis was diagnosed in psittacine birds that were negative for *S. falcatula* and positive for *S. calchasi.*[7] The psittacine birds included prince parrots, rose-breasted cockatoos, slender-billed cockatoo and a long-billed cockatoo. The main clinical signs in these psittacine birds were neurologic signs with encephalitis and intralesional protozoa. Experimentally cockatiels also have been shown to be susceptible to *S. calchasi.* Pigeons are more susceptible to *S. calchasi* resulting in encephalitis and myositis. Northern goshawk (*Accipiter gentilis*) has been determined to be the definitive host for *S. calchasi.* A pigeon, a dove and a flamingo with encephalitis negative for *S. falcatula* by IHC was probably infected with *S. calchasi* but was not confirmed. It is probable that infection caused by *S. calchasi* is more common than realized and that other species of birds may be susceptible.

Trichomonosis in Budgerigars

*Trichomonas gallin*ae causes upper digestive tract disease, most commonly in pigeons, doves and raptors.[8] It also has been reported in psittacine birds, most commonly in budgerigars and other species of birds including chickens, various species of passerine birds including canaries and finches.[8,9] Trichomonosis in psittacine birds is rare in the United States.

Endemic trichmonosis was diagnosed in approximately 15 budgerigars over two years from an aviary that housed about 320 budgerigars, 74 cockatiels and 25 parrots of various species. Only budgies were affected with clinical signs of lethargy, weakness, diarrhea, pasted vents with green feces and occasionally regurgitation. Necropsy revealed pale yellow plaques and/or necrotic exudate in the oral cavity and upper esophagus most commonly and lower esophagus and crop less frequently. Histopathology revealed severe ulcerative stomatitis, esophagitis and ingluvitis associated with numerous trichomonads. Trichomonas was confirmed by IHC.

Adenovirus

Avian adenoviruses cause various diseases in poultry such as inclusion body hepatitis and hydropericardium syndrome in chickens, bronchitis in quail, hemorrhagic enteritis in turkeys and marble spleen disease in pheasants.[10] Adenovirus infections have also been described in psittacine birds and the clinical signs are non-specific. They range from anorexia, lethargy, and diarrhea, to occasional ocular discharge, respiratory distress and neurologic signs.[10,11] Gross lesions include enlarged liver with white foci, hemorrhage in the intestine and pale kidneys.

Retrospective search for adenovirus infections in psittacine birds in the California Animal Health and Food Safety Laboratory System data base over a 20-year period yielded about 90 cases. Adenovirus was confirmed by transmission electron microscopy in some of these cases. Most common adenovirus infections were nephritis associated with intranuclear inclusion bodies in the renal tubular epithelial cells primarily in budgerigars and cockatiels but also in a few other psittacine birds. Systemic infection characterized by severe hepatitis, splenitis and enteritis, sometimes with hemorrhage associated with intranuclear inclusion bodies was observed primarily in African grey parrots, but also in other species of psittacines. One unique and interesting syndrome of adenovirus in the endothelial cells of the blood vessels associated with hemorrhage and minimal inflammation in the brain was observed in several parrotlets. These parrotlets had a short course of neurologic signs and hemorrhagic diarrhea.

Novel Herpesvirus

Psittacine birds are susceptible to two psittacine herpesviruses.[10] These include Pacheco's disease, often a fatal rapidly spreading disease of psittacine birds caused by psittacine herpesvirus -1 (PsHV-1) and psittacine herpesvirus-1 (PsHV-2) associated with cloacal and internal papillomatosis in African grey parrots. Recently a psittacine herpesvirus-3 (PsHV-3) has been described in Bourke parrots primarily associated with respiratory disease and mortality.[12] The disease is characterized by respiratory signs due to sinusitis, rhinitis, tracheitis, airsacculitis, otitis media and bronchopneumonia associated with syncytia and intranuclear inclusion bodies similar to infectious laryngotracheitis (ILT) in chickens but is different from ILT virus. Other species of psittacines in which this novel PsHV-3 vrus has been described include eclectus parrots, Indian ring-neck parakeets, yellow naped Amazon, and a cockatiel.[12,13]

References

1. Shivaprasad HL, Cadenas MB, Diab SS, et al. *Coxiella*-like infection in psittacines and a toucan. *Avian Dis.* 2008;52:426-432.

2. Woc-Colburn AM, Garner MM, Bradway D, et al. Fatal coxiellosis in Swainson's blue mountain rainbow lorikeets (*Trichoglossus haematodius moluccanus*). *Vet Pathol.* 2008;45:247-254.

3. Vapniarsky N, Barr BC, Murphy B. Systemic *Coxiella*-like infection with myocarditis and hepatitis in an eclectus parrot (*Ecluctus roratus*). *Vet Pathol.* 2012;49:717-722.

4. Stoute S, Cooper GL, Bickford AA, et al. *Yersinia pseudotuberculosis* in Eurasian collared doves (*Streptopelia decaocto*) and retrospective study of avian yersiniosis at the California Animal Health and Food Safety Laboratory System (1990-2015). *Avian Dis.* 2016;60:82-86.

5. Olias P, Maier K, Wuenschmann A, et al. *Sarcocystis calchasi* has an expanded host range and induces disease in cockatiels (*Nymphicus hollandicus*) and North American rock pigeons (*Columbia livia f. dom.*). *Vet Parasitol.* 2014;200:59-65.

6. Olias P, Gruber AD, Heydorn A, et al. A novel *Sarcocystis*-associated encephalitis and myositis in racing pigeons. *Avian Pathol.* 2009;38:121-128.

7. Rimoldi G, Speer B, Wellehan JFX. An outbreak of *Sarcocystis calchasi* encephalitis in multiple psittacine species within an enclosed zoological aviary. *J Vet Diagn Invest.* 2013;25:775-781.

8. Amini A Bilic I, Liebhart D, Hess M. Trichomonads in birds – a review. *Parasitol* 2014;141:733-747.

9. Mckeon T, Dunsmore J, Raidal S. *Trichomonas gallinae* in budgerigars and Columbid birds in Perth, Western Australia. *Aust Vet J.* 1997;7:652-655.

10. Gerlach H. Viruses. In: Ritchie BW, Harrison GJ, Harrison LH, eds. *Avian Medicine: Principles and Applications.* Lake Worth, FL: Wingers Publishing. 1994;862-948.

11. Droual R, Woolcock PR, Nordhausen RW, Fitzgerald SD. Inclusion body hepatitis and hemorrhagic enteritis in two African grey parrots (*Psittacus erithacus*) associated with adenovirus. *J Vet Diagn Invest.* 1995;7:150-154.

12. Shivaprasad HL, Phalen DN. A novel herpesvirus associated with respiratory disease in Bourke parrots (*Neopsephotus bourki*). *Avian Pathol.* 2012;41:531-539.

13. Gabor M, Gabor LJ, Peacock L *et al.* Psittacid herpesvirus infection in the eclectus parrot (*Eclectus roratus*) in Australia. *Vet Pathol.* 2013;50:1053-1057.

Herbal and Natural Health Supplements in Avian Medicine

Robert D. Ness, DVM

Session #332

Affiliation: From Ness Exotic Wellness Center, 1007 Maple Ave, Lisle, IL 60532, USA.

Integrative medicine involves the incorporation of holistic therapy modalities into practice with conventional allopathic medicine. Many terms are used to describe this type of health care, including alternative therapy, complementary medicine, holistic care and integrative medicine. Each term carries its own connotation, while none are entirely accurate. The most accurate term is integrative medicine, because it implies the integration of a variety of modalities into a more complete healthcare system. There are a number of modalities that can be integrated into avian medicine depending on the patient's need, including chiropractic, photobiomodulation, acupuncture, Chinese herbs, energy therapy, clinical nutrition, nutraceutical supplementation, medicinal herbs, flower essence and homeopathy.

Several forms of therapy involve giving the therapeutic product either orally or topically, similar to administration of a traditional medication. These modalities include nutraceuticals, herbal remedies, flower essence and homeopathy. Due to the relative ease of administration, these are the forms of integrative therapies commonly incorporated into avian practice.

Herbal Therapy

The use of herbal remedies for medicinal purposes dates back thousands of years. Herbs are mentioned in the Bible and excavated from ancient archeological sites. Herbs are used around the world, including Western herbal remedies in North America, Ayurvedic herbs in India and traditional Chinese herbal formulations in China.

Many common medications are made from herbs and other plants. Aspirin is a common example, generated from white willow bark. Approximately 25% of conventional drugs are directly derived from plants, usually by distilling down the plant material into a single active constituent and formulating it as a medication. A single isolated constituent may be more potent in effect for a particular action, but it is more likely to cause adverse side effects compared to when a whole herb is used. Herbs contain a wide array of compounds that tend to balance their effects on the body through synergistic actions of the herbal constituents. Herbal blends and formulations combine the benefits of multiple herbs, which further complement one another synergistically. Some herbs possess desired clinical effects that are often limited in allopathic medicine, such as immune-regulation, antiviral and anti-cancer properties.

Birds tend to be very responsive to herbal therapy for many common conditions, likely due to their ability to utilize nutrients from plant materials as well as their less domesticated nature. Herbal remedies are more effective than conventional therapies when treating metabolic conditions, such as liver or kidney disorders. Herbs also make reasonable alternatives to antibiotics in the treatment of infectious diseases, especially for viral and fungal infections. In addition, herbs support the immune system and help to build up resistance to recurrence of disease.

Not all herbal formulations are created equally. Numerous factors affect the potency and medicinal value of an herb, including method and time of harvest, the parts and preparation of the plant, and the handling and processing of the finished product. An herbal preparation can be manufactured that does not contain the medicinal portion of the herb or isn't processed properly to preserve the therapeutic benefits of it, yet still be labeled as that herbal remedy. Therefore, it is vitally important to utilize reputable and knowledgeable herbalist companies when purchasing these products. Whenever possible, fresh herbs or stable vegetable glycerin based extracts should be used.

Herbs are commonly used for support of the immune system. Although antibiotics and antifungal agents may be effective in battling invasive organisms, they do little to support the patient's immune system to ward off subsequent infections. Several herbs act to support the immune system by stimulating the patient's innate immunity, while others help modulate an immune response. Herbs to consider for an immune-stimulatory effect include Echinacea (*Echinacea angustifolia*) and Garlic (*Allium sativum*). Immune modulating herbs have a tonic or regulatory effect on the immune system, such as Astragalus (*Astragalus membranaceus*) or Siberian Ginseng (*Eleutherococcus senticosus*). Echinacea and astragalus are examples of herbs best used at the onset of an infection and lose their potency when administering later in disease or chronically. Olive Leaf (*Olea europaea*) is the quintessential herb shown to have strong antimicrobial effects as an antibiotic, antifungal and antiviral agent.

Liver disease is a very common condition in birds with numerous underlying etiologies, but conventional therapy has little to offer in its treatment. Hepatic effects that benefit from herbal therapy include Chlamydophila or bacterial hepatitis, hepatic lipidosis, hepatic encephalopathy or toxic hepatopathy. Certain herbs are well known and documented in treating these various hepatic related diseases, including milk thistle (*Silybum marianum*) and dandelion root (*Taraxacum officinalis*). Milk thistle seed is one of the most researched herbs with scientific documentation of the active constituent, silymarin, for many of the herbal claims, including being hepatoprotective, cholagogue and antioxidant. Although a very effective therapeutic herb for the liver, overuse can lead to false elevations in liver enzymes and potential slowing of liver function. Dandelion root is also used as strong liver tonic, while the leaves are a potent diuretic. Dandelion and milk thistle are complementary in supporting proper hepatic function, promoting proper bile flow and healing the liver.

Herbal remedies are also beneficial to the digestive system in birds. Everything from crop stasis to cloacal papilloma can be treated with herbs. Marshmallow root (*Althea officinalis*) is soothing and protective to internal tissues, nutritive and antimicrobial with an affinity to the digestive tract, as well as urinary and respiratory system. Slippery elm bark (*Ulmus fulva*) serves as a demulcent, by creating a protective film of the mucous membrane to reduce irritation, and astringent, causing constriction to minimize seepage of fluids in tissues. It is used in treatment of sore throats in the respiratory tract and for inflammation and irritation of the digestive tract. Chamomile (*Matricaria recutita*) is soothing and antiulcerogenic, in addition to reducing abdominal discomfort. Digestive herbal formulas often have at least one of these central herbs in addition to a blend of complimentary plants to soothe and heal the entire gastrointestinal tract.

Cardiac disease can be a challenge to diagnose and treat in avian medicine. However, there are safe herbal alternatives to the often more risky conventional cardiac drugs on the market. Hawthorn berry (*Crataegus sp*) has strong cardiovascular effects through various modes of action. It has been shown to modulate blood pressure, increase myocardial contractility and reduce peripheral vascular resistance through Na^+/K^+ ATPase and ACE inhibition. Gingko (*Gingko biloba*) leaves have a vasodilatory effect on peripheral circulation as well as a strong antioxidant effect. In addition, Gingko promotes vascular smooth muscle relaxation. These herbs can be adjuncts or even replacements to some of the more conventional cardiac medications for birds.

Nutraceuticals

Nutraceuticals are purified, chemically isolated compounds of nutritional substances that are used as therapeutic agents. These supplements include both micronutrients and macronutrients. Nutraceuticals take the form of vitamins, minerals, amino acids, proanthocyanidins, oils, probiotics and digestive enzymes. Some commonly used nutraceuticals in birds include aloe, apple cider vinegar, omega fatty acids and dimethylglycine.

Aloe is commonly known for its ability to heal burns when used topically, but it is also effective in treating wounds. Aloe squeezed directly from a freshly cut leaf or a gel extract is administered topically to burns, abrasions or lacerations. Aloe vera juice is very effective in boosting the immune system when taken orally. In addition, it has strong anti-inflammatory effects and is very soothing and nutritive to the digestive tract. Aloe juice is dosed orally to birds at 0.50mL/kg or added to the drinking water at 2 mL / 4 ounces of drinking water.

Apple cider vinegar is very nutritive to the digestive tract, providing numerous vitamins and minerals. Its greatest use however is as an acidifier of the intestinal tract, which is very effective in managing various disorders of the digestive tract, including bacterial dysbiosis, candidiasis, bacterial enteritis and chronic diarrhea. It is very effective in managing bacterial overgrowth causing foul-smelling stools from diseases such as proventricular dilatation disease and papillomatosis. A starting dose of 15 ml/8 ounces drinking water is usually effective in acidifying the GI tract in birds.

Dimethylglycine (DMG) is an amino acid derivative that has a wide range of benefits, including proper cardiovascular function, improved nerve transmission, boost immune system and metabolic support. DMG acts as an adaptogen that works in concert with other cofactors in the body to counteract effects of stress, promote healing and prevent degenerative diseases, thus maintaining homeostasis of the body. It increases oxygen utilization, modulates the immune response, aids in hepatic detoxification, regulates cholesterol metabolism and increases energy levels. DMG also serves as an antioxidant in preventing free radical damage caused by degenerative disorders.

Proanthocyanidins are a group of strong antioxidants including grape seed extract, pycnogenol, bilberry and citrus bioflavonoids. These are free radical scavengers that combat degenerative conditions and breakdown of tissues. Some beneficial effects include strengthening capillaries and regulating permeability as well as acting on the nervous system to nourish brain cells, thus affecting many aging disorders such as cardiovascular disease, cognitive dysfunction, and metabolic decline.

Omega fatty acids are beneficial to various tissues for the anti-inflammatory and immune-modulating effects. Birds utilize plant sources of fatty acids for their omega 3 and 6 requirements much more readily and effectively than animal sources, such as fish oils. Appropriate plant sources include flax seed, chia seed, pumpkin seed, red palm oil and coconut oil, depending on the ratio of omega 3 to 6 as well as omega 9 needed. Omega fatty acids are anti-inflammatory to joints and skin, as well as the kidneys and digestive tract. In addition, the proper ratio of fatty acids is important in cardiovascular function and nerve transmission.

Probiotics help balance the bacterial flora in a bird's gastrointestinal tract. This is most important after prolonged use of antibiotics or conditions disrupting the proper functioning of the digestive tract.

Birds differ from mammals in their normal bacterial flora, so a probiotic designed for mammals will not likely be fully beneficial to a bird's GI tract. In fact, even different species of birds are likely to have a different flora, but unfortunately research has not clearly distinguished which bacteria normally colonize the GI tract in different species. Therefore, most avian probiotics have a variety of presumed normal bacteria present for the typical parrot-type bird in hopes of reestablishing some of the normal flora in our avian patients.

Digestive enzymes in birds are usually derived from plant enzymes typically found in fresh foods that a bird would normally ingest, but are typically inactivated with the processing of a typical commercial avian diet. Pancreatic enzymes from animal sources are less effective nutritionally in birds. Plant source digestive enzymes have a broad pH range, which makes them more effective throughout the bird's GI tract. Birds that benefit from supplemental pant source digestive enzymes are those that may exhibit clinical signs of intestinal gas, voluminous stools, maldigestion, weight loss, chronic illness or intestinal bacterial overgrowth. Supplementation in the diet with probiotics aids in proper digestion.

Flower Essence

Flower essence therapy involves the art of using the energetics of flowers to affect the emotional state of the patient. The most recognized and original flower essence therapy was developed by Dr. Edward Bach in the late1920's, called Bach Flower Remedies. Based on the premise that the sight and smell of certain flowers elicit an emotional shift in people, Dr. Bach studied and characterized the emotional effects of thirty-eight flowers. There are flower essences identified to address anger, fear, anxiety, stress, depression, insecurity and many other core emotional states. Bach flowers work on the energetic level of the body, rather than a physical substance such as an herb or drug. As an energetic medicine, the effects are not dose dependent but rather frequency dependent.

Each of the thirty-eight flower essences are bottled as tinctures preserved in alcohol. These tinctures are used to make diluted formulas of combined essences in spring water, which also makes them safe to use in birds by minimizing the alcohol content. Although they can be used individually, typically several are combined in the same bottle to create an individualized formula for the complex emotional state of the patient. A flower essence formula is created by placing four drops of the tinctures from the chosen Bach flowers into a one ounce bottle of spring water. This diluted formulation is then administered to the patient directly by mouth or in the drinking water.

Rescue Remedy is the most common Bach flower formula. This is a combination of impatiens (impatience, stress), star of Bethlehem (trauma, shock), cherry plum (fear loss of control), rock rose (pain, terror) and clematis (sadness, withdrawn). It is the quintessential formula for stress and anxiety in people and animals alike. Rescue Remedy is particularly useful for birds that are stressed, anxious or frightened, so naturally useful in the veterinary clinic. Keeping Rescue Remedy diluted in a fine mist spray bottle for each exam room is an effective stress reducer for anxious patients and client. In addition, this formula can be used for controlling seizure activity, by lessening the severity or duration when used at onset of a seizure.

Lack of Regulation and Standardization

Since these products are not designated as medications, they are not regulated by the US Food and Drug Administration or any other governmental agency. Although this may initially appear to be a benefit, it is in fact a detriment in control of what can be marketed in the name of holistic therapy. There is no validation that a product contains the ingredients the manufacturer claims, or doesn't contain potentially fatal substances not included on the label. Regarding herbals and nutraceuticals, the wrong portion of a plant may be processed which may not have any therapeutic benefit, but still be presented as the desired product. In addition, there is no assurance that the preparation is properly harvested, handled and processed.

There is some hope in self-regulated validation of trustworthy holistic minded companies. Therefore, practitioners and consumers alike should only deal with reputable companies that are well known and bear the symbol of certain agencies or organizations. These include NSF International, National Animal Supplement Council

(NASC) and Consumer Lab. These organizations perform independent testing of products that are not otherwise tested or controlled. They then provide consumer reports on the products and grant approved companies their seal of validation. Although not a perfect system, this provides some credibility to properly run and ethically minded businesses.

NSF International is a public health and safety organization that establishes risk management solutions while serving interests of all the stakeholders, from the public, business community and government agencies. Its purpose is to develop uniform, consensus-based national standards, while bringing together regulators, industry, consumers and public health experts. This is an independent certification organization that lends its stamp of approval to products that meet or exceed its standards. The website for this organization is http://www.nsf.org.

The National Animal Supplement Council (NASC) consists of members that are committed to the quality and safety of animal health supplements for companion animal use. Members are self-regulated, but they must submit their products to independent audits to verify safety and compliance with specified standards in order to receive the NASC quality seal of approval. NASC's overriding goal is to promote the health and well-being of non-human food chain animals that are given animal health supplements by their owners, and to protect and enhance the integrity of the animal health product industry. The website for this council is http://nasc.cc. An associated website is http://www.pethealthlive.com , which is a resource for information regarding the health of companion animals. This site provides alerts on safety and recalls, lists company profiles and ratings, references an animal health library, and provides a resource for finding member products.

Consumer Lab is an independent laboratory dedicated to testing of supplements and nutritional products available in the human market. This group has been reporting on human product lines since 1999. Their mission statement is to identify the best quality health and nutritional products through independent testing. The website is http://www.consumerlab.com.

Summary

Integrative medicine for birds combines the art with science of practice by incorporating an assortment of modalities and methodologies in the treatment of our avian patients. This is not a cookbook approach to medicine, but rather a holistic endeavor to treat the entire patient to establish harmony in health on all levels and not just eradicate an organism from the body. It is by no means a replacement to conventional allopathic medicine, but rather a complement to quality medicine. The path to holistic medicine is an independent study with no two holistic practitioners quite the same, but rather following their own path in discovery and integration of knowledge in their repertoire that we refer to as medicine. As with any endeavor, the pursuit of holistic medicine is developed one step at a time and never really ends as we continue to learn and expand our knowledge base through continuing education, reading and studying, learning from our patients, and the art of practicing medicine.

Gross Lesion Recognition for the Avian Clinician

Michael M. Garner, DVM, Dipl ACVP

Session #006

Affiliation: From Northwest ZooPath, 654 West Main, Monroe, WA 98272, USA.

Gross lesion recognition, like other imaging modalities, is a bit of an art form. Vast zones of grey may confront the investigator when considering what the lesion is, what it could be, and what to do with it. And nobody is good at all of it: A seasoned livestock pathologist could possibly clear an entire days' necropsies without the help of a single histology slide. A bee keeper would recognize immediately the maggot stages that parasitize his colony. A dog-cat pathologist in a busy private laboratory or a companion animal clinician may not have a clue about gross lesions in livestock or bees. Historically, our attempts at accurately diagnosing disease by gross lesion recognition have been a humbling experience, and that has been the impetus for all further diagnostic specialties.

So now that we have put gross pathology in its proper place, why even bother with such an imperfect science? Obviously, gross lesions, or lack thereof, are the first visual indication of what may be wrong with the patient. These are the lumps, the effusions, the asymmetrical oddities, the discolorations, the odiferous clues to disease that (hopefully) stimulate a scientific-thought process culminating in a list of differential diagnoses. The purpose of this workshop is to present images of common and not-so-common avian gross lesions in live and necropsy specimens, and in a participatory manner, establish a differential diagnosis and means for establishing a definitive diagnosis.

Recommended Bibliography

1. King JM, Dodd DC, Roth L, Newson ME. *The Necropsy Book*. Charles Louis Davis Foundation; 2002. (Mammals)

2. Reimschuessel R. Postmortem examination. In: Stoskopf MK, ed. *Fish Medicine*. Philadelphia, PA: W.B. Saunders:160-165. (Fish)

3. Garner, MM. Overview of biopsy and necropsy techniques. In: Mader DR, ed. *Reptile Medicine and Surgery*. 2nd ed. St. Louis, MO: Elsevier Science;2006:569-580. (Reptiles)

4. Nichols DK. Necropsy. In: Wright KM, Whitaker BR, eds. *Amphibian Medicine and Captive Husbandry*. Malabar, FL: Krieger Publishing Co.;2001:331-332. (Amphibians)

5. Latimer KS, Rakich PM. Necropsy examination. In: Ritchie BW, Harrison GJ, Harrison LR, eds. *Avian Medicine: Principals and Application*. Lake Worth, FL: Wingers Publishing, Inc.;1994:356-376. (Birds)

6. Campbell TW, Ellis C. *Avian and Exotic Animal Hematology and Cytology*. Ames, IA: Iowa State University Press;2007.

West Nile Virus: Diagnosis and Differentiation from Other Diseases in Raptors

David E. Scott, DVM

Session #043

Affiliation: From Carolina Raptor Center, 6000 Sample Road, Charlotte, NC 28078, USA.

Abstract: West Nile virus (WNV) infection is very common and can lead to devastating disease. Unfortunately, it can be difficult to diagnose and there are many causes of neurologic disease with similar clinical findings. This paper will summarize the main differentials and provide a decision-making guide designed to help diagnose WNV as well as several other common causes of neurologic disease.

Introduction

Infections with West Nile virus (WNV) were first seen in the western hemisphere in 1999. The infection is due to a *Flavivirus* and the clinical signs were initially quite severe. Many species of birds were affected as the epidemic spread from the east to the west coast. Presently, the disease can be characterized as endemic, however, there have been several outbreaks of higher virulence and mortality over the years.

The virus is spread via mosquitoes and birds can become reservoir hosts if viremia is present. Non-avian species can become clinically ill but typically do not become viremic, and are therefore characterized as "dead-end hosts." Clinical disease, when present, ranges from mild to severe and birds literally "falling from the sky" have been reported.[1,2] In less severe cases, neurologic signs of varying degrees are the hallmark of infection. Many species are affected, but corvids are particularly susceptible. Unfortunately, there is a great deal of overlap with other common diseases and conditions which makes diagnosis a challenge.

Differential Diagnoses

There are many differentials for neurologic disease in birds of prey (Table 1). They fall into three general categories: trauma, infectious and toxic.

Table 1. Summary of differential diagnoses.

Category	Cause	Diagnosis
Trauma	Vehicular trauma	Physical exam
	Window strike	Physical exam
Infectious	West Nile virus	CBC, Serology, PCR, histopathology with special stains
	Paramyxovirus (PMV-1, Newcastle's)	Serology, PCR
	Avian influenza	Serology, PCR
	Avian vacuolar myelinopathy (AVM)	History, histopathology
Toxic	Lead	Blood level (can be done in-house)
	Mercury	Blood level
	Pesticides/insecticides (4-aminopyridine, organophosphates, carbamates, metaldyhyde, organochlorines)	Toxicology screen on liver or stomach contents Response to atropine ACHE blood levels

Trauma is usually easy to detect on physical exam. Besides an altered mental state, there may often be fresh or dried blood in the nares, oral cavity, and ears. It is critically important to complete a thorough ophthalmologic exam as extensive amounts of blood and fibrin may be present in the posterior chamber and this is easily missed unless the fundus is examined carefully. The history is also important as vehicular traumas are often found on the roadside and window strikes are often found down below a large window. Necropsy may reveal large intracranial hematomas (Fig 1) as well as contusions in the lungs and free blood in the coelomic cavity.

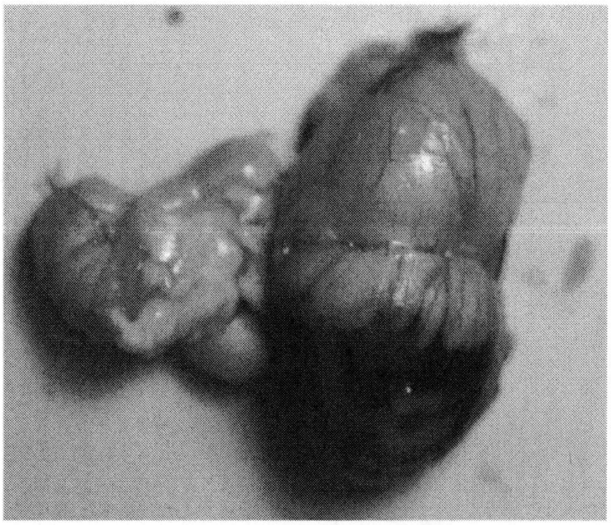

Figure 1. Severe intracranial hemorrhage.

There are many infectious agents that can be confused with WNV. *Paramyxovirus* (PMV-1) is the causative agent in Newcastle's disease commonly seen in poultry. Raptors can be affected after eating infected birds and it is known to cause severe torticollis. Geographic location can provide a clue, especially in areas such as the Middle East where the disease is endemic.

Avian influenza (AI) can cause lethargy and neurologic signs and should also be considered. As with PMV-1, this is typically a poultry disease but many species can be affected as was seen in the North American outbreak in 2015. Avian vacuolar myelinopathy (AVM) is another disease that causes neurologic signs and lethargy. It affects mostly short birds such as coots, but also raptors that feed on these birds such as bald eagles. The disease is not fully understood but it is believed to be caused by a cyanobacteria[3] that grows on aquatic plants often ingested by waterfowl. This disease should be suspected in the appropriate species and in areas such as the southeastern United States where the proper habitat exists.

There are many toxins that can cause neurologic signs in raptors.[6] These compounds are in widespread use so exposure is probably quite common. Unfortunately, with a few exceptions, diagnosis is difficult, expensive and often times impossible.[7] Intoxication with lead is very common and widely reported. Symptoms occur when lead is ingested[4] and does not generally occur from gunshot wounds with lead fragments embedded soft tissue.[5] Intoxication is much more prevalent in species such as eagles and vultures that feed on carcasses (i.e., deer) that may contain gunshot. Mercury is also a concern and can be bio-accumulated in various fish species so this may be of particular concern in osprey (*Pandion haliaetus*) and bald eagles (*Haliaeetus leucocephalus*).

There a literally hundreds of different chemicals used as insecticides and pesticides. Metaldehyde is a common snail bait and 4-aminopyridine (Avitrol) can induce seizures by affecting $K+$ channels. The organophosphates (malathion and dichlorvos) and the carbamates inhibit acetylcholine esterase (ACHE). The organochlorines such as DDT and chlordane are often used in termite control and they interfere with $Na+$ channels in the peripheral nervous system.

Diagnosis

Accurate diagnosis is critical in order to treat the cases appropriately. However, the various tests can be prohibitively expensive, some can take several days to a week for results and some are generally only done post-mortem. To maximize chances of success, to save money and to get an answer as quickly as possible, it makes sense to do the quick, easy and least expensive tests first. Then, if a diagnosis is not obtained, continue with the slower, more expensive tests later. The following diagnostic protocol is summarized in Fig 2.

1. Physical exam; look for signs of trauma. If none found, continue to Step 2.

2. Collect blood for a lead level and save more for future tests. A blood lead level can be obtained in 3 minutes using a Lead Care II unit (Magellan Diagnostics, North Billerica, MA, USA). If the lead level is above the toxic limit (typically 20 µg/dl), begin chelation therapy. If not, continue to Step 3.

3. Administer atropine. If there is a response, continue treatment for organophosphorus toxicosis. If not, continue to Step 4.

4. Begin treatment with meloxicam in case the cause is an infectious agent such as WNV.

5. Submit blood samples for PCR testing and serology for WNV, AI, PMV-1. Combined choanal/chloacal swabs can also be submitted for WNV antigen testing. These tests can take some time and the results can be difficult to interpret. The window of viremia can be short and false negatives are not uncommon. In addition, a single positive result on any serology test only demonstrates exposure but does not necessarily indicate an active infection. Paired serology testing is much more helpful but clearly takes more time and is more expensive.

6. Depending on how the case progresses, supportive care may be indicated. In other cases, euthanasia may be the best choice. If the bird is euthanatized, histopathology of the brain is the next step. Samples of the liver, brain and stomach contents should also be saved and frozen. WNV typically causes a perivascular, lymphoplasmacytic encephalitis. If present, continue to Step 7. If not, continue to Step 8.

7. In order to definitively diagnose WNV, virus-specific immune-histochemical stains must be applied. A positive result appears as in Fig 2.

8. If WNV has been ruled out, the next step is a toxicology workup. Typically, these tests are run on liver, brain, and/or stomach contents. The results can be somewhat difficult to interpret as the toxic levels of the multitude of possible chemicals in the various tissues of any number of species may not be known. However, the results can be extrapolated and a tentative diagnosis can often be made.

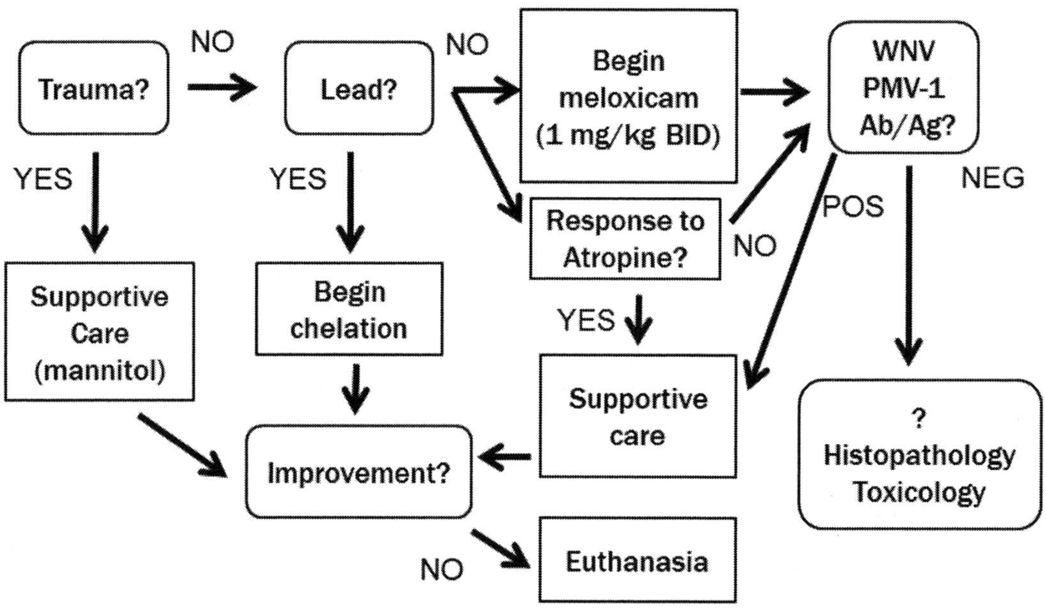

Figure 2. Diagnostic flowchart.

Case Study

A hatch-year red-shouldered hawk (*Buteo lineatus*) presented with severe neurologic signs including ataxia, tremors and inappropriate vocalization. The physical examination did not reveal any obvious signs of trauma. Initial care was purely supportive and included fluids and meloxicam (1 mg/kg IM). A blood sample was collected and the lead level was within normal limits. The bird's condition deteriorated rapidly and it was euthanized within 24 hours of admission.

This species is known to hunt insects and grubs in lawns so, besides WNV, we were concerned about ingestion of an insecticide or pesticide. The gross necropsy, including the brain, was normal. Stomach contents and liver were submitted for a pesticide screen. Those results were negative. The tissue samples were then submitted for histopathology and a characteristic perivascular inflammatory pattern was evident. Follow up with viral-specific immuno-histochemical stains was positive for WNV (Fig 3.)

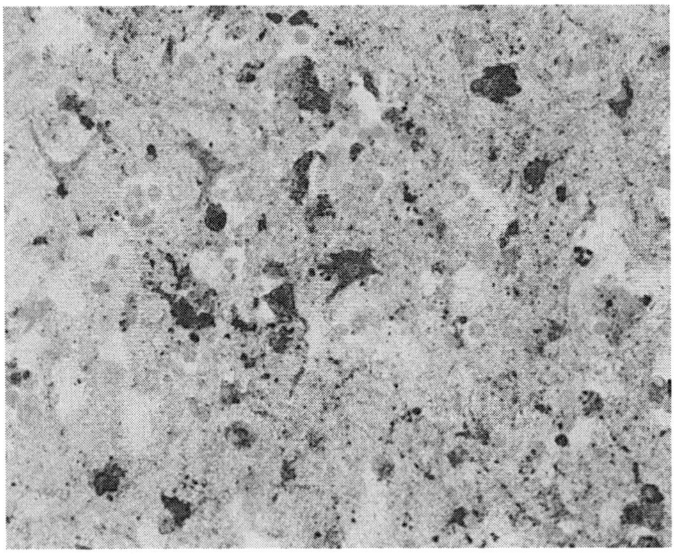

Figure 3. Positive result for WNV using immuno-histochemical stains on brain tissue in a red-shoulder hawk (*Buteo lineatus*).

Conclusion

Neurologic signs are very common in raptor patients presented to rehabilitation centers. Unfortunately, they can be result of any number of toxins, infectious agents or simply the result of trauma. More importantly, effective treatment and prognosis absolutely require an accurate diagnosis.

Since the vast majority of wildlife patient admissions are for traumatic reasons, trauma should be and can be easily ruled in or out with a thorough physical examination. Following that, a step-by-step diagnostic protocol can be followed in order to maximize the chances of making a diagnosis without wasting money and time on unnecessary tests. In general, WNV cases are very difficult to diagnose ante-mortem and the gold-standard relies on histopathology of the brain using special stains.

References

1. Nemeth N, Gould D, Bowen R, Komar N. Natural and experimental West Nile virus infection in five raptor species. *J Wildl Dis*. 2006;42(1):1-13.

2. Nemeth N, Kratz GE, Bates R, Sherpelz JA, et al. Clinical evaluation and outcomes of naturally acquired West Nile virus infection in raptors. *J Zoo Wildl Med*. 2009;40(1):51-63.

3. Wilde SB, Murphy TM, Hope CP, Habrun SK, et al. Avian vacuolar myelinopathy linked to exotic aquatic plants and a novel cyanobacterial species. *Environ Toxicol*. 2005;20(3):348-353.

4. Redig P. Lead from spent ammunition: a source of poisoning in bald eagles. *Proc Annu Conf Assoc Avian Vet*. 2009;427-428.

5. LaDouceur EE, Kagan R, Scanlan, M, Viner T. Chronically embedded lead projectiles in wildlife: a case series investigating the potential for lead toxicosis. *J Zoo Wildl Med*. 2015;46(2):438-442.

6. Mineau P, Fletcher MR, Glaser LC, Thomas NJ, et al. Poisoning of raptors with organophosphorus and carbamate pesticides with emphasis on Canada, U.S. and U.K. *J Raptor Res*. 1999;33(1):1-37.

7. Mineau P, Tucker KR. Improving detection of pesticide poisoning in birds. *J Wildl Rehab*. 2002;25(2):4-13.

Avian Respiratory System: Function, Disease, and Treatment

Thomas N. Tully, Jr., DVM, MS, Dipl ABVP (Avian), Dipl ECZM (Avian)

Session #051

Affiliation: From the Zoological Medicine Service, Department of Veterinary Clinical Sciences, School of Veterinary Medicine, Louisiana State University, Baton Rouge, Louisiana, LA 70803, USA.

Abstract: Respiratory diseases as a group are one of the most common clinical presentations that an avian veterinarian will diagnose. The complex anatomic structure of the avian respiratory system demands that the astute practitioner correctly diagnose and treat a respiratory disease condition. Distinguishing between upper and lower repiratory disease is essential so that the focus of the treatment protocal be directed to the affected area. Proper diagnostic testing will aid in the diagnosis of microbiologic infections, foreign body blockage, viral disease, reproductive tract problems, tumors, trauma, nutritional deficits/toxicity, cardiovascular disease, toxins and parasites. Only a complete physical examination, which includes an external overview of the patient plus diagnostic testing, will allow the veterinarian to make the correct diagnosis. A complete understanding of the avian respiratory system is necessary to correctly treat many respiratory diseases. Avian respiratory anatomy dictates special aggressive treatment for many infectious organisms and disease conditions that may infect/affect the avian patient. A review of the basic anatomy of the upper and lower respiratory systems is the critical foundation to undertand avian respiratory disease. Correct diagnosis and treatment of a respiratory problem will prevent secondary disease, chronic health problems and possibly death.

Upper Respiratory System Anatomy

The upper respiratory system of the bird includes the external nares, operculum, nasal concha, infraorbital sinus and choanal slit.[1,2] The first anatomic structure of the upper respiratory system is the naris. Depending on the species, the nares should have symmetrical clear openings, with the operculum clearly visible within the opening.[3] Soiling or matting of the small feathers around the nasal opening may indicate nasal discharge even if this is not evident on the beak or within the nares, especially if the bird has been sneezing or has mild rhinorrhea. Obvious nasal discharge (dried mucoid material) on the beak, disfigurement, or malformation of the nares are overt clincal signs of upper respiratory disease.

The nares communicate directly with the nasal cavity, which is divided by the septum. Each side of the nasal cavity contains a rostral, middle and caudal nasal concha in most avian species.[2] After the inspired air is warmed, filtered and moistened by the nasal concha, it enters the extensive infraorbital sinus. Birds have one respiratory sinus that is divided into five diverticula: the rostral, preorbital, infraorbital, mandibular, and postorbital diverticula.[2] The extensive nature of the infraorbital sinus and its diverticula often make upper respiratory infections difficult to treat and prevent. Infraorbital sinuses communicate with the nasal cavity by only 2 openings, both in the dorsal wall: one into the caudal nasal concha and the other into the nasal cavity.[2] The infraorbital sinus in most birds opens caudally into the cranial and cervical portions of the cervicocephalic air sac that extends and terminates along the lateral cervical region. The treatment and presentation of infraorbital sinus infections depend on the species of bird affected. In Passeriformes, Anseriformes and Psittaciformes, the right and left infraorbital sinues conmmunicate.[2] In noninsectivorous passerines, the infraorbital sinuses are separate.

The upper respiratory tract ends at the choana. The choanal slit may be viewed on the dorsal aspect of the oral cavity forming a V, with the apex at the rostral position. Eptithelial projections (papilla) extend to the center of the choanal slit in most avian species. The appearance of the papilla at the time of physical examination can give the veterinarian an idea of the chronicity of the condition or nutritional status of the bird. Blunted, edematous papilla often indicate chronic irritation or inflammation. Foreign bodies and abscesses can occur in the rostral aspect of the choana and are associated with dyspnea and/or infection.

Lower Respiratory System Anatomy

The lower respiratory system includes the glottis (laryngeal mound), trachea, syrinx, bronchi, lungs, parabronchi and air sacs.[2] The laryngeal mound is composed of muscles that dilate and constrict the cartilages of the glottis and are innervated by cranial nerve IX.[3] Dyspnea may occur if neurologic damage affects the *dilator glottidis* and *constrictor glottidis* muscles or if abnormal tissue growth occludes the glottis.[3,4] Complete cartilaginous rings support the avian trachea.[3] The rings may ossify with age in many avian species, making the rings more resistant to collapse. Although ratities have complete tracheal rings, the large size of the rings predisposes the trachea in these large birds to collapse when captured.

The syrinx is modified cartlaginous tissue found at the caudal aspect of the trachea. The syrinx is the vocal organ of birds and is a common site for fungal lesions in raptors, psittacformes and waterfowl. The syrinx bifurcates into incomplete cartilaginous rings called the primary bronchi.[3] In most avian species, the primary bronci transform into four pairs of secondary mesobronchi.[4] Primary bronchi, secondary bronchi, and tertiary bronchi leading to air capillaries together are consideredthe lung architecture of most avian species. Primary bronchi have both extrapulmonary and intrapulmonary components and begin at the bifurcation of the trachea. The secondary bronchi are commonly divided into four groups, commonly named for the area of the lungs they supply. Mediodorsal and medioventral secondary bronchi are part of the paleopulmonic system, a unidirectional system that promotes efficient movement of air through the lungs. Tertiary bronchi (parabronchi) are a large number of pathways that branch off the secondary bronchi. Their primary structures including the atria, infundibula and air capillaries. Atria are channels that allow air to move from the parabronchi into the infundibula. The infundibula lead to air capillaries, the primary site where gas exchange occurs. Air capillaries can anastomose with each other and the blood-gas barrier is much thinner in birds than in mammals, allowing greater diffusion of oxygen. The lung air passageways communicate with the air sac system, which in most avian species includes four paired air sacs and one unpaired air sac. The clavicular, paried cervical, cranial, and caudal thoracic and abdominal air sacs commonly comprise the avian air sac system.[2] The clavicular air sac is located in the cranial mediastinum. Depending on the species, the air sac extends into the bones of the thoracic girdle, humerus, sternum and sternal ribs.[5] The blood vessels, nerves, esophagus and trachea are found within pleural folds of the clavicular air sac.[5] Surrounding the cervical vertebrae is the appropriately named cervical air sac. The cranial and caudal thoracic air sacs surround the majority of major organs in the avian celomic cavity. The abdominal air sac surrounds the intestinal loops and, depending on species, extends into the pelvic bones and femur.[5] Pneumatic bones including the humerus, vertebrae, coracoid, femur and diverticula around the heart and cranial kidney are diverticula of air sac epithelium.[3]

Diagnostic Techniques: Upper Respiratory System

A diagnostic work-up on a bird with an upper respiratory infection should include a thorough history, external examination of the nares and choana, and one or more of the following techniques: choanal cytology, infraorbital sinus lavage and bacterial culture, radiography, endoscopy of the choana and infraorbital sinus diverticula, complete blood count and blood chemistry panel. Rhinography and sinography are useful radiographic techniques for examination of the upper repiratory tract when an anatomic abnormality, mass or foreign body is suspected.[6] Case history and presentation will help the veterinary surgeon to determine which tests are indicated. Because of the transitory nature of microorganisms within the upper respiratory system, bacteria identifided by culture or Gram's stain from samples taken from a bird with no clinical abnormalities have questionable interpretive significance. If clincal signs are present, or abnormalities such as nasal granulomas, choanal edema, hyperemia, or abscesses are noted on physical examination, isolation of potentially pathogenic organisms may be significant.[2] Lavage of the infraorbital sinus and air sacs may increase the likelihood of identifying the organism responsible for the presenting problem.[7]

Endoscopic examination of the upper respiratory tract is an effective means for diagnosing disease processes. This technique is especially usedful to detect foreign bodies and fungal granulomas at the level of the lower nasal septum. Often when examinaing the choanal area with an endoscope suction is needed to remove nasal discharge that may occlude the field of vision.

Computed tomographic (CT) and magnetic resonance imaging (MRI) are advanced diagnostic imaging techniques that may be used for a more detailed evaluation of the upper respiratory system.

Diagnostic Techniques – Lower Respiratory System

Asculatation, radiography, endoscopy and lavage can be used to diagnose a bird exhibiting clinical signs of lower respiratory system disease.[2] Aucultation through a thin towel after a small amount of exercise may demonstrate abnormal lung or air sac sounds. Although a definitive diagnosis cannot be made by auscultation, detection of abnormal sounds will indicate the need for further diagnostic tests, such as radiographic and endoscopic examination.[8] Foreign bodies, granulomas, chronic irritation and inflammation can be diagnosed by radiology or endoscopy. Radiology is a very useful diagnostic tool in evaluation of lower respiratory disease. Anesthesia of the patient with isoflurane is recommended, and full-body radiographs should be obtained during inspiration. Two views are required for proper radiographic evaluation. The trachea, heart, lungs, and air sacs should be closely examined on the radiographic images.[3] One of the most useful diagnostic tests for respriatory disease in companion avian species is an endoscopic examination of the lungs and air sacs.

An air sac breathing tube can be placed at the same location described for a respiratory endoscopic examination of the body cavity. To maintain a bird on a surgical plane of anesthesia with an "abdominal" air sac tube it is recommended to increase the oxygen flow and percentage of isoflurane anesthesia at a higher level than commonly practiced when using an endotracheal tube. If the common oxygen flow to maintain a surgical plane is 0.75-1 L then it should be increased to a 2 L flow. If the common isoflurane percentage to maintain a surgical plane is 2-2.5% then it should be increased to 3.5-4% with an air sac breathing tube. An endoscopic sheath is not commonly used when examining the tracheal lumen. Instead the tracheal lumen compensates as a "physiologic sheath" to direct biopsy forceps toward the site of interest or aspirate microbiologic samples. The bird should be monitored as is normally performed when an endotracheal tube is used.

Fluid accumulation or a space-occupying mass may be evaluated through abdominocentesis, ultrasound, contrast gastrointestinal radiology, flouroscopy, or exploratory endoscopy. Even with the limitations posed by the avian air sac system, ultrasonography may be helpful in evaluating cardiac disease, space occupying masses or fluid densities. Electrocardiography my be helpful in diagnosis of degenerative, metabolic or inflammatory cardiac disese.[3] The paper speed should be fast (50-100 mm/sec) to detect abnormal heart rhythm and chamber enlargement.[3]

Evidence of infectious or noninfectious disease and disease chronicity may be obtained by complete blood count and blood biochemistry panels. The total leukocyte count may help determine whether an infectous disease is preseent. In birds with total leukocyte counts over 50,000 cells/µl, four primary differtial diagnoses must be considered: *Chlamydia psittaci*, *Aspergillus* spp., *Mycobacterium* spp. infections or leukemia. White blood cell counts less than 3,000 cells/µl, or pancytopenia indicate bacterial septicemia or viral infection. The differential cell count may alert the clinician to the nature of the respiratory problem; for example, monocytosis is indicative of chronic infection and eosinophilia may be indicative of parasitism.

Other tests that may be helpful in the diagnosis of respiratory disease include fecal flotation and direct fecal examination. Fecal examinations may reveal respiratory tract parasites, including *Syngamus trachea*, *Cyathostoma bronchialis*, *Trichomonas*, *Capillaria*, *Atoxoplasma*, *Cryptosporidium*, and *Sternostoma* spp.[9]

Chlamydia psittaci and *Mycoplasma* spp. infections may be diagnosed through serologic antibody testing, fecal ELISA tests and DNA polymerase chain reaction (PCR) technology through blood and choanal/cloacal swabs.[10] A negative test result for these intracellular organisms does not rule out infection; results of other diagnostic tests and the patient's history must be considered.

Clinical Considerations

Prior to examining a bird exhibiting abnormal respiratory signs, the clinician must decide whether clinical signs indicate critical or life-threatening disease. If the bird is in severe respiratory distress, emergency measures must be taken to ensure a patent airway and to stabilze the patient. The client must be informed when respiratory signs are potentially life-threatening prior to examination or treatment of the bird. Open-mouth breathing, tail bobbing, and abdominal distention may indicate respiratory compromise, and even minimal restraint for examination may cause death. There has been criteria established to leave the bird alone and perform only minor critical life saving procedures if any of the criteria are observed. If the patient has an increased respiratory rate, excessive or unusual vocalization and/or has difficulty breathing during the physical examination, the bird may have to be left alone and placed in a critical care unit until it is able to withstand the rigors of the evaluation process.[11] Along with oxygen, the placement of an air sac breathing tube may be necessary for patient stabilization.[9] A quick thorough history, clinical signs and prior knowledge of common respiratory problems associated with species being examined will aid in forming a list of differential diagnoses.

When performing the external physical examination the respiratory system covers a significant part of evaluating the avian patient's overall condition. The external physical examination should follow the same protocol for each patient regardless of the presenting complaint. If the presenting complaint is life threatening then that condition should be treated and the patient stabilized before continuing the examination. The upper respiratory system that can be examined by an external examination includes the external nares, infraorbital sinus, choanal slit and glottis. Clinical signs associated with upper respiratory disease include open-mouth breathing, change in vocalization, odor from sinus, rhinorrhea, nasal granulomas, exercise intolerance, dyspnea, head-shaking, mucopurulent nasal discharge, inflammed swollen cere stretching of the neck, yawning, epiphora, periophthalmic swellings and plugged nares.[2] Clincial signs associated with lower respiratory disease include tail-bobbing, change or loss of vocalization, labored respiration, wings out, exercise intolerance, open beak breathing, coughing and abnormal sounds on auscultation.[2] The external physical examination should begin with the head and a frontal examination viewing the beak and eyes straight on. Every bird should have its nares observed for symmetry, matted feathers and nasal discharge or granulomatous plugs.

The oral cavity and choanal slit is the next area examined. To examine this area an avian speculum is recommended to open the beak of the bird. Most parrot species will not open their beaks willingly for an examination. To open the beak an avian speculum is recommended. The avian speculum must be placed in the rostral notches of the beak to prevent iatrogenic trauma to the fragile lateral beak rhinotheca and gnatotheca. Once the beak is open by sliding the narrow part of the speculum to the wider area the choanal slit, oral cavity, tongue and glottis can be observed. The choanal slit should be smooth and thin with papillary projections extending toward the medial aspect of the choana. Blunting of the choanal papilla, edema and hyperemia of the choanal edges suggest irritation and inflammation. A wooden applicator stick can aid in lowing the tongue within the oral cavity to gain a better view of the glottis is many species, particularly budgerigars. White plaques can develop in the oropharynx as a result of fungal infections, bacterial granulomas, mycobacterial granulomas, trichomoniasis, pox lesions and neoplasia.[5] It is not unusual to view granulomas or abscesses in the oral cavity affecting the choana and glottis due to hypovitaminosis A. This usually develops in older Amazon parrots that have been on a seed diet most if not all of their lives. Treating birds with oral abscesses begins with removal of the abscesses, vitamin supplementation followed by a diet change. Most Amazon parrots, even older birds, will readily switch to a pellet based diet following oral granuloma treatment.

Infraobital or periorbital swelling or asymmetry associated with the postion of the globe may indicate disease of the infraorbital sinus. There may be swelling or asymmetry in the area between the medial canthus of the eye and the nares. Fluid or an inspissated abscess may be present causing this abnormal clinical presetnation. Since there is a single layer of epithelium covering the sinus in this area fluid may be aspirated or a simple incision over the swelling will reveal the abscess, which then can be removed. Culture and sensitivity of the affected area will aid in selecting the proper antimicrobial treatment.

Prior to examining the bird or before the lower respiratory tract is examined one may gain an appreciation of the bird's respriatory condition. Avian patients having difficulty breathing will show specific signs that correspond directly to the degree of distress the bird is having when breathing. The degree of respiratory distress ranges from almost nondiscernable tail bobbing to open beak breathing, tongue protruding, wings out and significant tail bobbing. When a bird breathes normally the tail does not move. If there is difficulty breathing and the bird wants to expand its respiratory capacity the tail moves up and down, commonly called "tail-bobbing." The tail bobs because the bird breathes with its intracostal muscles, moving the keel in and out in a bellows action. Birds do not have a diaphram and need to mechanically move air in and out of their respiratory system and the more difficulty they have in breathing the greater degree of movement they have on their keel and intracostal muscles. If the tail is bobbing there is a clinical condition that must be treated.

Although auscultation of avian patients is not emphasised by many practitioners there is a proper way to examine a bird with a stethoscope. The stethoscope should be placed dorsally and ventrally over the thorax and the trachea. Abnormal sounds that may be heard include clicks, wheezes and squeaks. With a normal ambient temperature and only brief handling, the respiratory recovery time should be less than 3 minutes.[5] A prolonged recovery time suggests primary or secondary respiratory involvement or cardiovascular disease or both. Auscultation also is helpful in determining heart health. Irregular heart beats or murmurs that may be associated with congenital heart problems and lead to dyspnea especially in cockatoo species. A pediatric stethoscope is the best instrument for auscultation of the sinuses, trachea and lungs, as well as the thoracic and abdominal air sacs in avian patients. Increased inspiratory sounds are associated with upper respiratory tract infections while increased expiratory sounds are associated with lower respiratory tract disease.[5] Normally no alveolar sounds are noted in avian species because birds have nonexpandable air capillaries rather than expandable alveloi. Auscultaion through a thin towel may help in detecting abnormal respiratory sounds.

The flexible trachea may be observed through the apterylae (the unfeathered areas of the cervical neck region). This inspection is particularly useful for detecting airsac mites in passerines by transillumination or, in larger birds, by using a flexible or rigid endoscope intralumninally.

As one moves down the bird's body during an external physical examination the caudal celomic cavity should be palpated. Any reduction in a bird's respriatory capacity will affect its ability to properly breathe. By palpating the caudal celomic area for space occupying lesions, this may give an indication why the patient may be having difficulty breathing. Common space occupying lesions of the body cavity that can lead to dyspnea include hepatomegly, tumors, eggs within the reproductive tract or internal ovulation that leads to nonseptic celomitis. The celomitis associated with internal ovulation often results in significant amounts of fluid produced within the body cavity. This fluid may be pale yellow in color with few if any cells found in sample smears submitted for clinical pathology evaluation.

During the physical examination one may hear sounds that may be associated with respiratory distress. Sneezing is often a condition that owners are concerned about when bringing their bird in for evaluation. Birds will sneeze for the same reasons that other animals and humans sneeze, irritation to the upper respiratory mucosa, in many instances to clear the respiratory tract of debris. Birds often shed feather dander and live in dusty environments predisposing themselves to sneezing and clearing the upper respiratory system. Often bird sneezes come in intervals of 3 or more. This is normal, but if the sneezing occurs on a regular basis other clinical signs (e.g., nasal discharge) may be indicative of a more serious illness.

Many parrot species, particularly Amazon parrots and grey parrots, have an uncanny ability to mimic their owner's habits including sneezing and coughing. A veterinarian must determine if the coughing, sneezing and or wheezing is a mimick trait or real. A good thorough history from the owner or caretaker will help determine if the condition is real or a learned behavior. Certain species have normal sounds that may be mistaken for abnormal respiratory conditions, grey parrots growl, cockatoos and cockatiels hiss, and pionus parrots will elicit moist respiratory sounds when they become excited.

Conditions Affecting the Upper Respiratory System

Companion birds are commonly exposed to infectious bacterial and fungal organisms through the environment, handfeeding techniques, foods and contact with other birds or owners who have handled unfamiliar birds.[12] In an immune-compromised state (e.g., neonatal or handfed birds, young birds, or birds with nutritional deficiencies or concurrent infectious processes), avian species become susceptible to these ubiquitous infectious organisms.

Gram-negative bacteria frequently isolated from birds with upper respiratory tract infections include *Escherichia coli*, *Klebsiella pneumoniae*, *Pseudomonas aeruginosa*, *Pasteruella mutocida*, *Yersina pseudotuberculosis*, and *Salmonella* spp.[2] Strains of *Streptococcus* spp. and *Staphylococcus* spp., *Mycobacterium tuberculosis*, and *Norcardia asteroides* are gram-positive bacteria that have been isolated from birds exhibiting rhinitis and sinusitis.[2] *Bacillus* spp., *Corynebacterium* spp., and *Lactobacillus* spp. are common nonpathogenic bacteria isolated from the upper respiratory system.[2] Spirochetes have been isolated from cockatiels diagnosed with upper respiratory infections, but this is uncommon.[2]

Aspergillus spp. is a common environmental fungal pathogen that infects the respiratory system of susceptible birds.[2] Grey parrots are commonly infected with this fungal organism. In the upper respiratory tract grey parrots usually develop fungal granulomas that result in enlargement of the nares. Immature immune systems, stress, growth, poor management, malnutrition, toxin exposure are common conditions that predispose birds to

Aspergillus spp. infections. Early diagnosis of aspergillosis is essential for treatment success and reduction of adverse anatomical changes associated with granuloma formation within the nares. Not only is *Aspergillus* spp. implicated as a causitive agent of proliferative nasal granulomas but E. *coli*, *Enterobacter* spp., *Pseudomonas* spp., *Klebsiella* spp. and *Candida* spp. have been isolated from birds showing signs consistent with this disease. Similar to *Aspergillus* spp. organisms causing this condition, malnutrition and poor air quality are important factors predisposing the nasal mucosa to these lesions.[2]

Candida spp. infections that originate in the oropharyngeal cavity may spread from the choana to the infraorbital sinus, resulting in varying degrees of dyspnea.[2] *Candida* spp. are usually opportunistic pathogens that infect birds suffering from primary infectious diseases or malnutrition, or young birds with immature immune systems. Lesions usually appear as white plaques in the oral cavity, sometimes extending into the choana or coexisting with a primary bacterial sinusitis.[13]

Gelatinous myxomatous material found in the nasal cavity and infraorbital sinus may indicate respiratoary cryptococcosis infection.[9] Birds with cryptococcosis will be depressed, dyspneic and unresponsive to treatment.[9]

There are several less common bacterial and fungal pathogens that may cause signs of upper respiratory disease including *Proteus* spp., *Haemophilus* spp., *Bordetella avium*, and *Rhinosporidium* spp.[9] Since these are uncommon causes of upper respiratory infection, microbial culture and isolation of the causative agent is indicated in any bird exhibiting rhinitis or sinusitis.

The intracellular bacteria *Chlamydia psittaci* and *Mycoplasma* spp. are often isolated from birds with clear nasal discharge (rhinorrhea), sneezing and conjunctivitis.[12] *Chlamydia psittaci* causes the disease avian chlamydiosis (AC) in pet bird species. *Chlamydia psittaci* is arguably the most important infectious disease that veterinarians treat because it is a zoonotic infection that can lead to death in improperly treated humans. This organism can also be present in an infected bird that does not exhibit any overt clinical signs. The organism is shed in its infectious form, the elementary body, through respiratory and fecal secretions. Recommendations for controlling infection among humans and birds includes protect persons at risk, maintian accurate reords of all bird-related trasactions for not less than one year to aid in identifying sources of infected birds and potentially exposed persons, avoid purchasing or selling birds that have signs of AC, isolate newly acquired, ill or exposed birds, test birds before they are to be boarded or sold on consignment still housing them under quarantine in a room separate from other birds, practice preventive husbandry, control the spread of infection and use disinfection measures. *Chlamydia psittaci* is susceptible to most disinfectants and detergents as well as heat, but it is resistant to acid and alkali pH. Recommended disinfectants includes Roccal® at a 1:1000 diluation or a 1:100 dilution (2.5 tablespoons/gallon of water) of household bleach.[10] Doxycycline is the treatment of choice and can be administered in the water, food, manufactured pellets, orally and intramuscularly.[10]

Although *Mycoplasma* spp. infections are not as serioius as AC in pet birds, infected birds show similar clinical signs.[14] There is currently a PCR antigen test available to detect *Mycoplasma* spp. in avian species.[14] *Mycoplasma* spp. are relatively low in infectivity but may affect high-density, poorly managed cockatiel aviaries.[9] Many known *Mycoplasma* spp. infect waterfowl, but unidentified species may be associated with psittacine infections.[9]

Members of the Poxviridae family are large DNA viruses with lipophilic inclusion bodies called Bollinger bodies.[15] The cutaneous form, characterized by papular lesions, is the most common form of *Avipoxvirus* infections in raptors and passrines but not in parrot species.[15] Papular lesions usually form on the featherless areas of the body due to the main form of transmission of the virus, the mosquito vector.[9] If the lesions interfere with air flow through the nares, respiratory distress may occur. Oral lesions associated with the wet or diphtheroid form may hinder respiration and cause dyspnea in psittaciformes and columbiformes.[15] Diagnosis of poxvirus infections may be made through lesions biopsy with histologic demostration of Bollinger bodies.[15] Good management techniques in young birds, prevention of vector (mosquito) exposure, separation of affected birds, and vaccination help control poxvirus outbreaks.

Certain forms of Newcastle disease (*Paramyxovirus*) virus manifest as acute upper respiratory infections, with mucoid nasal discharge and necrotic foci at the lateral aspect of the choanal cleft.[15] Isolation of the virus in respiratory discharge or fecal samples can identify affected birds during quarantine period. Live vaccines are manufactured for chickens and specified gallinacious birds. It is recommended to use Newcastle disease vaccines only on the species for which they have been approved.

The Orthomyxoviridae family of viruses contains the avian influenza virus and all other influenza virus and all other influenza virus taxons.[15] Influenza-group viruses are classified as A or C, with the A-group viruses being isolated from birds.[9] Purulent rhinitis may be noted in birds infected with influenza A virus.[9] virual isolation from swab samples from the cloacal and choanal cleft is the best way to identify the organism in live birds.[15] The importance to aviculturalists of psittacine birds is dependant on the amount of importantation occurring within the region or country.

Secondary physiologic reaction to scaly face mites (*Knemidocoptes pilae*) may occlude the nares, thereby causing upper respiratory distress.[9] Budgerigars are the most common avian species affected by this parasite. The mite can be identified from a skin/beak scraping of the affected areas. Reduction of stress through good nutrition and management techniques will help reduce mite infestations in budgerigars and other susceptible avian species.[9] Treatment includes systemic ivermectin and topical mineral oil.

The diet of many pet birds consists primarily of seeds, thus predisposing them to vitamin A deficiency. Vitamin A is essential in maintaining proper epithelial growth and health. If the lining of the respiratory tract is damaged or unhealthy, birds are inherently more susceptible to microbial infections. Epithelial integrity is lost and clincial signsappear when hepatic storage concentrations of vitamin A drop to less than 10% of normal.[16] Clinical respiratory signs associated with vitamin A deficiency may take the form of chronic upper respiratory infection, irritation, or hyperkeratotic plugs of the nares.[9]

Birds are sensitive to inhaled toxins and have been used historically as sentinel animal s to detect toxic levels of poisonous gases.[2] Clients should be educated about the adverse effects that fumes emitted from common household compounds have on companion birds.[12] Similar upper respiratory signs can occur with possible allergic reactions, seasonal occurrrences (low humidity), or when central heating or air conditioning systems are used. Clinical changes following inhalation of household fumes (e.g., Teflon®, air fresheners) may include irritation of mucous membranes, conjunctivitis, rhinitis, dyspnea, ataxia or peracute death.[2] One of the most common inhaled toxins is cigarette smoke. Exposure to passive cigarette smoke and associated irritated, inflamed respiratory epithelium may increase susceptibility to infectious diseases.

Tumors are rarely diagnosed in the upper respiratory tract of birds.[9] Squamous cell carcinomas have been found in the sinuses of birds. Recently a older Amazon parrot was diagnosed with a very aggressive squamous cell carnioma in the choanal cleft and a blue and gold macaw diagnosed with a proliferative lymphosarcoma mass in the same area. Fibromas, fibrosarcomas, and basal cell carcinomas are associated with the beak and fibrosarcomas and papillomas can occur in the oral cavity.[17] Budgerigars are the most common avian species reported with upper respiratory tumors.[17] Fine-needle aspirates or biopsies of the masses will aid in diagnosis of tumors.

Papilloma-like growths occasionally occur within the oral cavity, sometimes occluding the glottis or choanal cleft. The casuative agent has not be identified, but a virus is suspected, possibly a herpes virus. These masses are often very proliferative within the area in which they are found and extremely vascular. The vascular nature of these growths makes it difficult to remove a significant amount of affected tissue without causing considerable blood loss. Cryosurgery is a treatment technique that may be used with these oral/nasal growths.

Trauma of the upper respiratory tract occurs commonly in all avian speices. Flight concussion or trauma by a cagemate or other animal may affect the nares and/or sinuses. Treatment is based on location and severity of the injury. The first treatment concern is focused on cessation of the hemorrhage and pain management. After an assessment of beak trauma the owner should be informed on prognosis for healing and return to normal appearance. Even in the most extreme cases of upper beak trauma, birds can breathe normally and learn how to eat. Beak prosthesis is only a temporary resolution of the problem. There is only one guarantee, the beak prosthesis will fail.

Choanal or sinus atresia has been commonly identified in grey parrots, although other psittacine species may be affected.[2] Clinical signs of choanal or sinus atresia usually begin druing the first week of life with bilateral serous nasal discharge. The inability of the serous fluid to drain from the upper respiratory tract causes oculonasal discharge and swelling in the ventral areas under the eye.[2] Contrast media placed in the nares will stop abruptly at the level of a thickened palate, confirming the diagnosis of atresia.

Infections that cause irritation and subsequent blockage of the sinus communications may casue a "sunken-eye" sinusitis, whereby the skin covering the infraorbital sinus collapses against the frontal bone and suborbital arch. There is no skeletal involvement, but bacterial infections have be associated with this presentation in macaws and emus.[2]

The inhalation of foreign bodies (e.g., seeds, splinters and toys) into the external nares or choanal cleft occasionally occurs in companion birds, causing dyspnea and secondary irritation (rhinorrhea).[2] As mentioned earlier, endoscopic examination or contrast radiography will help determine whether a foreign body is present.

Conditions Affecting the Lower Respiratory System

Many of the same infectious organisms and problems that affect the upper respiratory tract also cause disease of the lower respiratory system. Unfortunately the increased surface area, environment and unique anatomy of the avian lower respiratory system allows for susceptibility to many more parasitic and viral diseases.[9] The bacterial species discussed as etiologic agents of upper respiratory tract infections can also casue tracheitis, laryngitis, pneumonia, respiratory abscesses and airsacculitis.[2] Biopsy and culture by endoscopy may reveal the causative microorganism.

Lower respiratory aspergillosis is more commonly diaganosed than upper respiratory fungal granulomas. Unfortunately lower respiratory aspergillosis is more difficult to diagnose than the overt nasal granuloma. The bird infected with aspergillosis involving the lungs and air sacs may be depressed and present with mild dyspnea. A complete blood count, plasma chemistry panel and radiographs are the basic diagnostic protocol for grey parrots presenting with the clincial signs described above. Many times the complete blood count will show a severe, chronic inflammatory leukogram. The total leukocyte count can range as high as 70,000 to 80,000 or it can be much lower. Radiographic evidence can range from "cloudy" air sacs, obvious opaque granulomas to no abnormalities noted, even in the most severe cases. At this time the recommended dignostic test for lower respiratory aspergillosisis is celomic endoscopy. Serologic antigen and antibody testing has not been perfected to the point of offering reliable results for antemortem diagnosis. Once the diagnosis is confirmed or a differential diagnosis list is formulated for a patient, treatment can be initiated. The proper treatment may be aided by specific methods that are effective treating the avian respiratory system. Voriconazole (20 mg/kg orally q24h for 21 days) or itraconazole (5-10 mg/kg orally q12-24h) is the drug of choice for avian aspergillosis.[18] Caution should be used when contemplating the use of this drug in grey parrots. Grey parrots are extremely sensitive to itraconazole and may become anorexic and develop a depressed state when treated with this drug. Terbinafine is alternative to itraconazole, dosed at 5-15 mg/kg. Due to the concern that terbinafine has questionable therapeutic effect in treating aspergillosis in avian species as a singlar medication it can be used in combination, at higher dosages and also nebulized, with itraconazole and voriconizole. For concurrent upper respiratory therapy or sinular upper respiratory therapy there is an effective sinus flush for parrot species. The sinus flush is comprised of 3.5 mg neomycin, 124.5 mg trypsin, 10 mg amphoteracin B added to 30 ml of a water soluble base to form a stock solution. For application of the sinus flush, 0.1-0.4 ml of the stock solution is added to 10-20 ml of saline for direct flushing through the nares once a day (Dr. B. Speer, written communication, 2016). Humidification, vaporization and nebulization are three types of aerosol therapy that have been used successfully to treat birds suffering from respiratory problems.[12] Humidification is the administration through nebulization of a saline solution into a treatment enclosure. The humidification of inspired air can improve the efficiency of the mucociliary blanket of the patient's respiratory tract.[12] Vaporization uses cool or warm mist to deliver topical medications to the mucous membranes. Volitile oil-based products added to human vaporizer formulations (e.g., eucalyptus oil) should not be used on avian species. These oil-based products can cause irritation and complicate and already compromised respiratory system. Nebulization is the exposure of a bird's respiratory tract to a finer mist than vaporizers. Nebulization helps maintain proper hydration of the respiratory epithelium, breaks up necrotic debris, and delivers antimicrobial agents to the upper and lower respiratory tract.[2] Nebulization is the most common form of aerosol therapy used in avian medicine. Depending on the agents being delivered, nebulization can be used 3-4 times per day for 10-15 minutes per session. Recommended equipment includes a nebulizer compressor (Sportneb®, model 3050, Medical Industries America, Adel, IA), infant nebulizer, nebulizing chamber. Therapeutic agent formulation for treatment of systemic aspergillosis, 5-10 ml saline, 3-5 drops mucomyst and amphotericin B, 7 mg/ml. Clotrimazole can be substituted for amphotericin B, 10 mg/ml for 30-60 minutes. Normally nebulization intervals are 10-15 minutes per treatment.[18]

Fungal hyphae of *Candida* spp. may extend into the proximal trachea from the oral cavity, resulting in varying degrees of dyspena.[2] Fungal hyphae may be seen in cytologic specimens taken from the trachea or other affected areas. Crytococcosis, trichosporonosis, mucormycosis and nocardiosis have been isolated from the lungs and air sacs of birds showing signs of respiratory compromise.

Respiratory disease outbreaks caused by viral infections are not uncommon in psittacine aviaries. Although most avipox infections affect the upper respiratory system, some forms of this virus affect the lower respiratory system. The diphtheroid form or "wet pox" may grow on the larynx or rarely, in the bronchi, causeing dyspnea.[15] A desquamative pneumonia with air capillary involvement may be found in canaries suffering from the septicemic form of canary pox. Diagnosis of avipox in the lower respiratory tract is made through histologic demonstration of Bollinger bodies in samples of suspect lesions. Diagnosis of poxvirus involving the lower respiratory system is usually made at necropsy because of the difficulty in obtaining biopsy samples and the life-threatening nature of the infection.

Infectious laryngotracheitis is caused by a herpesvirus that affects adult and growing birds older than 8 weeks of age resulting in acute severe respiratory disease. Canaries and several gallinaceous species are susceptible to this respiratory disease.[15] Birds usually die from asphyxiation because the cellular epithelial reaction in the infected trachea obstructs normal respiration. Bloody mucus emanating from the glottis, swelling of the infraorbital sinus and respiratory distress are clinical signs associated with infectious laryngotracheitis. Diagnosis can be made through ELISA, immunofluorescence, virus neutralization and immunodiffusion testing in suspect cases.[15] Amazon tracheitis virus is considered an adapted form of infectious laryngotracheitis virus.[15] Affected psittacine species have similar signs to birds suffering from infectious laryngotracheitis virus and usually die approximately 7 days after infection.[15]

Adenoviruses are separated into 3 groups by common group antigens, with group one, fowl adenovirus, causing lower respiratory infections in many avian species.[9] Although group one adenoviruses also affect other organ systems in psittacines, guinea fowl, quail, and pigeons, clincial respiratory signs, including tracheal rales, coughing and respiratory distress, indicate that this is primarily a respiratory infection. Viral isolation from fecal, pharyngeal, kidney or liver tissue samples will give a definitive diagnosis of an adenoviral infection.[15]

Virulence of Newcastle disease virus (*Paramyxovirus*) is host-specific and varies considerably among avian species. The virus enters the host through the oral or respiratory epithelium and disseminates through the body by infected erthrocytes.[15] Depending on the virulence of the Newcastle disease virus, acute respiratory death can occur, or chronic neurologic signs develop. Clinical signs associated with the viral infection usually involve the neurologic, gastrointestinal and/or respiratory systems. Diagnosis is made by serologic testing or by viral isolation from swab samplesof the feces or respiratoary tract.

Parasites more commonly affect the lower respiratory system of birds than the upper respiratory system. The lower respiratoary tract has a large surface area and possibly a better environment for these organisms to live and multiply.

Trichomonas gallinae is a protozoan parasite usually restricted to certainareas of the gastrointestinal system, but respiratory tissues may also be affected. Infections are manifested by white plaques on the gastrointestinal mucosa, glottis or trachea.[9] Diagnosis may be made by direct microscopic examination of a swab of the affected tissue and identifying the organism. Infection in young birds may reduce growth rate, while adults are usually emaciated and show respiratory distress.[8] Wild raptors are exposed through ingestion of infected pigeons.

Cryptosporidia are coccidian parasites that affect the mucosal epithelial cells of the respiratory tract, causing sinusitis, tracheitis, air sacculitis, coughing, sneezing and dyspnea.[9] Many cryptosporidial infections are considered opportunisitic. Infected birds should be evaluated for other primary disease problems. The organism is small and fecal smears stained with Giemsa, carbolfuchsin or periodic acid-Schiff stain can demonstrate the oocyst.[9]

Toxoplasma gondii are common parasites that have caused death in many avian species, including psittacines and raptors.[9] Respiratory distress may be noted in affected birds caused by congestion or necrotic foci within the lung tissue.[9] A definitive diagnosis can be made by histologic examination of the affected lung tissue and idenfication of the organism.

Sarcocystis appears to be isolated to North America. This coccidian parasite can cause respiratory distress in infected birds. Diagnosis can be made through histologic examination of affected tissue samples.

Gapeworms, *Syngamus trachea* and *Cyanthostoma* spp. cause severe dypsnea, frothing, and a bloody mucous discharge from the trachea and glottis.[2] Examining the discharge under a microscope or fecal flotation will reveal the worm or eggs. The copulating adults can be identified as a red "Y," with the female adult the larger of the 2 worms.

Filarid nemades are transmitted to birds by blood-feeding diptera.[9] The adults may be found in the air sacs but do not appear to cause tissue reaction. Passerines, especially Goulidan finches , are extermely sensitive to the tracheal mite *Sternostoma trachacolum*. The usual presenation of a finch or canary suffering from a tracheal mite infestation is a characterisitic clicking sound made during respiration..

Airborne toxins are inherently more serious in birds than in mammals because of the larger surface area and efficiency of the avian respiratory system. Polytetrafluoroethylene (PTFE) gas is released when nonstick cooking surfaces overheat or burn. Birds should never be housed or maintained in a kitchen because of potential exposure to this and other airborne contaminants generated from cooking. The lungs are the target organ of PTFE gas; hemorrhage and congestion are noted in exposed birds at necropsy.[9]

Disinfectants such as chlorine-based products may cause topical irritation to the respiratory tract when birds are exposed to low concentrations and death when high concentrations are used. Birds should be removed from cleaning areas if high concentrations of volatile compounds (e.g., carpet cleaner, aerosol disinfectants) are being used. Carbon monoxide is fatal at lower concentration in birds than in mammals because of the efficient avian respiratory system. Clinical signs may be acute death with bright red blood and pink to red tinged mucosal tissue. Tobacco smoke has already been described as a respiratory toxin and birds should be placed in areas where they are not exposed to this environmental toxin.

Reported tumors that involve the lower respiratory system of avian species include fibrosarcomas, papillomas, adenomas, adenocarcinomas and mixed epithelial and connective tissue tumors.[17] Clinical signs associated with tumors of the lower respiratory tract are dyspnea, wheezing and tail bobbing. Endoscopy and biopsy of the air sacs and lungs help identify the tumor type. Pulmonary hypersensitivity of macaws has been associated with poor ventilation and birds that produce powder down.[9] This syndrome is characterized by severe dyspnea and polycythemia because of hypoxia and can can result in right ventricular failure.[9] A definitive diagnosis can be made from a lung biopsy sample that shows interstitial infiltrate and fibrosis.[2]

Aspiration pneumonia can occur in young birds being hand fed, sick birds being tube fed and/or treated with oral medication. When treating birds orally, the last procedure should be oral administration of a therapeutic agent. If the bird regurgitates it should but in the cage or incubator without being held thereby giving it the best chance to clear its glottis. A struggling bird being held will cause aspiration of the ingested material to occurr. As mentioned earlier heart failure induced by disease or congential deformities will casue respiratory signs and often death.

Conclusion

Education of the owner regarding treatment protocol and basic respiratory anatomy is important to prevent chronic problems and reduce treatment noncompliance. If a bird with signs of dyspnea has no abnormalities on the diagnostic workup of the respiratory system, other factors must be examined (e.g., environment, allergies, cardiac disease). Only through knowledge of the anatomy and physiology, as well as understanding of disease processes of the avian respiratory system, can a veterinarian be successful in treating the avisn respiratory patient.

References

1. AAV Practice Survey. AAV Today. 1987; 1(3):98-99.

2. Tully TN, Harrison GJ. Pneumonology. In: Ritchie BW, Harrison GJ, Harrison LR. eds. *Avian Medicine: Principles and Application.* Lake Worth, FL: Wingers;1994:556-581.

3. Fudge AM, Reavill DR, Rosskopf WJ. Diagnosis and management of avian dyspnea: a review. *Proc Annu Conf Assoc Avian Vet.* 1993;187-195.

4. Fedde MR. Respiration. In: Sturkie PD. ed. *Avian Physiology.* 3rd ed. New York, NY: Springer-Verlag;1976:122-143.

5. Tully TN. Dyspnea and other respiratory signs. In: Olsen GH, Orosz SE. eds. *Manual of Avian Medicine*. St. Louis, MO: Mosby;2000;47-61.

6. Greenacre CB, Watson E, Ritchie BW. Choanal atresia in an African grey parrot (*Psittacus erithacus*) and an umbrella cockatoo (*Cacatua alba*). *J Assoc Avian Vet*. 1993;7:19-22.

7. Graham A. Nasal flushes for diagnosis and treatment. *Proc Annu Conf Assoc Avian Vet*. 1994;359-361.

8. McMillan MC. Radiology of avian respiratory diseases. *Compend Contin Educ Vet*. 1986;(8)8:551-558.

9. Tully TN. Avian respiratory diseases: clinical overview. *J Avian Med Surg*. 1995;9:162-174.

10. National Association of State Public Health Veterinarians. Compendium of measures to control *Chlamydia psittaci* infection among humans (psittacosis) and pet birds (avian chlamydiosis), 2010. Available at: http://www.nasphv.org/Documents/Psittacosis.pdf . Accessed July 2, 2016.

11. Lightfoot T. Avoiding disaster in the critical patient. *Proc Assoc Avian Vet*. 1998;265-271.

12. Gerlach H. Bacteria. In: Ritchie BW, Harrison GJ, Harrison LR. eds. *Avian Medicine: Principles and Application*. Lake Worth, FL: Wingers;1994:949-983.

13. Morrisey JK. Diseases of the upper respiratory tract of companion birds. *Semin Avian Exot Pet Med*. 1997;6:195-200.

14. Phalen D. Respiratory medicine of cage and aviary birds. *Vet Clin North Am Exot Anim Pract*. 2000;423-452.

15. Gerlach H. Viruses. In: Ritchie BW, Harrison GJ, Harrison LR. eds. *Avian Medicine: Principles and Application*. Lake Worth, FL: Wingers;1994:862-948.

16. Dorrestein GM, Zwart P, van der Hage MH, Schrijver J. Metaplastic alterations in the salivary glands of parrots in relation to liver vitamin A levels. *Proc First Int Conf Zoo Avian Med*. 1987;69-74.

17. Leach MW. A survey of neoplasia in pet birds. *Semin Avian Exot Pet Med*. 1992;1:52-64.

18. Carpenter JW: *Exotic Animal Formulary*. 4th ed. St. Louis, MO: Elsevier/Saunders;2013.

Handling Respiratory Emergencies

Susan E. Orosz, PhD, DVM, Dipl ABVP (Avian), Dipl ECZM (Avian), and
Marla Lichtenberger, DVM, Dipl ACVECC

Session #276

Affiliation: From Bird and Exotic Pet Wellness Center, 5166 Monroe St, Toledo, OH 43623, USA (Orosz), and Milwaukee Emergency Center for Animals, 3670 S 108th St, Greenfield, WI 53228, USA (Lichtenberger).

Abstract: Acute respiratory distress is usually a life-threatening emergency. In the emergency room, observation of the patient's breathing pattern, respiratory sounds and a brief physical exam are the most important tools for diagnosis and treatment of respiratory distress. Anatomic and physiologic understanding of the 5 different areas of the respiratory system are important in understanding the diagnosis and treatment protocols for each area.

Introduction

Acute respiratory distress is usually a life-threatening emergency. In the emergency room, observation of the patient's breathing pattern, respiratory sounds and a brief physical exam are the most important tools for diagnosis and treatment of respiratory distress.[1-5] These tools will help the clinician localize the lesion which will facilitate immediate steps to stabilize the patient. This presentation will give an anatomic and physiologic overview of the 5 different areas of the respiratory system and describe diagnosis and treatment protocols for diseases occurring in each area. Every bird requires immediate stabilization in a warmed oxygenated incubator when there is respiratory distress (oxygen flow at 5 L/min delivers an oxygen concentration of 78%-85% in the Lyons cage). A bronchial dilator, terbutaline (e.g., 0.01 mg/kg IM q6-8h; Brethine, Novartis Pharmaceuticals Corporation, East Hanover, NJ, USA) and an anti-anxiety analgesic, butorphanol (1-2 mg/kg IM q2-3h; Torbutrol, Intervet Inc, Millsboro, DE, USA) are usually given to birds in respiratory distress prior to placing in the oxygen enriched incubator.

Five Areas of the Respiratory System

Upper airway and infraorbital sinus

History: The owner will often describe nasal discharge, sneezing or yawning, redness and/or swelling of the sinus area around the eye, and ocular discharge, and some birds will rub their beak on the perch or on the cranial margin of the carpus.

Pattern of breathing, respiratory sounds, and physical exam: There will be a soft nasal sound with open beak breathing only with complete, bilateral nasal obstruction. There will be an increased respiratory rate without increased effort. There may be exudate or plugging of the opening of the nares. The area around the eye may be swollen. The bird may flick his head.

Etiology: Causes include infraorbital infection (bacterial, fungal, avian TB), foreign body, neoplasia, abscess, granuloma, fibrinous plug,[6] and avipox virus.[7]

Diagnostic tests: The material from a nasal flush or infraorbital sinus aspiration can be used in a number of diagnostic tests. These may include: PCR for *Aspergillus* sp, *Chlamydophila psittaci*, avipox virus and mycobacterium sp. (TB), gram stains, and culture and sensitivity of the aspirate. Endoscopy of the choanal area and biopsy of any lesions is the next diagnostic procedure that could be used to diagnose cases that are nonresponsive to initial treatment. Radiographs, ultrasound and CT scan of sinuses may be useful.

Treatment: Many times a nasal flush will help remove mucous and discharge so that the bird can breathe easier. Initial cytology of the fluid can help the clinician decide on treatment (antibiotics, antifungals) pending cultures and sensitivities. Eye wash saline that is pH balanced can be used for the diagnostic flush and/or for treatment (can be 2 times a week as needed). F 10 has also been used for a flush in our clinic as well. We use the 1 part to 250 part dilution and mix with eye wash saline. This is important as the pH should be balanced close to 7.2. Saline from the bag for injection often has a low pH and will cause additional edema to the mucosa. Therapeutic laser use combined with flushes have also been successful in our clinic.

Large Airway (Glottis, Trachea to the Syrinx)

Anatomy and physiology: The rima glottis and trachea are larger in diameter on a per weight basis than they are in mammals. This principle provides one reason why it is easier to intubate birds than mammals. This increased diameter directly reflects one physiologic adaptation of birds—the use of their beaks for the manipulation of objects. The increased length of the trachea is part of an adaptation in birds whereby the increased length of their necks allows use of their beak to manipulate objects. However, the increase in length causes an increased air resistance. A wider rima glottis, as well as a wider trachea, is necessary to reduce this resistance. To compensate for the wider and longer trachea and its corresponding increase in dead-space, the respiratory rate of birds is less than that of mammals, with a larger tidal volume. These anatomic and physiologic differences are important when selecting endotracheal tubes for anesthetic procedures. In addition, the tracheal rings of birds are complete. They are shaped like a signet ring and overlap to form a more rigid trachea than that of a comparably sized mammal. These two facts are important to reduce the possibility of tracheal kinking when moving the neck. Overinflation of a cuffed endotracheal tube is potentially easier in a bird because of the less expansive nature of the trachea. Overinflation can lead to mucosal edema and difficulty breathing. Whenever possible, uncuffed endotracheal tubes should be used.

The trachea often bifurcates immediately after entering the thoracic inlet. At this bifurcation is the syrinx, which represents the voice box of birds. This narrowed internal diameter of the syrinx is a common site for tracheal granulomas and foreign bodies.

History: The bird may have demonstrated inspiratory stridor that has increased in intensity or the respiratory distress could be acute due to complete obstruction. If this happens after extubation, there is tracheal swelling from the endotracheal tube. This will require administering non-steroidal anti-inflammatories (NSAIDs) by IM injection. The NSAIDs can include meloxicam, robenacoxib or carprofen. An air sac tube will need to be placed in the left caudal thoracic air sac with possible ventilation support until the swelling subsides. The left caudal thoracic air sac is larger than the right. Patency of the trachea can be tested by clamping off the air sac tube and monitoring breathing.

Pattern of breathing, respiratory sounds, and physical exam: There is loud inspiratory stridor with large airway blockage. There is an increase rate and effort with open beak breathing. The bird may have no other physical exam findings. This is often acute and so there will be no other findings. Since this amount of respiratory effort is demanding, it will require that the bird use a considerable amount of energy to maintain this respiratory effort. This will cause rapid reduction of the pectoral muscle mass. If there is a complete obstruction the bird will present gasping for air.

Etiology: Causes of large airway disease include: obstruction by a foreign body (ie, seed in cockatiels or squames), and masses that fall into the trachea and these can include: *Aspergillus* granuloma or oropharyngeal granulomas. The squames can include plaques of yeasts from infection in the upper airway and from the hyperplasia that accompanies hypovitaminosis A. In addition, there can be granulomas from *Mycobacterium* infection of which *genavense* [8] is associated with transmission from humans to birds. A tracheal fibrinous seal[9] will also close off the airway.

Diagnostic tests: Radiographs, bloodwork, trans-tracheal wash, and endoscopy can help to diagnose the cause of airway obstruction.

Treatment: Large airway obstruction requires rapid anesthetic induction with sevoflurane/isoflurane anesthesia, rapid intubation and ventilation with 100% oxygen. If there is a total upper airway obstruction, emergency air sac tube placement is performed. The bird will relax and breathe easier under inhalant anesthesia. Endoscopy can be used to remove an obstruction as a granuloma or seed caught at the syrinx. With material that is soft and flexible stuck in the trachea, blowing into the air sac tube to send the material into the lungs, particularly small birds where the endoscope does not fit is an option to try.

Due to difficulty breathing most birds improve at least slightly with anxiolytic drugs and/or the use of an opioid. Midazolam at 0.2-2 mg/kg IM or intranasally and/or butorphanol at 1-2 mg/kg IM relieves stress and allows them to breathe to enhance perfusion.

Small Airway

Anatomy and physiology: The main stem or primary bronchi represent the continuation of the syrinx. Each perforates the septal surface of the lung and continues into the lung parenchyma as the intrapulmonary primary bronchus. The portion of the bronchus is long as it continues to the caudal extremity of the lung.

History: In blue and gold macaws, they often present with the history of periodic respiratory distress and commonly are raised with a cockatoo or African grey. This feather dander provides the basis of presumed allergic responses in the small airways. Other historical information for the patient can include exposure to an offending toxin such as smoke, aerosol, or cigarettes.

Pattern of breathing, respiratory sounds, and physical exam: There will be a very soft (heard by placing ear next to bird's beak) expiratory wheeze. The bird will be in extreme respiratory distress with open beak breathing and may even fall on its side gasping.

Etiology: Small airway disease includes inhalant respiratory irritants as smoke, aerosols, cigarette smoke, candles, heated polytetrafluoroethylene,[10] overheated foods or other items in the microwave.[11]

Diagnostic tests: Stabilize the bird for 12-24 hours with bronchodilators and oxygen. Radiographs and CBCs are useful in the diagnosis for an infectious disease. Since it is often not purely a small airway disease process an IM injection of Lasix may improve lung performance. Gram stains of the choanal slit will also help to rule in a disease process.

Treatment: The most effective bronchodilators are pharmaceutical agents that stimulate β-adrenergic receptors in bronchial smooth muscle and promote smooth muscle relaxation. For this reason, they have become the first line of treatment for bronchoconstriction causing acute respiratory distress. These β-receptor agonists are most effective and least toxic when they are given as an aerosol that is inhaled. B-agonists, such as terbutaline can be given IM in birds (0.01 mg/kg q 6-8 h) initially and then continued by nebulization.

Nebulizers are designed to convert liquids to appropriately sized aerosol particles for inhalation. The nebulizer consists of a disposable or reusable nebulizer and a pressurized gas (air or oxygen) source. A small volume of medication and a larger volume of diluent are placed into the nebulizer chamber.

Ultrasonic nebulizers generate an aerosol through ultra-high-frequency vibration of a piezoelectric crystal at the bottom of a liquid. The advantage of using a ultrasonic nebulizer include faster nebulization time, longer product life, smaller particle size, and quieter operation compared with a jet nebulizer. The veterinary nebulizer used most commonly is an ultrasonic nebulizer (DVM pharmaceuticals, Inc, IVAX Corp, Miami, FL, USA), delivers particle size of 2-7 microns, and can be used by an owner in the home. It is not known if the particle size will reach the atria of the lung but use of this nebulizer will deliver particles to the bronchi for treatment of bronchoconstriction.

With disease or irritants to the mucosal surface, one of the first lines of defense for the bird is the use of the mucociliary escalator. This beating of the cilia on the pseudostratified epithelium becomes exhausted as the mucus becomes thicker. It thickens with reduced humidity in the environment, and dehydration of the patient. The epithelium will be modified with all seed or poor diets. Hypovitaminosis A causes squamous metaplasia of the epithelium making it ineffective for removal of foreign material including infectious particles. For this reason, nebulization is an important part of therapy for most respiratory problems.

Coelomic Cavity Disease (Mass, Fluid, or Organomegaly)

Anatomy and physiology: The coelomic cavities of birds are similar between species and in general there are 16 distinct and separate cavities in the adult bird. Eight of the cavities are air sacs and the remaining 8 are cavities of the coelom proper with 5 as peritoneal cavities and the remaining 3 represented in mammals as well. The mammalian cavities include: the right and left pleural cavities and the pericardial cavity. The 5 peritoneal cavities of birds include the left and right ventral hepatic peritoneal cavities, the left and right dorsal hepatic cavities, and the intestinal peritoneal cavity.

History: The bird will likely have a history of lethargy, decreased appetite and not doing well prior to the respiratory distress.

Pattern of breathing, respiratory sounds, and physical exam: The bird will have increase respiratory rate and effort which will become extreme. This will lead to exertional dyspnea with open beak breathing on handling. There will be no respiratory sounds. A distended abdominal lift will be found on physical exam. The abdomen often is distended and may be firm.

The respiratory recovery time will be prolonged. Respiratory recovery time should normally be less than 5 minutes after brief handling during the physical exam. Birds with lower respiratory tract disease will have prolonged times before normal recovery or may not completely recover based on the degree of involvement.

Etiology: Causes of coelom space disease include: Fluid accumulation (heart disease; hypoalbuminemia; liver disease; egg-related peritonitis[12]; organomegaly (hepatic lipidosis), mass (neoplasia, abscess, egg).

Diagnostic tests and treatment:

Ascites: Ascites causing coelomic distension will require immediate coelomocentesis. Coelomocentesis is done on the right lateral coelom just cranial to the cloaca. Use a 25-22 gauge angiocatheter and syringe. The size of the needle and syringe is based on the size of the bird. However, thick material like pus or yolk will not move through the catheter smaller than at 22 g. It is better to use a smaller syringe as it produces less suction so that the organs do not shut off the opening as you are drawing off the fluid. Often, ultrasound can help guide thoracentesis if the fluid is flocculated and difficult to aspirate.

Save some of the fluid recovered for culture/sensitivity, measurement of total solids and cytology. Transudate is characterized by a clear to pale-yellow color, with a low specific gravity (<1.020), low protein (1 g/dl) and a low cellularity. The most common causes of a transudate are liver failure (eg, cirrhosis) or heart failure. An exudate is characterized by a high specific gravity (>1.020), a high protein content (>3 g/dl) and presence of inflammatory cells and/or mesothelial cells. Septic exudates may contain intracellular bacteria.

Septic peritonitis is differentiated from nonseptic peritonitis by identifying a lower coelomic glucose than serum glucose in septic peritonitis. Turbid yellow, green or brown yolky fluid of inspissated yolk material in the coelomic cavity is indicative of ectopic ovulation or a ruptured oviduct. Cytology reveals yolk material and fat particles.

Free yolk in the abdomen may be absorbed and inflammation is treated with NSAIDs. If bacteria are found on the gram stain of the material or cytology, antibiotics are used systemically along with an antifungal drug. Culture and sensitivity testing for bacterial and fungal elements should be considered as well but therapy can be initiated after gram staining of the material. The inflammatory aspect of this condition will require that the patient is checked frequently as fluid build-up with respiratory compromise can occur quickly. As the abdomen is aspirated frequently it is important to monitor the protein in the fluid and the protein in the blood. Often this will result in protein depletion and further complicate therapy. Feeding foods higher in protein and/or providing enteral nutrition may be helpful.

In neoplastic disease, there is usually bloody fluid on cystocentesis. Occasionally exfoliated neoplastic cells can be identified on cytology. Abdominal ultrasound is performed to identify presence of an abdominal mass. A 22-gauge needle or angiocatheter is used for ultrasound guided aspiration of the mass and cytology evaluation is then performed.

Organomegaly, egg binding, or mass lesions: Organomegaly, egg binding or a mass lesion is suspected when the coelom is firm on palpation. Ultrasonography is useful for identification of the egg (eg, laminated egg, ectopic egg, or normal egg) or ultrasound guided aspiration of the mass. Endoscopy may be used to biopsy the mass or liver. Medical care for an abnormal egg is attempted first with fluids provided by SC injection, warmth, enteral nutrition by gavage feeding and calcium injection. Pain medications such as butorphanol or tramadol with or without midazolam are helpful to get the egg to pass normally or by using abdominal pressure. Non-steroidal anti-inflammatories are also important. Surgical exploratory may be required if the egg does not pass in 24 hours or if the egg is within the coelomic cavity.

Parenchymal Disease (Lungs and/or Air Sacs)

Anatomy and physiology: Birds do not have a diaphragm and therefore rely on changes in pressure between their air sacs relative to the atmospheric pressure to move air through a non-expandable lung. The pressure changes within the air sacs are the result of volume changes in the thoracoabdominal cavity. On inspiration, the ribs move cranioventrally, thereby increasing the thoracoabdominal space. This results in a lowering of the pressure in the air sacs. On expiration, the ribs move caudodorsally, thereby reducing the space in the "chest" and increasing the pressure in the air sacs. Expiration is an active process requiring skeletal muscle contraction. For this reason, it is extremely important when holding a bird that the sternum can move freely or the bird will suffocate.

Gas exchange occurs in the walls of the parabronchi, the atria, and more importantly in the air capillaries, which are the avian equivalent of alveoli. The blood-gas barrier of birds is similar to that of mammals in that it consists of an endothelial capillary cell, a common basal lamina, and an air capillary epithelium of squamous cells. The difference is that the blood-gas barrier is much thinner in birds than in mammals. The diameter of the air capillaries of birds is much smaller than that of a mammalian alveolus, allowing for a much larger number of air capillaries in a given space when compared with mammals.

The epithelial cells of the lung act as fixed macrophages and transfer engulfed material to interstitial macrophages. Air that enters the caudal air sacs is filtered less than that of the cranial air sacs, making the caudal air sacs more susceptible to disease.

History: There is usually a history of not doing well, lethargy, and anorexia. Birds that are having difficulty breathing are unable to spend time eating as that further complicates their ability to breathe.

Pattern of breathing, respiratory sounds and physical exam: The bird shows signs of an increased respiratory rate and or effort. There may be no respiratory sounds or there are harsh sounds from the chest. The bird is usually in poor body condition with reduced pectoral mass. There feather quality is often dull as well. The heart may be loud or bounding due to the increased resistance of blood flow in the parenchyma of the lungs.

Etiology: Causes of parenchymal disease include: cardiogenic pulmonary edema; smoke inhalation; aspiration pneumonia; *Aspergillus*, *Chlamydophila psittaci* infection, bacteria, avipoxvirus, reovirus, Encephalitozoonosis,[13] and neoplasia.[14] Acute trauma to the lungs can also occur.

Diagnostic tests and treatment: Cardiac disease will require blood pressure measurements, radiographs, and an echocardiographic exam. Other diseases on the differential list require a variety of testing, endoscopic biopsies, and/or cultures and sensitivities. Gram stains of the choanal slit may give some indication of a bacterial or fungal infection. A CBC with a profile should be considered as well.

Parenchymal disease: Parenchymal disease may benefit from use of parental or nebulization of bronchodilators and antibiotics. An air sac tube may partially improve respiration in birds with primary lung and air sac disease. Each tube can be left in the patient approximately 3 days.

- Careful auscultation is indicated for a heart murmur or gallop or persistent arrhythmia suggestive of cardiogenic edema. If heart failure is suspected, administer furosemide (2-4 mg/kg IV) and nitroglycerine ointment on the back and allow the bird to stabilize prior to further diagnostics.

- Birds with a history of regurgitation or vomiting may benefit from broad-spectrum antibiotics (eg, enrofloxacin combined with clavamox or clavamox alone, trimethoprim sulfadiazine) and bronchodilators when aspiration pneumonia is suspected. Since these birds are compromised, it is important to administer an antifungal agent as well to reduce the incidence of secondary fungal infections.

- Parrots and other birds may benefit from furosemide to reduce poor performance of the blood-air barrier.

- Non-steroidal inflammatory medications may also reduce the inflammatory process in the lung parenchyma.

- Birds suspected of aspergillosis (leukocytosis and pulmonary nodules) should be treated with terbinafine hydrochloride (Lamisil, Norvartis Pharmaceuticals, East Hanover, NJ, USA) at 10-15 mg/kg q12h for a minimum of 4 weeks or an azole that may include itraconazole or voriconazole. Itraconazole should not be used in African grey parrots as they have an idiosyncratic reaction possibly based on plasma protein binding. Monitoring with CBC and TP at 3-4 week intervals will help determine when the antifungal drug can be withdrawn.

- Birds suspected of psittacosis (leukocytosis, clinical signs varying from chronic unthriftiness to acute anorexia, diarrhea and respiratory distress, and "lime green feces") should be treated with drugs inhibiting the growth of *Chlamydophyla*. Doxycycline is the main agent. Treat with doxycycline for 45 days. Testing by combined PCR testing using samples of the oropharynx, blood, and fresh feces is most accurate and should be performed before the start of therapy. Many protocols are recommended.

- Birds with pulmonary hypertension should be administered sildenafil TID.

References

1. Hillyer H. Clinical manifestations of respiratory disorders. In: Altman R, Clubb S, Dorrestein G. Quesenberry K, eds. *Avian Medicine and Surgery*. Philadelphia, PA: WB Saunders; 1997:419-454.

2. Rupley A. Critical care of pet birds. Procedures, therapeutics, and patient support. *Vet Clin North Am Exotic Anim Pract.* 1998;1:21-22.

3. Quesenberry K, Hillyer E. Supportive care and emergency therapy. In: Ritchie B, Harrison G, Harrison L, eds. *Avian Medicine: Principles and Application.* Lake Worth, FL: Wingers Publishing; 1994:406-407.

4. Doneley B, Harrison GJ, Lightfoot TL. Maximizing information from the physical examination. In: Harrison GJ, Lightfoot TL, eds. *Clinical Avian Medicine.* 2nd ed. Palm Beach, FL: Spix Publishing; 2006:153-212.

5. Graham JE. Approach to the dyspneic avian patient. *Semin Avian Exotic Pet Med.* 2004;13(3):154-159.

6. Burr EW. Intranasal caseous fibrinous plug causing upper respiratory distress in two African grey parrots (*Psittacus erithacus*). *Avian Dis.* 1981;25(2):542-544.

7. McDonald DE, Lowenstine LJ, Ardans AA. Avian pox in blue-fronted Amazon parrots. *J Am Vet Med Assoc.* 1981;179(11):1218-1222.

8. Kiehn TE, Hoefer H, Bottger EC, Ross F, Wong M, Edwards F, Antinoff N, Armstrong D. Mycobacterium fenavense infections in pet animals. *J Clin Microbiol.* 1996;34(7):1840-1842.

9. Good DA, Heatley TN, Tully Jr, JA Smith. Anesthesia case of the month. Partial obstruction of the trachea was a differential diagnosis for the bird's respiratory problems. *J Am Vet Med Assoc.* 2001;219(11):1529-1531.

10. Wells RE, Slocombe RF, Trapp AL. Acute toxicosis of budgerigars (*Melopsittacus undulates*) caused by pyrolysis products from heated polytetrafluoroethylene: clinical study. *Am J Vet Res.* 1982;43(7):1238-1242.

11. Zaen AL, Rietveld. Inhalation trauma due to overheating in a microwave oven. *Thorax*. 1993;48(3):300-302.

12. Gorham SL, Akins M, Carter B. Ectopic egg yolk in the abdominal cavity of a cockatiel. *Avian Dis*. 1992;36(3):816-817.

13. Poonacha KB, William PD, Stamper RD. Encephalitozoonosis in a parrot. *J Am Vet Med Assoc*. 1985;186(7):700-702.

14. Andre J-P, Delverdier M. Primary bronchial carcinoma with osseous metastasis in an African grey parrot (*Psittacus erithacus*). *J Avian Med Surg*. 1999;13(3):180-186.

Section 7

AAV Master Classes Track 2

Backyard Poultry Practical Information You Can Use Presented by Systems

Cheryl B. Greenacre, DVM, Dipl ABVP (Avian), Dipl ABVP (Exotic Companion Mammal)

Session #076

Affiliation: From the College of Veterinary Medicine, Department of Small Animal Clinical Sciences, 2407 River Drive, C-247, University of Tennessee, Knoxville, TN 37996, USA.

Abstract: Backyard poultry are becoming popular and therefore are presenting to veterinary practices for individualized care more frequently. Some of the more common diseases of backyard poultry are discussed below, as well as diagnostic and treatment options.

Respiratory Diseases

Mycoplasmosis (most common)

There are three different species of *Mycoplasma* that can infect chickens. *Mycoplasma gallisepticum* causes respiratory disease in chickens, but an infectious sinusitis in turkeys. *Mycoplasma meleagridis* causes an air sacculitis and skeletal deformities in turkeys. *Mycoplasma synoviae* causes air sacculitis and synovitis /lameness in chickens. *Mycoplasma gallisepticum* (MG) is seen in backyard flocks and is of concern because it can easily spread to nearby commercial flocks and cause economic devastation for that commercial flock. Most commercial flocks are MG free. To participate in the National Poultry Improvement Plan (NPIP) a flock needs to be MG free. Transmission is through fomites. Clinical signs of MG in chickens includes an upper respiratory disease with swelling of the infraorbital diverticulum of the infraorbital sinus with caseated pus. The best prevention is to depopulate and repopulate with clean stock. Treatment can be attempted with antibiotics (spectinomycin, lincomycin, erythromycin, or tylosin), but birds remain carriers for life.

Infectious bronchitis virus

Infectious bronchitis virus, caused by a coronavirus, affects only chickens and has a worldwide distribution. Younger, immunosuppressed chickens show worse clinical signs than older immunocompetent chickens. Clinical signs include upper respiratory signs including sneezing, gasping, rales, and nasal discharge. Young are affected worse than adults showing gasping and labored breathing. In an affected flock of chicks, all birds will typically develop clinical sings within 36-48 hours and the clinical disease will typically last approximately 4 days (longer if secondary infections develop). Older chickens show a 5-10% drop in egg production for about 10-14 days. The reproductive tract can also be affected by this virus resulting in irregular and roughened eggshells with watery albumin and decreased egg production. Also newer strains may affect the kidneys. Tests include virus neutralization, hemagglutination inhibition or ELISA. The best method to control is to disinfect, repopulate and use live vaccine. This disease is highly contagious that easily spreads via airborne particulate matter and via fomites. There is no treatment, but antibiotics can be given to prevent secondary bacterial infection, especially with infectious coryza. The virus is easily destroyed by disinfectants, sunlight and heat. Increasing the environmental brooder temperature by 5 degrees Fahrenheit helps chicks to recover. There is a vaccine available, but it is not used in backyard flocks since there are numerous serotypes.

Infectious coryza

Infectious coryza affects mainly chickens (it can affect pheasants and guinea fowl as well). The causative agent is *Haemophilus paragallinarium*, a Gram negative rod. The incubation period is 1-3 days and the course of the disease is approximately 4-12 weeks. Clinical signs include upper respiratory signs and swelling of the face with foul smelling (hallmark) and sticky nasal and ocular discharge, dyspnea and rales. Clinical signs also include decreased egg production. Birds that recover are lifetime carriers. Transmission is through direct or indirect contact. Mortality is about 20-50%. Treatment consists of giving antibiotics such as sulfa drugs, erythromycin or tetracycline, but since affected birds are lifetime carriers, the only way to control the infection is through depopulation and leaving the premises vacant for at least 30, preferably 60 days after cleaning and disinfecting before repopulating. Avoid mixing flocks or mixing different ages and sources of birds. There is a vaccine available, but it is seldom used in backyard flocks.

Infectious laryngotracheitis (ILT)

Infectious laryngotracheitis affects only chickens (and pheasants). The causative agent is a Herpesvirus. Chickens older than 14 weeks are more affected than younger chickens, so the disease is usually seen in mature chickens. There is a mild form in the US that is associated with decreased egg production, conjunctivitis, nasal discharge, swollen infraorbital sinus and in more severe cases moist rales. Shaking of the head and flinging necrohemorrhagic material from the trachea is a hallmark of this disease including an inspiratory dyspnea and death. Milder strains in the US may not present with hemorrhagic tracheitis. At gross necropsy a mucoid to necrohemorrhagic tracheitis is present. Diagnosis is confirmed via virus isolation, ELISA, or indirect fluorescent antibody test. Prevention is through the use of a live vaccine. The disease can be spread by fomites. Properly dispose of dead birds to prevent spread (incinerate). Antibiotics can be used, but it is better to depopulate and then vaccinate new birds.

Avian influenza – need to know about this one

Avian influenza affects many species. The causative agent is an orthomyxovirus. It is also known as "Fowl Plague". The clinical signs are variable since there are mild and highly pathogenic forms, but include anorexia, decreased egg production and respiratory disease with the mild form to respiratory distress, facial swelling, diarrhea and neurological signs with the highly pathogenic form. Usually the mild form of the disease is associated with high morbidity and low mortality. Basically coughing, sneezing and sudden death are the typical signs of this disease. This a highly contagious disease associated with high mortality of both domestic and wild birds. There is no treatment. Testing usually involves virus isolation. This is a reportable disease. A killed vaccine is available, but not used in backyard flocks. Depopulation is usually the recommended action during an outbreak and the disease should be reported to the state veterinarian.

Newcastle's disease – need to know about this one

Newcastle's disease affects many species of birds. The causative agent is a Paramyxovirus. There are 4 different forms that vary in severity. The least pathogenic form is the Lentogenic form which causes a mild upper respiratory disease and usually only affects the young. The mesogenic form also causes a mild upper respiratory disease, decreased egg laying, and has a low mortality. The Neurologic/velogenic form causes a sudden onset of upper respiratory disease followed by neurological signs with approximately a 50 to 90% mortality. The worst form, which is a foreign animal disease for the US, is the "Exotic Newcastle's Disease" END or also known as the viscerotrpic velogenic form that is associated with neurological signs and high mortality. In the US there is a vaccine for the first three forms listed here, but in the US we do not vaccinate against END, we test and eradicate since it is a foreign animal disease. There is a serology test available at California's San Bernardino County Laboratory for PMV 1,2 and 3. A paramyxovirus; many different kinds of Newcastle disease; mostly mild disease in poultry and they are vaccinated at one day of age; causes mild conjunctivitis in people; test at US Quarantine stations. Exotic Newcastle's Disease (aka VVND or viscerotropic-velogenic Newcastle's Disease) is a foreign animal disease and millions are spent to eradicate all exposed birds when there is an outbreak in the US (last one was in California, 2002/2003 – 3.5 million birds)

Musculoskeletal Diseases

Marek's disease (most common)

Marek's disease affects only chickens. Causative agent is a Herpervirus. Clinical signs are generally seen in birds that are 12-20 weeks of age, but can be seen as young as 3 to 4 weeks of age. This highly contagious disease is associated with a high morbidity and low mortality. The incubation period is typically 4-12 weeks, clinical signs are seen in 10 -12 weeks chickens. Transmission is mainly via virus shed in skin and feathers. The virus can persist in the environment indefinitely (feathers and dander in poultry houses and yards). Clinical signs of the paralytic form include an assymetrical paralysis since the virus affects the "sciatic" (technically the ischiadic) nerve and presents as a bird that cannot stand because one leg is pointed forward while the other leg is pointed backward. A clinical sign of the ocular form is "gray eye" where the color of the iris gradually changes from brown to gray and the pupil becomes irregularly shaped. There is no treatment. Prevention is by administering a polyvalent vaccine in the egg or after hatching at 1 day of age. Understand that no vaccine is100% protective and that vaccine administration may not be 100% at the hatchery since it is typically a spray on vaccine as the chicks go by on the conveyor belt. Marek's disease must be differentiated from lymphoid leucosis based on age of clinical signs, and gross and histopathological lesions (ischiadic nerve enlargement).

Lymphoid leukosis

Technically LL is an infectious disease, but is basically a neoplasia and presents as such. Affects only chickens. Clinical signs usually occur in older chickens at approximately 24 to 40 weeks of age. The clinical signs are non-specific and include anorexia, weakness, pale comb and distended abdomen. At gross necropsy grey to white tumors are observed in the liver and other organs. The clinical signs are sometimes difficult to differentiate from those of Marek's disease, but lymphoid leukosis does not occur before 14 weeks of age, whereas Marek's disease usually occurs at 10-12 weeks of age. The best prevention is to test and cull positive breeder birds.

Pododermatitis

Ulcerative or non-ulcerative pododermatitis tends to occur if the chicken is overweight, on a roughened surface, or if one leg and foot bears more of the body weight than the other, or a combination of all these factors. There are varying grades of ulcerative pododermatitis from mild with hyperemia of the skin, to severe with osteomyelitis of underlying bone. A radiograph is the best method to determine if there is underlying osteomyelitis, a condition that requires long term antibiotic therapy and probably debridement of necrotic tissue. Most cases of ulcerative pododermatitis are somewhere in between mild and severe and consist of a thickened area of skin on the plantar surface of the foot usually over the metatarsal pad, but can also be seen on the phalangeal pads. There are differing opinions as to how aggressive to be with debriding the tissue or not – usually if there is necrosis, then debridement is necessary. Soaking the foot will greatly soften the tissue and ease surgery. Surgery should be performed under general anesthesia with administration of a pain reliever such as butorphanol since this is inherently a painful procedure. An aspirate or tissue sample is often needed to culture the area and prescribe the appropriate antibiotics. The substrate should be made as soft as possible and kept clean. Underlying lameness should be corrected. Pain relief should be addressed.

Egg-related peritonitis (most common)

Commonly at necropsy hens can have egg related peritonitis. Generally it is mild and not as severe as we are used to seeing in parrots. Diagnosis is based on typical clinical signs of lethargy, radiographs, or endoscopy. Occasionally antibiotics and/or surgery is necessary.

Intestinal/Hepatic Parasites

Coccidiosis

There are many different species of coccidia in domestic fowl: 9 in chickens, 7 in turkeys, and at least 4 in quail. They are host specific and not zoonotic. A flock may develop resistance to one species only to be infected with another species. The typical clinical signs of coccidiosis are diarrhea, unthriftiness, and variable levels of mortality. Cecal coccidiosis is worse in that it typically causes bloody droppings and is associated with higher mortality, whereas intestinal coccidiosis is typically more chronic in nature and is associated with a lower mortality. The clinical signs of coccidiosis are more severe in young (4-16 weeks of age) chickens by having bloody diarrhea, pale combs, lethargy, tendency to huddle, partial anorexia, weight loss, dehydration, and death. As chickens get older they become more resistant and show little to no clinical signs, but can act as carriers to later expose young chicks. Transmission is through direct or indirect contact with droppings from infected birds (fomites, free-flying birds, insects and rodents). The oocysts shed in feces are not immediately infective, they have to first go through a maturation phase (sporulation) which can take as little as 1-3 days in warm, damp litter. The disease is most common in the springtime. Diagnosis is based on a fecal float. At necropsy, if *Eimeria tenella*, the cecal coccidian, is the cause, then hemorrhage of the ceca is seen. The other species (*E. acervulina and E. necatrix*) affect the intestine and are less severe. The key is prevention. Wet litter, poor sanitation, poor nutrition, and concurrent immunosuppressive diseases are the most common triggers of a coccidiosis outbreak. Treatment is with a coccidiostat such as amprolium or sulfamethazine. The best recommendation is to prevent the disease by feeding medicated feed between the ages of 0 and 16 weeks. There is a coccidia vaccine available for use in 1-3 days old chicks, but it is only useful in certain poultry operations, since it uses live organisms and re-ingestion at 4-25 days is necessary as a booster.

Dermatologic Parasites

External parasites

External parasites such as mites and lice are common in poultry. Prevention, checking your flock periodically for external parasites and treating early helps prevent a larger flock outbreak. Fall and winter are the most common times to see a louse problem. The lice are big enough to see with the naked eye and they move fast in comparison to mites. Nits (louse eggs) are laid in clumps at the base of feathers along the ventrum. Treatment of lice and mites consists of either carbaryl powder, pyrethrin powder or ivermectin. Follow label directions for withdrawal times in food producing poultry. If selling the eggs commercially, sometimes the only allowed treatment is diatomaceous earth. Poultry lice and mites are host specific and are not zoonotic.

The Northern fowl mite, also known as the feather mite (*Ornithonyssus sylvarium)* is the most common external parasite in poultry, especially in cool weather climates. This mite spends its entire life cycle on the chicken. Clinical signs include soiled feathers around the vent, tail and rear legs.

The chicken mite, also known as the red roost mite (*Dermanyssus gallinae*) is associated with worse clinical signs than the Northern fowl mite and is more difficult to diagnose because this mite feeds on poultry at night and then remains secluded during the day within the poultry house. This mite can live off the bird for two to three weeks. Clinical signs include mild weight loss and decreased egg production.

The body louse (*Menacanthus stramineus)* is very common. Clinical signs include hyperemia and irritation of the skin of affected birds with small scabs. They are chewing ice, chewing on skin fragments and feather debris on the surface of the skin. They spend their entire life cycle on the chicken.

References

1. Saif YM. *Diseases of Poultry*, 12th ed. Oxford, UK: Blackwell Publishing; 2008.
2. Carpenter JW. *Exotic Animal Formulary*, 4th ed. St. Louis, MO: Elsevier Saunders; 2012.
3. Damerow G. *The Chicken Health Handbook.* North Adams, MA: Storey Publishing; 1994.
4. Damerow G. *A Guide to Raising Chickens*. North Adams, MA: Storey Publishing; 1995.
5. Ussery H. *The Small-Scale Poultry Flock.* White River Junction, VT: Chelsea Green Publishing; 2011.
6. Cattoli G, Susta L, Terregino C, Brown C. Newcastle disease: a review of field recognition and current methods of laboratory detection. *J Vet Diagn Invest.* 2011;23(4):637-56.
7. Greenacre CB, Morishita TY. *Backyard Poultry Medicine and Surgery: A Guide for the Veterinary Practitioner*. Ames, IA: John Wiley & Sons; 2015.

Avian Anesthesia—Wild Style

Heather Wilson Barron, DVM, Dipl ABVP (Avian), and
Michelle Hawkins, VMD, Dipl ABVP (Avian)

Session #087

Affiliation: From Clinic for the Rehabilitation of Wildlife, 3883 Sanibel-Captiva Rd, Sanibel, FL 33957, USA (Barron), Department of Medicine and Epidemiology, School of Veterinary Medicine, 2108 Tupper Hall, University of CA, Davis, Davis, CA 95616, USA (Hawkins).

Abstract: This 2-hour master class reviewed anesthetic considerations for a wide variety of wild birds. It detailed clinically relevant anatomic and physiologic differences, pre-anesthetic practices, intubation, ventilation, monitoring, and patient support. Additionally, the class covered many of the anesthetic drugs and appropriate combinations for balanced anesthesia.

Introduction

Anesthesia is used routinely and successfully in wild birds for a wide variety of procedures from the relatively mundane, such as radiographs or venipuncture, to the complex, such as orthopedic surgeries. Understanding the anatomy and physiology of the myriad of wild avian patients that may present for care and being fully prepared prior to any procedure can increase the success of anesthetic procedures considerably. Recent advances in avian anesthesia and analgesia continue to improve the health and welfare of avian patients.

Anatomic and Physiologic Considerations

The respiratory system of birds has some important differences from that of mammals and this necessarily produces some unique considerations for avian anesthesia. The avian respiratory system is composed of groups of air sacs connected to each other and also to the lungs by bronchi. Gas exchange in the lungs does not occur in alveoli, as it does in mammals, but rather through a mechanism of capillary exchange of oxygen, and thus anesthetic gases, in the parabronchi. Since birds lack a diaphragm, air is pumped through their respiratory system by movement of the ribs and sternum. Anything that prevents sternal movement, even something that seems lightweight and innocuous like a warm towel, may prevent breathing. During respiration, air flows into and out of the air sacs which serve as a sort of bellows to push air into the lungs. Compared to that of mammals, the avian lung has a higher surface to volume ratio and a thinner blood-air barrier making it more efficient at gas exchange. Therefore, birds have very rapid induction times as compared to mammals with inhalant anesthetics. While birds do have a smaller lung volume than mammals, because of their air sacs they actually have a greater total respiratory volume. However, they have far less functional residual capacity in their parabronchi than mammals do in their alveoli. Blood oxygenation can only occur if a constant flow of gas is maintained in the parabronchi. Therefore, apnea results in death much more quickly in birds than in mammals. Generally, birds under anesthesia will go into respiratory arrest first and then cardiac arrest if breathing is not resumed quickly. Some researchers feel that the large amount of gas stored in the air sacs makes birds relatively inefficient at eliminating inhaled anesthetics. Oxygen should be continued for a few minutes after the anesthetic gas has been discontinued to aid the bird in clearing the air sacs of residual anesthetic.

Because the trachea is relatively longer in birds than in mammals, it is also relatively wider in diameter in order to decrease resistance. Birds under anesthesia for more than ~ 5 minutes should, in most cases, be intubated. Birds, unlike mammals generally have complete tracheal rings, and cuffed tubes are rarely used; although the author occasionally finds that the anatomy of some wild birds, such as pelicans, may lends itself well to cuffed tubes gently inflated. Other unique anatomic variations that may have clinical implications for anesthesia (such as increasing dead space) include an elongated and often very complex trachea in some 50 different species of birds; a septum that divides the cranial trachea into two tubes in some penguins, mallards and petrels; a bulbous ossified intrathoracic tracheal expansion in male Anseriformes; and an inflatable sac-like tracheal diverticulum in emus and ruddy ducks.[1] On the other hand, because of their highly efficient respiration, birds generally have a resting respiratory rate that is relatively slow compared to that of a comparably sized mammal. The 'normal' respiratory rate for an awake resting bird can be calculated at $17.2 \times M_b^{0.31}$ (M_b is body weight in Kg).[2] Intermittent partial pressure ventilation is important in birds as in mammals and ventilators are often used. It is important not to over-ventilate birds as it is CO2 levels that stimulate respiration and over ventilation can result in longer recovery times.

Pre-Anesthetic Considerations

While analgesia is not the focus of this class, appropriate pain medications should be considered prior to procedures that would be expected to produce discomfort. Generally, most species should be fasted a minimum 4-6 hours prior to induction. Consideration should be given to performing anesthetic procedures early in the morning in diurnal birds after overnight fast. If a crop is present in that species, palpate crop to be sure it is empty. If not, the crop may be flushed with a gavage tube. Ideally, avian patients should be pre-oxygenated for roughly 5 minutes prior to induction.

Intubation and Ventilation

Birds are easily intubated as the laryngeal mound is large and visible at the base of the tongue. Some species of birds, such as pelicans, have a cartilaginous protuberance, called the crista ventralis, at the base of the glottis. This may complicate intubation and care should be taken not to damage this structure which is reported in *Anas* (mallards), *Apteryx* (kiwi), Spheniscidae (penguins), *Gallus* (fowl) and *Corvus* (crows) species.[3,4]

Non-rebreathing systems work well on birds. Whether administered by facemask or through an endotracheal tube, the recommended gas flow rate is at least 660 ml/kg/min with a non-rebreathing system. It is recommended to minimize air dead space by having appropriately sized tubes and bags used in the non-rebreathing system. Generally, rates of 1-3 L/min are used, depending on if the patient is intubated or not.[5]

In some cases, birds may be unable to breathe through their trachea or disease may necessitate surgery in this area and an air sac cannula or tube may be placed. Research has shown that oxygen and anesthesia delivered in this way gives comparable blood gas levels compared to endotracheal intubation.[6] Gas can be instilled into an air sac through a tube and then the anesthetic passes from the air sac through the parabronchi and out through the trachea. The caudal thoracic or abdominal air sacs are generally utilized. Gas is still delivered at 660 ml-1 liter/kg/min.

Patient Monitoring

Perhaps the most important factor in monitoring is a dedicated anesthetist. A good indicator of the depth of anesthesia is respiratory rate and character. Respiration becomes slower and shallower as the plane of anesthesia deepens. Since apnea rapidly results in death, immediate measures must be taken to lighten anesthesia and provide ventilation should the bird stop breathing on its own. Average respiratory rates generally vary inversely with size (i.e. the smaller the bird the higher the rate).

Heart rate should also be closely monitored. This can be done with a stethoscope (external or esophageal) or an ECG monitor. ECG leads can be placed on the wings and legs. Heart rate also varies inversely with size. Small patients may have a rate of >400 bmp which may be difficult to follow on an oscilloscope. The normal lead II QRS complexes in birds are negative. One of the best ways to monitor heart rate and character is with a Doppler probe. Direct or indirect blood pressure may also be monitored although few normal reference ranges are established.

Several reflexes are useful for monitoring the depth of anesthesia in birds. The palpebral reflex is present only in very light planes, as is reflex movement in response to cere stimulation. The corneal reflex (movement of the 3rd eyelid in response to corneal stimulation) and the pedal reflex (withdrawal of the foot in response to toe stimulation) remain longer and are slow but present at a surgical plane of anesthesia.

The most commonly encountered problem in avian anesthesia is hypothermia. Normal avian body temperature is 104°-110° F. Being small, birds have a high surface to volume ratio which predisposes them to heat loss when they are not moving. Hypothermia markedly decreases avian respiratory rate and greatly prolongs recovery time. All birds undergoing anesthesia should be kept warm throughout the procedure and the recovery period. This can be accomplished with BAIR huggers (forced warm air), rice bags, circulating warm water blankets, and warm room temperatures.

Hypothermia can be compounded by hypoglycemia, which may occur in debilitated birds. Administration of ½ strength LRS + 2.5% dextrose or 5% dextrose IV or SC may help to maintain blood glucose. Daily maintenance in birds is 75-150 ml/kg, depending on size, age, and species. However, while under anesthesia, the current convention is to use10 ml/kg/hour.

Patient Support

Catheters

Placing intravenous (IV) or intraosseous (IO) catheters is a simple, quick procedure that every clinician should be able to perform. Most patients expected to undergo anesthesia for more than brief, simple procedures will probably warrant having venous access. For avian patients, there are a variety of IV or IO sites, depending on the species and size of the patient. For wild birds, an IV catheter is generally preferred to avoid potential joint damage, which might complicate later release. If an IO catheter is to be placed, pneumatic bones (such as the humerus or femur) should be avoided. Generally the proximal tibiotarsus or distal ulna works well. These IO catheters are easy to place, stable, and offer nearly instantaneous access to general circulation. There are several different considerations that may dictate which choice is used.

Fluid Therapy

Crystalloids: Use of crystalloid fluids in birds and reptiles has been based primarily on data extrapolated from other species as there have been few studies to evaluate different fluid types in exotic species. While there are many species for which normal biochemical reference ranges have yet to be established, a profile may help to determine appropriate fluid therapy. Some species of birds have a higher glucose requirement than mammals and may frequently be hypoglycemic on emergency presentation. If bloodwork has yet to be acquired, a safe fluid choice in most circumstances is Lactated Ringers Solution (LRS) or ½ strength LRS + 2.5% dextrose, both of which may be given IV, IO, SQ, or PO. LRS may help correct metabolic acidosis when lactate is metabolized in the liver to liberate bicarbonate. It is best to warm the fluids (100-103°F) so that they more closely approximate normal avian body temperature (104-110°F). Generally this can be done by placing the capped syringe of fluids in a warm water bath, or by microwaving the bag of fluids or by having a fluid warmer that is kept stocked with fluids. Fluids may be given by a variety or a combination of routes (IV, IO, PO, and SQ) and administration method may depend on the stability of the patient, the type of fluids used, the species, the age, and the temperament of the animal. When maintaining intravenous or intraosseous catheters, it is important to minimize the amount of flush used in very small patients so that the animal is not over-heparinized (recommend a flush solution of heparinized saline 10 units/ml). Fluids may be bolused IV or IO in 10-25 ml/kg increments over 5-7 minutes (shock doses of 90 ml/kg/hr) in birds. For emergency stabilization in acute shock, rapid fluid administration has been safely given at a rate of 1-2x blood volume/hour. General recommendations are to make up fluid deficits over 24-48 hours. This may be calculated by: Fluid deficit (ml) = Body weight (g) x % dehydration. Only about 25% of a crystalloid solution remains in the vasculature 30 minutes after administration due to fluid shifts into the interstitial space.

Colloids: Colloids, such as plasma, dextrans and hetastarch, should be used in patients in shock, those suffering acute blood loss, and in cases of severe hypoalbunemia. The use of synthetic colloids in humans and dogs has been associated with coagulopathies, but this has not been reported in birds. Hetastarch may be given at a bolus dose of 10-15 ml/kg IV or IO (bolus over 10-15 minutes) q8h up to 4 treatments. Colloidal therapy may be used in conjunction with crystalloid therapy.

Blood transfusions: After acute blood loss, mammals are dependent on red blood cell (RBC) regeneration to maintain oxygen delivery to the tissues. In response to tissue hypoxia, erythropoietin stimulates RBC production by the bone marrow. Reticulocyte release from the bone marrow in mammals rarely occurs before 2 to 4 days and most commonly longer than 5 days after acute blood loss. Reticulocytosis, or polychromasia, is the hallmark of intensified erythropoiesis in mammals and birds, allowing classification of anemias into regenerative or nonregenerative types. A previous study on acute blood loss in ducks documented an early regenerative response shown by presence of increased polychromasia starting at 12 hours after blood loss. The relatively short live span of the avian red blood cell (28-45 days) and presence of a nucleated red blood cell (RBC) may account for a birds' ability to mount a very early regenerative response. The use of early supportive care with fluid therapy in avian shock may help bridge the gap for the first 24 hours, after which birds can mount their own RBC regenerative response.

Trauma or disease resulting in hemorrhage is a common emergency presentation in avian wildlife. Total blood volume in birds is estimated to be approximately 7-10% of BW (i.e. a 100 gm bird would have 10 ml blood volume). Interestingly, birds are less susceptible to shock from blood loss than are mammals. An otherwise healthy bird can lose up to 30% of blood volume with no apparent ill effects. This is due to a bird's ability to quickly mobilize large numbers of RBC's from the bone marrow and also probably from an ability to rapidly replace vascular fluid loss from the extravascular space. The pathophysiology of hemorrhagic shock is poorly understood in avian species. Acute blood loss of 30–40 % of blood volume has been shown to result in 50%

mortality (LD50) in mammals. Blood loss is better tolerated in birds than in mammals. The LD50 for acute blood loss in ducks has been shown in a recent study to be 60% of the total blood volume. A recent hemorrhagic shock study in mallard ducks (*Anas platyrhynchos*) documented an increase in heart rate and decrease in blood pressure following acute blood loss. That study may show that birds have a baroreceptor response to shock similar to that seen in mammals.

Whether to transfuse an avian patient with acute blood loss should be given careful consideration. One study showed that pigeons suffering from acute blood loss had a better response to IV fluids than to heterologous or homologous blood transfusions. The same study showed a single iron dextran given IM (10 mg/kg) to pigeons was associated with a significant increase in packed cell volume within 48 hours. If the PCV from acute hemorrhage is less than 20% or in more chronic cases less than 15% (normal PCV for most birds is 35-50%) and the patient is demonstrating clinical signs of anemia, a whole blood transfusion may be beneficial. Transfusions should ideally be from same species. However, if a homologous donor is not available, a single heterologous transfusion from a donor of the same genus (i.e. Accipiter to Accipiter) may be of some benefit.

There is no information on blood grouping in wild birds. In chickens, at least 28 different blood group antigens have been found. A rough cross match (major and minor) can be performed mixing donor and recipient red cells and serum on a glass slide. An absence of gross agglutination or hemolysis suggests compatibility. The recommended anticoagulant is sodium citrate, although acid citrate dextrose (ACD), citrate phosphate dextrose (CPD), or heparin can also be used. The recommended ratio is 0.1 ml of citrate per 0.9 ml of blood. The amount transfused is 10-20% of the patient's blood volume (1-2% BW). This amount would be expected to increase the PCV by 2-5%. Donor blood should be used within 12-24 hours. Long term storage of avian blood is not recommended as mammalian storage media appears to be inadequate to support avian erythrocyte metabolism. An 18 μm blood filter can be used. Anaphylactic reactions following a single transfusion are rare.

Cardiopulmonary-Cerebral Resuscitation

CPCR is a comprehensive term used to describe both the basic principles of cardiopulmonary resuscitation (CPR) as well as advanced life support and post-resuscitation care. The word cerebral resuscitation was added to cardiopulmonary resuscitation to identify the importance of being alive with complete neurologic function. The prognosis for respiratory arrest in birds, especially when caused by inhalant anesthetic overdose is fairly good. Cardiac arrest in avian species carries a poor prognosis, since direct compression of the heart is not possible because of the overlying sternum. Because birds do not have a diaphragm, closed-chest compressions in birds cannot use the thoracic pump mechanism to increase overall negative intrathoracic pressure.[7]

The goal is to use monitoring devices to identify cardiovascular instability early. Early recognition of cardiovascular instability is the key to success in avian anesthesia. If the bird arrests on anesthesia, stop the anesthetic gas administration. Place an endotracheal tube and start positive pressure ventilation with 100% oxygen. The bird can be given an injection of dopram (0.2 ml for large birds and 0.1 ml for small birds intramuscularly) to stimulate the respiratory center. Epinephrine (0.01 mg/kg) and atropine (0.04 mg/kg) can be given IV, IO, or via the endotracheal route (using a tom cat catheter inserted down the endotracheal tube and doubling the dose used for IV). Intracardiac injections should be attempted only in extreme cases and with care, because of risk for laceration of the coronary vessels. An electrocardiogram, Doppler blood pressure, end-tidal CO_2 monitor and reflexes can be used to evaluate the effectiveness of CPR resuscitation.

Balanced Anesthesia

Inhalant anesthetics have the advantage of allowing rapid changes in anesthetic depth, and rapid recoveries due to their low solubility in tissues and elimination via the respiratory tract. But inhalants also induce dose-dependent cardiopulmonary depression. "Balanced anesthesia" uses multiple drugs at lower overall doses, to minimize the adverse effects of each drug while still providing appropriate anesthesia and analgesia. This approach reduces the magnitude of cardiovascular depression and promotes hemodynamic stability. Pre-anesthetic medications, constant rate infusions and local anesthetic use can reduce the inhalant anesthetic concentrations necessary.

Even with balanced anesthetic protocols, morbidity and mortality increase with increased duration of anesthesia. Minimize anesthesia time by preparing the patient, all drugs and equipment in advance and provide close patient monitoring so that anesthetic adjustments are identified and corrected immediately. Morbidity and mortality also occur during recovery so it is important to monitor animals until they are fully recovered.

Pre-anesthetic Medications

The necessity of routine medication of birds with parasympatholytics is questionable. Atropine and glycopyrrolate reduce mucus production and decrease mucus plug formation, but also increase mucus viscosity, making it harder to clear secretions from the tube and sometimes seem to increase the development of mucus plugs. Additionally, some practitioners avoid using anticholinergics in birds because they increase heart rate, cardiac work and myocardial oxygen demand.

Sedatives and tranquilizers are used to decrease anxiety and fear during induction. However, most tranquilizers lack analgesia. For balanced anesthetic protocols, often a benzodiazepine such as midazolam can be combined with an opioid medication for sedation and pre-emptive analgesia. Historically the most commonly used opioid medication in birds has been butorphanol, and pre-operative butorphanol administration decreases the inhalation anesthetic concentration needed for general anesthesia with minimal cardiopulmonary changes in many species. In contrast, 1- 6 mg/kg butorphanol did not provide analgesia or sedation for American kestrels; hyperaesthesia or hyperalgesia and agitation was seen in males receiving 6 mg/kg.[13] Hydromorphone administered 0.1-0.6 mg/kg IM to American kestrels has a dose responsive thermal antinociception effect, suggesting antinociception in this species for up to 6 hours.[14] At the higher dosage, some birds were sedated and when handled they appeared mildly agitated. In a recent study in red-tailed hawks, fentanyl as a CRI produced a dose-related decrease of isoflurane MAC with minimal cardiovascular effects.[15] If a fentanyl CRI is to be utilized, a loading dose is administered as part of the pre-medication protocol.

Injectable anesthetics

Most injectable anesthetic agents are associated with pronounced cardiopulmonary depression and prolonged induction and recovery times. Because of this, it is recommended to use balanced anesthetic techniques instead of high dosages of single agents. Additionally, reversible anesthetic agents permit faster recovery. Ultimately, the general dosages published for different protocols are adjusted based on individual factors such as species, gender, age, health status and others. If injectable anesthetics are to be used, it is vital to have accurate patient body weights to ensure appropriate drug dosing. Emergency and supportive drug dosages should be calculated and prepared in advance of anesthetic induction anyway, but especially if using injectables. Finally, cardiopulmonary parameters should be monitored closely throughout the anesthetic period and ventilatory equipment, including endotracheal tubes and a supplemental oxygen supply, should be available if needed.

Alpha$_2$ agonists: The use of medetomidine alone or combined with ketamine, ketamine/butorphanol, or with a benzodiazepine have been evaluated in pigeons, rock partridges, ducks and falcons. Alpha$_2$ agonists are not recommended for short-term anesthesia in birds because in several studies there was inconsistent immobilization, adverse cardiovascular and respiratory effects, unreliable sedative effects and excitatory reactions. To date there are few dexmedetomidine studies alone or in combination with midazolam in raptors and pigeons.[8,9]

Ketamine: Medetomidine/ketamine or xylazine/ketamine combinations have been evaluated in several avian species and often provide a suboptimal plane of anesthesia for painful procedures. Ketamine dosages tend to be higher for smaller birds. When effective, anesthesia occurs within 5-10 minutes of IM injection and may last 5-20 minutes depending on the dose and size of the bird. Ketamine has also been reported to cause salivation, excitation and convulsions when given to vultures, but these signs are rare in other avian species.

Propofol: Propofol has a narrow safety margin in pigeons however, low dosages were effective and safe in turkeys and mallard ducks. In mute swans, propofol CRI was comparable with multiple boluses for short procedures, but about 50% of the birds from both groups demonstrated short periods of excitement during recovery.[10] Prolonged recovery, with or without moderate-to-severe CNS excitatory signs, has been reported after the use of a CRI in red-tailed hawks and great horned owls.

Alfaxalone: Alfaxalone is a new drug to us in the United States that in mammals has a relatively wide margin of safety, little cardiovascular or respiratory depression, and minimal induction and recovery excitement. It has a general anesthetic effect by binding to GABA receptors in the CNS. GABA is a major inhibitory neurotransmitter in the CNS. Alfaxalone is rapidly metabolized by the liver with no cumulative effect, making it suitable for continuous rate infusion. There is currently one clinical study using the newly formulated alfaxalone in birds. In flamingos, sedation was rapid and induction quality was smooth, lowering isoflurane concentrations needed for maintenance anesthesia. Alfaxalone produced moderate cardiorespiratory effects not seen in the isoflurane induction group.

Inhalant anesthetics

Inhalant anesthetics offer rapid induction and recovery, the ability to rapidly change anesthetic depth, and their use does not require accurate body weight. Very little is metabolized, reducing hepatic and renal impact, and recovery is independent of either. But inhalant anesthetics can induce life-threatening cardiovascular and respiratory depression. Inhalants are potent negative inotropes therefore a dose-related decrease in cardiac output is expected, reflected as a decrease in blood pressure. Thus, the primary goal is to maintain the lightest plane of anesthesia possible allowing for minimal cardiopulmonary depression and completion of the procedure. In some cases these goals may not be achieved by simply reducing the inhalant, and balanced anesthetic/analgesic protocols, or the addition of dopamine or dobutamine (only in the volume-loaded patient) may be necessary to offset inhalant-induced hypotension.

Isoflurane, sevoflurane, and desflurane are the currently available inhalants. Their speed of onset and offset is largely determined by their solubility and isoflurane>sevoflurane>desflurane. Despite these physiochemical differences, the clinical differences overall may be fairly small. All three agents have similar dose-dependent cardiovascular effects. Sevoflurane has a less pungent odor than other inhalants and is generally more tolerated during mask induction. Many practitioners induce using sevoflurane then maintain anesthesia with the more cost-effective isoflurane.

Local and regional anesthetics

Concerns about avian sensitivity to local anesthetics have been attributed to rapid absorption and delayed metabolism. Toxic effects have been previously reported in birds at lidocaine doses of 2.7-3.3 mg/kg. In contrast, chickens survive much higher doses of lidocaine compared to mammals, with the convulsive dose of lidocaine approximately 10 times the reported toxic dose.

If a surgical procedure such as fracture repair or amputation is being performed, local anesthetics can be used for nerve blocks to minimize pain associated with these procedures. Using these blocks has reduced the perioperative bradycardia seen when manipulating fractured wing bones in birds. High doses of lidocaine, bupivacaine and ropivacaine have been studied for brachial plexus blocks in ducks without adverse effects.[12] Brachial plexus blocks using palpation, ultrasound or nerve locator have been used with variable degrees of success in avian species.[12] The safe use of transdermal patches, epidural infusions, spinal blocks, and intravenous blocks have not yet been reported in birds.

References

1. Edling TM. Anesthesia and monitoring. In: Harrison GJ, Lightfoot TL, eds. *Clinical Avian Medicine*. Vol. 1. Palm Beach, FL: Spix Publishing, Inc.;2006:747-760.

2. Heard DJ. Critical care monitoring. *Vet Clin N Am Exot Anim Pract.* 1998;1(1):1-10.

3. Wilson D, Pettifer GR. Anesthesia case of the month. *J Am Vet Med Assoc.* 1994;225:685-688.

4. King AS. Apparatus respiratorius [systema respiratorium]. In: Baumel JJ, et al., eds. *Handbook of Avian Anatomy Nomina Anatomica Avium* 2 ed. Cambridge, MA: Nuttall Ornithological Club;1993:257-299.

5. Bennett RA. Avian anesthesia. *Proc Annu Symp Am Board Vet Pract.* 1998;21-24.

6. Jaensch SM, Cullen L, Raidal SR. Comparison of endotracheal, caudal thoracic air sac, and clavicular air sac administration of isoflurane in sulphur-crested cockatoos (*Cacatua galerita*). *J Avian Med Surg.* 2001;15(3):170-177.

7. Curro TG. Anesthesia of pet birds. *Sem Avian Exot Pet Med.* 1998;7(1):10-21.

8. Hornak S, Liptak T, Ledecky V, et al. A preliminary trial of the sedation induced by intranasal administration of midazolam alone or in combination with dexmedetomidine and reversal by atipamezole for a short-term immobilization in pigeons. *Vet Anaesth Analg.* 2015;42:192-196.

9. Santangelo B, Ferrari D, Di Martino I, et al. Dexmedetomidine chemical restraint in two rator species undergoing inhalation anaesthesia. *Vet Res Commun.* 2009;33:S209-S211.

10. Muller K, Holzapfel J, Brunnberg L. Total intravenous anaesthesia by boluses or by continuous rate infusion of propofol in mute swans (*Cygnus olor*). *Vet Anaesth Analg.* 2011;38:286-291.

11. Villaverde-Morcillo S, Benito J, Garcia-Sanchez R, et al. Comparison of isoflurane and alfaxalone (Alfaxan) for the induction of anesthesia in flamingos (*Phoenicopterus roseus*) undergoing orthopedic surgery. *J Zoo Wildl Med.* 2014;45:361-366.

12. Brenner DJ, Larsen RS, Dickinson PJ, et al. Development of an avian brachial plexus nerve block technique for perioperative analgesia in mallard ducks (*Anas platyrhynchos*). *J Avian Med Surg*. 2010;24:24-34.

13. Sanchez-Migallon Guzman D, Drazenovich T, Olsen G, et al. Evaluation of the thermal antinociceptive effects of intramuscular butorphanol tartrate in American kestrels (*Falco sparverius*). *Proc Annu Conf Am Assoc Zoo Vet*. 2012;198-199.

14. Sanchez-Migallon Guzman D, Drazenovich T, Olsen G, et al. Evaluation of the thermal antinociceptive effects of intramuscular hydromorphone in American kestrels (*Falco sparverius*). *Proc Annu Conf Assoc Avian Vet*. 2012;343-344.

15. Pavez JC, Hawkins MG, Pascoe PJ, et al. Effect of fentanyl target-controlled infusions on isoflurane minimum anaesthetic concentration and cardiovascular function in red-tailed hawks (*Buteo jamaicensis*). *Vet Anaesth Analg*. 2011;38:344-351.

Interpreting a Pharmacokinetic Study: Applying Research in the Clinic

Marike Visser, DVM

Session #092
AAV 2016 Donald W. Zantop Emerging Leader Award Recipient

Affiliation: From the Department of Anatomy, Physiology and Pharmacology, Auburn University College of Veterinary Medicine, 1500 Wire Rd, Auburn AL 36849, USA.

Abstract: Principle components of a pharmacokinetic study include area under the curve, maximum concentration, time to maximum concentration, clearance, volume of distribution, elimination half-life and bioavailability. This session defines each pharmacokinetic parameter, its clinical significance and limitations. Furthermore, the use of pharmacokinetic parameters to determine bioequivalence of generic and compounded drugs are discussed. Finally, cases highlight the use of PK data when selecting antimicrobials and determining the dosing interval for seizure patients.

Introduction

Pharmacokinetics (PK) of various drugs are frequently reported for multiple species, offering dosage ranges and intervals. PK data is frequently included on the package insert for FDA-approved drugs in the approved species. However, this information is limited in exotic and avian medicine, where extra-label drug use is common due to the lack of approved products. Understanding the purpose and limitations of the parameters reported in PK studies, allows the clinician to critically evaluate and apply research results in the patient.

Pharmacokinetic Parameters

Pharmacokinetics can be defined as the modeling of the time course of a drug in the body. These studies seek to generate mathematical models to quantify physiologic processes in the body, tailoring a dose for a given population, optimize therapeutic effectiveness and minimize toxicity. A PK study is a measures the concentration of a given drug or its metabolites in blood, tissue, feces or urine over a set period of time. The most common and accessible site is blood, leading to the majority of PK study reporting this compartment. The most convenient method of modeling this data is a non-compartmental model, which allows for the mathematical modeling of multiple parameters. The draw-back of this model is specific statements about sites of absorption, metabolism and elimination cannot be made.[1]

Pharmacokinetics differs from pharmacodynamics (PD), which measures the effects of a drug at a given dose in the body. Both PK and PD are necessary to help predict and integrate clinical efficacy of a drug and also forms the basis of physiologic-based modeling. Examples of PD would be studies regarding the efficacy of pain medication or whether a NSAID is COX-1 or COX-2 selective.[2]

The study design should be scrutinized before evaluation of the data. In a cross-over, the same animals are given two or more formulations, route or dose of a drug (e.g., intravenous (IV) vs oral (PO) meloxicam).[3] This removes individual variability when comparing parameters such as area under the curve (AUC) and bioavailability. It is also important to note that an appropriate wash-out period if the same animals are repeatedly used in the same study, there is. This wash-out period refers to the amount of time that has elapsed before the animal is again used for the same study but given a different dose, formulation or route. This wash-out time is dependent on the half-life of the drug, with some drug only requiring a 1-2 day wash-out period while others may require weeks. If a drug is given while the previous dose is present this will not only interfere with quantification of the drug but if this is a PD study, there may be misleading results since the previous drug is still present and exerting an effect. To determine whether a study applied an appropriate wash-out period, the elimination half-life can be used. In general, 5 half-lives must have passed for a drug to have been eliminated.[4]

Intravenous administration of drugs, results in direct administration into the systemic circulation without having to penetrate barriers such as the GI tract and prevents the effect of first-pass metabolism by the intestinal tract and liver. Pharmacokinetic parameters (Table 1) usually reported include: maximum concentration (C_{max}), time at which peak plasma concentration is reached (T_{max}), the area under the curve (AUC), apparent volume of distribution (Vd), the elimination half-life ($T_{1/2}$) and clearance (Cl). The three primary pharmacokinetic parameters describing drug disposition in the body is: Cl, Vd and $T_{1/2}$.[5] Combined with the bioavailability (F), a dosing regimen can be determined.

Table 1. Definition and clinical application of parameters.

Parameter	Definition	Application
AUC	Drug concentration in plasma over a time period	Determining the AUC/MIC Calculate the bioavailability
Cmax	Peak plasma concentration at a given dose.	At least 10x MIC for concentration dependent antimicrobials Concentrations which may pose a risk for adverse events
Tmax	Time to maximum concentration.	Time at which to collect peak sample for therapeutic drug monitoring Help determine time before adverse events will be seen
Vd	Distribution of a drug through the body: intravascular, extravascular, cellular	Calculate a loading dose Determine drug distribution (e.g., antimicrobials)
Cl	Volume of plasma completely cleared by a drug over a given period of time	Adjust dose based on renal/hepatic disease
T1/2	Time required for 50% of drug plasma levels to decrease	Time to steady state Determination of dosing interval
F	Area under the curve of an extravascular route compared to intravascular drug administration.	Alteration in dose based on route of administration

The area under the curve (AUC) indicates the duration and extent of exposure of a drug in the body at a specific dose. The AUC is used to determine the bioavailability across different routes and is also very useful when integrating the minimum inhibitory concentration (MIC) of culture and susceptibility reports, especially for fluoroquinolones and aminoglycosides.[1,6]

The maxiumum concentation (C_{max}) is the peak plasma concentration at a given dose and reflects the rate of absorption compared to the rate of elimination. For IV, the Cmax is considered the time when a bolus of the drug is given and therefore usually at t=0. For other routes, the C_{max} can vary based on the bioavailability at the site of absorption. C_{max} can be used to predict the risk of toxicities, achieving therapeutic efficacy when using concentration dependent antimicrobials and targeting therapeutic doses for anticonvulsants and immunosuppressant. C_{max} is linked with T_{max}, which the time of C_{max}. This can help the clinician when to anticipate the time before side effects are seen and when to collect peak samples for therapeutic drug monitoring. Both C_{max} and T_{max} are dependent upon the rate of absorption, which is dependent on drug administration (e.g., IM vs PO). These two factors can further be impacted by the presence of absence of food, which can either increase or decrease absorption.

Clearance (Cl) is the volume of plasma that is completely cleared of drug over a given time period. Clearance measures the overall ability of the organism to eliminate drug, including hepatic biotransformation, biliary and renal excretion. Clearance is calculated as Dose/AUC, and is one of the three parameters used to determine the dosing rate and can be affected by hepatic or renal disease, thereby requiring a change in the dose. Clearance can vary between species based on blood flow, but if the clearance rate is known, this can provide a tentative mechanism for dosage extrapolation.[7,8] Whether these recommendations (based on human medicine) can be extrapolated to avian species is unknown and further research is required.

Volume of distribution (Vd) is a mathematically concept relating the distribution of the drug to the dose administered. This concept can help determine whether a drug is distributed to plasma space or extensively distributed to fat and muscle. Volumne of distribution can help determine if a drug is protein bound, usually with a low Vd but a prolonged elimination half-life or if this drug is widely distributed and can penetrate. Volumne of distribution can clinically be used to determine a patients loading dose. Equation 1 (Eqn 1) indicates a loading dose if given IV, while the loading dose needs to be corrected for F if given by another route (Eqn 2).

(Eqn 1) $$Loading\ Dose\ (mg/kg) = (desired\ plasma\ concentration\ (mg/L) x\ Vd(L/kg)$$

(Eqn 2) $$Loading\ Dose\ \left(\frac{mg}{kg}\right) = \frac{desired\ plasma\ concentration\ \left(\frac{mg}{L}\right) x\ Vd\left(\frac{L}{kg}\right)}{F}$$

The elimination half-life ($T_{1/2}$) is the time it takes for 50% of drug plasma levels to decrease. It is generally accepted that 5 half-lives are required to remove 97% of a drug. Alternatively, since 50% of a dose is cleared during one $T_{1/2}$ time period, doubling the dose will result add one half-life. Elimination half-life is important when determining steady state concentrations, the dosing interval and to minimize the risk of toxicity.[4] Elimination half-life is dependent upon the Cl and the Vd of a given drug, which can change based upon disease state. Examples of clinical application includes determining steady-state concentration in cyclosporine, dosing frequency with levetiracetam and zonisamide and wash-out period for drugs such as clomipramine and amikacin.[8]

Bioavailability (F) refers to the percentage of a drug that appears in the bloodstream and is dependent upon the route of information. When a drug is given intravenously, the bioavailability is a 100% (F=1) and other routes (oral, subcutaneous, intramuscular) are compared to the IV route. Bioavailability is a ratio of the AUC of two routes, thereby allowing comparison regardless of the dose being used. If the bioavailability is 25% compared to the IV dose, then the oral dose needs to be four times higher than the IV dose given. There are multiple reasons for low bioavailability, including large molecular size, highly polar drug, drug binding to cationic minerals, degradation by the GI tract, transporter efflux and biotransformation by both the intestine and liver before the drug reaches systemic distribution.[8] While IV is used to give true bioavailability, any two routes of drug administration can be compared to see what the comparable bioavailability would be and allowed changes in dose as necessary, this is known as relative systemic bioavailability. For example, comparing the F of meloxicam when given IV vs PO indicates that the F in Hispaniolan Amazon parrots is 49-75%, great horned owls 62% and red-tailed hawks 74%.[3,9] Therefore Amazon Hispaniolan parrots may require at least a 50% dose increase when the dose is given orally, compared to red-tailed hawks, which only needs a 25% dose increase.

Bioequivalence can be used to compare the pharmacokinetics of two products containing the same active pharmaceutical ingredient (API) and given at the same dose/molar concentration by the same route. Examples of bioequivalence studies would be comparing a manufactured to a compounded product, or a pilot drug (e.g., Rimadyl®) to a generic product (e.g., Novox®). The majority of differences is when two products are given orally, when the products are subject to potential degradation by the stomach and must traverse the small intestine. Bioequivalence differences have been reported when comparing compounded or generic products to pilot products in penguins[10] and horses.[11]

Principles in Practice

Case 1: Applying culture and susceptibility data in osteomyelitis

An adult bald eagle (BAEA), presented for an open fracture of the right metacarpus. After re-fracturing the wing twice, BAEA was taken to surgery where the bone was noted to appear necrotic. During his treatment, the patient received both TMS and enrofloxacin since this was an open fracture. A sample was collected for aerobic bacterial culture and susceptibility. This patient was extremely fractious and stopped eating if handled too often. Two organisms were isolated, *Escherichia coli* and *Enterococcus faecalis* (group D).

Table 2. Aerobic bacterial culture and susceptibility isolates.

	Escherichia coli		*Enterococcus faecalis*	
	Interpretation	MIC	Interpretation	MIC
Amikacin	S	<2		
Amoxicillin/Ca	S	4		
Ampicillin	R	>32		
Cefovecin	S	<0.5		
Cefpodoxime	S	<0.25		
Ceftiofur	S	<1		
Cephalexin	S	16		
Chloramphenicol	R	>64	S	<4
Enrofloxacin	R	>4	S	<0.5
Erythromycin			I	2
Gentamicin	R	>16		
Imipenem	S	<1		
Marbofloxacin	R	>4	S	1
Penicillin			S	2
Tetracycline	R	>16	S	<1
Tobramycin	I	8		
Trimethoprim/Sulfa	R	>320		

The patient required a systemic antibiotic that has a large Vd because it needed to penetrate bone. PK reports of cephalosporins indicate a very short $T_{1/2}$ in avian species or requires daily injections; therefore, this class would not be effective due to the need for frequent handling.[12] Amikacin would be effective, but does not have a Vd and will not distribute to bone.[13] In this case, amikacin and vancomycin beads were created and embedded at the surgical site, allowing for localized therapy.[14] Additionally, the patient was placed on doxycycline to treat *E faecalis*.

Case 2: Managing seizure medication in a patient

When managing seizing patients, options include phenobarbital, levetiracetam, zonisamide and bromide. Pharmacokinetics report that phenobarbital has unpredictable absorption,[15] but levetiracetam has shown to have predictable absorption and PK parameters.[16,17] In this case, knowing that the Cmax is at 1 hour, therefore a peak concentration sample should be taken 1 hour after giving the dose and then before the next dose for the trough value. This will then allow for determination of the half-life for the individual patient.

References

1. Martinez M. Noncompartmental methods of drug characterization: statistical moment theory. *J Am Vet Med Assoc.*1998;213:974.

2. Cole GA, Paul-Murphy J, Krugner-Higby L, et al. Analgesic effects of intramuscular administration of meloxicam in Hispaniolan parrots (*Amazona ventralis*) with experimentally induced arthritis. *Am J Vet Res.* 2009;70:1471-1476.

3. Lacasse C, Gamble KC, Boothe DM. Pharmacokinetics of a single dose of intravenous and oral meloxicam in red-tailed hawks (*Buteo jamaicensis*) and great horned owls (*Bubo virginianus*). *J Avian Med Surg.* 2013:204-210.

4. Toutain P-L, Bousquet-Mélou A. Plasma terminal half-life. *J Vet Pharmacol Ther.* 2004;27:427-439.

5. Riviere JE. *Comparative Pharmacokinetics: Principles, Techniques and Applications.* Ames, IA: John Wiley & Sons;2011.

6. Papich MG. Pharmacokinetic–pharmacodynamic (PK–PD) modeling and the rational selection of dosage regimes for the prudent use of antimicrobial drugs. *Vet Microbiol.* 2014;171:480-486.

7. Toutain P-L, Bousquet-Mélou A. Plasma clearance. *J Vet Pharmacol Ther.* 2004;27:415-425.

8. Trepanier LA. Applying pharmacokinetics to veterinary clinical practice. *Vet Clin North Am Sm Anim Pract.* 2013;43:1013-1026.

9. Molter CM, Cole GA, Gagnon DJ, et al. Pharmacokinetics of meloxicam after intravenous, intramuscular, and oral administration of a single dose to Hispaniolan Amazon parrots (*Amazona ventralis*). Am J Vet Res. 2013;74:375-380.

10. Smith JA, Papich MG, Russell G, et al. Effects of compounding pharmacokinetics of itraconazole in black-footed penguins (*Sphenicus demersus*). *J Zoo Wildl Med.* 2010;41:487-495.

11. Merritt A, Sanchez L, Burrow J, et al. Effect of GastroGard and three compounded oral omeprazole preparations on 24 h intragastric pH in gastrically cannulated mature horses. *Equine Vet J.* 2003;35:691-695.

12. Sadar MJ, Hawkins MG, Byrne BA, et al. Pharmacokinetics of a single intramuscular injection of ceftiofur crystalline-free acid in red-tailed hawks (*Buteo jamaicensis*). *Am J Vet Res.* 2015;76:1077-1084.

13. Bloomfield RB, Brooks D, Vulliet R. The pharmacokinetics of a single intramuscular dose of amikacin in red-tailed hawks (*Buteo jamaicensis*). *J Zoo Wildl Med.* 1997:55-61.

14. Atilla A, Boothe HW, Tollett M, et al. In vitro elution of amikacin and vancomycin from impregnated plaster of Paris beads. *Vet Surg.* 2010;39:715-721.

15. Powers LV, Papich MG. Pharmacokinetics of orally administered phenobarbital in African grey parrots (*Psittacus erithacus erithacus*). *J Vet Pharmacol Ther.* 2011;34:615–617.

16. Schnellbacher R, Beaufrere H, Arnold RD, et al. Pharmacokinetics of levetiracetam in healthy Hispaniolan Amazon parrots (*Amazona ventralis*) after oral administration of a single dose. *J Avian Med Surg.* 2014;28:193-200.

17. Visser M, Boothe DM. Population pharmacokinetics of levetiracetam and zonisamide in the African grey parrot (*Psittacus erithacus*). *Proc ExoticsCon.* 2015;7.

Avian Ophthalmology: Principles and Application

R. T. Korbel, Prof., Dr. med. vet. Dr. med. vet. habil., Cert. Spec. Avian Med., Cert. Spec. Vet. Ophthalmol., Dipl ECZM (Avian)

Session #181

Affiliation: Department for Clinical Veterinary Medicine, Clinic for Birds, Reptiles, Amphibians and Ornamental Fish, and Reptiles, University Ludwig-Maximilian of Munich, Sonnenstrasse 18, D – 85764 Oberschleissheim, Germany.

Abstract: Ocular disorders in birds are of great importance to vision as well as the fact that they commonly hint at numerous systemic disorders, which may even be transferable to man (e.g., salmonellosis, mycobacteriosis). The lab will provide the practitioner with a practical review of suitable instrumentation following a step-by-step tutorial through the complete avian ophthalmologic examination procedure as a reliable basis for diagnoses and therapy of ocular disorders in birds. Within the proposed practical lab, diurnal and nocturnal raptors with healthy and diseased eyes will be used to demonstrate physiologic and pathologic situations. Certain avian-specific capacities, like flying, are closely related to fully functional vision, underlining the importance of avian ophthalmology. In addition, ocular lesions in birds are an expression of systemic disorders more than in mammals and therefore represent an important diagnostic criterion. Ocular symptomatology frequently enables specific conclusions to be drawn on suspected disorders or it may even be pathognomonic for a certain disease. Thus the avian eye may be seen, to a much larger extent than in mammals, as a diagnostic window. On average, more than 30% of all traumatized birds (incidence higher in raptors than in pet birds) are suffering from ocular lesions, which are most often hidden within the inner structures of the eye with hemorrhages arising from the pectin oculi. Ophthalmoscopy (i.e., examination of the posterior eye segment) is obligatory in traumatized birds. Avian ophthalmology is not a highly specific working field within avian medicine but it should be an integral part of the general examination procedure. After a short review of anatomic and physiologic peculiarities, commonly and newly developed routine ophthalmologic examination procedures will be described followed by a description of clinical pictures in the eye of various patient groups based on an etiologic schematization. Avian ophthalmology is increasing in importance in terms of avian visual perception, which strikingly differs from that of man (including avian visual perception within the UV and the ability to resolve high-frequency flicker rates). These aspects are animal welfare issues because commercial poultry are commonly kept under artificial light sources, which meet human visual perception requirements but lack those of avian vision.

Introduction

In many birds the eye is the most important sensory organ. The capacities of the avian eye which exceed those of the mammalian eye in part are an adaptation to the specific way of life and habitats as well as physical activities that are closely bound to perfectly functioning vision (e.g., flying). Even partial impairment of vision that can be caused by any one of many known eye diseases, always has far-reaching consequences because compensation by other senses (including olfactory and acoustic sensory perceptions) is usually insufficient, if possible at all. Hence ornitho-ophthalmology occupies an important position in avian medicine.

Part I. Anatomic and Physiologic Basics

Basics principles in avian ophthalmology are similar to mammalian ophthalmology; however, a broad variety of ophtalmologic, physiologic, and anatomic peculiarities have to be considered within the class of Aves representing

8,800 species and more than 28,000 subspecies. Some of the most important aspects include the striated rather than smooth intraocular musculature, making it necessary to use bird specific techniques for inducing mydriasis. Additionally the fundus oculi includes numerous peculiarities such as the pecten, an anangiotic retina, none, one or bifoveate retina, lack of the retinal pigment epithelium (RPE) and more. The presentation will include a section on avian visual capacities, which are of greatest importance of keeping both pet birds, raptors as well as commercial kept poultry under artificial light source circumstances and thus are of animal welfare importance.

Significance of visual capacities

Visual function in raptors is i. a. essential for flying, surviving in the wild and reproduction. The eye as the main sense organ in birds and its visual capacities have no general superiority compared to mammals but shows a highly specialisation as an adaption to living conditions. Thus visual acuity is 2-8 times higher compared to mammals, visual fields are up to 360°, stereopsis ranges from 0-70°, maximum spatial frequency (the ability to dissolve a certain movement into single frames) is up to 160 frames/sec (10-15 in man) and minimum detection of movements is up to 15°/h (very slow movements). It should be recognized that ultraviolet-perception (UV-perception between 320 and 680 nm), an ability common in diurnal birds bound to special UV-sensitive rods within the retina and an aspect not investigated very well to date, plays a probably plays a very important role in inter- and intraspecific communication based on plumage-UV-reflection, even in birds which appear monomorphic for the human eye, for the identification/assessment of fruit ripeness based on varying UV-reflection of fruit wax layers, for phenomena of camouflage, orientation, and others. On the other hand, even complete blindness is not a reason for euthanasia in pet birds. Thus, canaries, suffering from blindness due to cataract formation (a condition that occurs often in this species) act normal, as long as the interior of the cage or aviary is not modified.

Anatomic peculiarities

Despite numerous anatomic and physiologic differences, such as striated rather than smooth intraocular musculature, anangiotic fundus oculi, and pecten oculi, the approach to avian ophthalmology is similar to that of mammalian ophthalmology. Some anatomic peculiarities with relevance for the ophthalmologist are:

Size and weight of the avian eye:

- Axial length-8 mm (kiwi, aptery species) up to 50 mm (ostrich, *Struthio camelus*).
- Weight-oculus sinister (OS) and oculus dexter (OD) in man 1%, fowl 7% (adult), fowl 12% (juvenile) compared to overall head weight.

Shape:

- Determination by annulus ossicularis sclerae (single bony platelets).
- Flat type (diurnal birds with narrow heads, for example Columbiformes).
- Conical type (diurnal birds with broad heads, for example Falconiformes).
- Tubular eye (nocturnal birds, for example Strigiformes).

Adnexal structures:

Eyelids

- Palpebra inferior larger than palpebra superior (exclusive owls).
- Lower eyelid with tarsus, no meibomian glands.
- Membrana nictitans highly motile, regulation of precorneal tear film, protective function for cornea, snorkel mask effect (compensation for loss of refractive power during underwater vision).

Lacrimal glands
- Small or absent glandulae lacrimalis.
- Large glandulae lacrimalis, membrana nictitans (vel. Harderian gland).
- Replacement by nasal salt gland (aquatic species).

- No musculus retractor bulbi.
- Complete decussation of chiasma opticus (no consensual pupillary reflex).

Anterior eye segment
- Striated intraocular musculature (highly motile iris corresponding to external stimuli, no sensitivity to parasympatholytics/sympathomimetics, no reliable pupillary reflex corresponding to light stimuli).
- Fundus diameter much larger than pupil diameter.
- Ligamentum pectinatum/ciliary cleft with species specific peculiarities.
- Lens with annulus pulvinus (Ringwulst), accommodation range 2 (owls) up to 80 (waterfowl) D.

Posterior eye segment
- Avascular (anangiotic) retina.
- Afoveate, uni- or bi-foveate retina.
- Rods and cones (inclusive special UV-cones) in functional units.
- UV-sensitive cones in most diurnal birds.
- Slight retinal pigmentation in nocturnal, heavily pigmented retina in diurnal birds.
- No retinal tapetum lucidum.
- Pecten oculi.
- Choroidal, heavily pigmented structure.
- Protruding into the vitreous.
- Obscuring papilla nervi optici.
- Pleated, vaned and conical type.
- 32 functional theories: nutritive, thermo- and pressoregulative function most obviously.

Part II. Instrumentation

This chapter on avian ophthalmology will provide the practitioner with detailed informations on specific instrumentation needed going through a routine ophthalmologic examination procedure in exotic animals. Specified instrumentation includes a Finoff transilluminator allowing examination of the anterior and posterior eye segment (using monocular direct ophthalmoscopy), a panoptic monocular ophthalmoscope, suiting specific needs for avian ophthalmology or a combination of a head-band ophthalmoscope in combination with each 30, 40, 78 and 90 diopter lenses. Measuring suitable artificial light conditions for birds, with suitable luxmeter and spectralanalytic devices will be provided.

General examination

Minimum requirement for basic and general purposes:
- Focused light source with magnification lens (Finoff transilluminator).
- Instrumentation for manipulation of the eye lids (Graefe hook).
- Lacrimal cannula (Anel).
- Topical anesthetics: Proxametacain, Oxybuprocain (duration of action approximately 7-8 minutes) or lidocaine (duration of action approximately 17 minutes).

Advanced equipment:

- Slit lamp (magnification 5-15x, better 20x).
- Monocular direct ophthalmoscope with 15 D lens or even better.
- Head band ophthalmoscope with 30 and 78 D lens (additional aspherical 40, 60, 90 D).

General examination procedure (adnexal structures and anterior eye segment):

- Without restraint - Assessment of visus via food intake, reluctance to fly, orientation.
- With restraint - Examination of the ear opening, pupillary reflex.
- Examination of the anterior eye chamber with lateral illumination.
- Examination of the anterior eye chamber with lateral transillumination.
- Examination of the anterior eye chamber with retroillumination.

Equipment and procedures for specific examinations:

- Slit lamp biomicrography.
- Gonioscopy (Lovac lens). Examination of the angulus iridocornealis with the pectinate ligamentum. Etiologic assessment of primary/secondary glaucoma status.
- Tonometry. Estimation of the intraocular pressure (IOP). Use electronic short time acting tonometer or Schioetz tonometer in raptors. Standard reference values measured with an electronic tonometer calibrated for avian eyes, range from 9-22 mgHg and are available for 42 species from 7 orders. Standard reference values for psittacine birds range from 12-15 mmHg for nocturnal birds from 9-12 mmHg, for various poultry species intraocular IOP values are as high as 22 mmHg. Minimum corneal diameter for reliable value is 9 mm. Reference values for newly developed tonometers (Tonovet^{R} and Tonolab^{R}) will be available soon.
- Schirmer's test. Test for the estimation of the lacrimal function. Use standardized filter strips of 2, 3 and 5 mm width. Standard reference values using filter strips of various width for 42 species from 7 orders showing a wide range of interspecific variations are available. Strigiformes show conspicuously low values.
- Electroretinography. Measurement of retinal function by recording electrical potentials after light stimulation. This technique gives no information about the visus, only on retinal function. Basic principles of electroretinography for routine examination have been established. Indications are retinal disorders and diotric apparatus opacities.
- Laboratory examinations include bacteriologic examination of the conjunctival flora. Physiologic bacterial flora contain gram-positive bacteria, while gram-negative bacteria are an indicator for pathologic conditions. Standard reference values have been worked out for 42 different bird species from 8 orders.

Part III. Examination Procedures

Basic examination principles are equal to those in mammalian ophthalmology; however, the generally small size of avian eyes plus many anatomic and physiologic peculiarities are demanding for numerous specific adaptations. These will be demonstrated on a comparative basis used to techniques in mammals including basics of patient restraint techniques, examing adnexal structures and anterior eye segment, induction of mydriasis using curariforme drugs, mon- and binocular ophthalmoscopic techniques and visualizing the fundus. The lecture will include a review on most recent and sophisticated diagnostic imaging techniques including ultra sonography, optical coherence tomogryphy, digitial scanning ophthalmoscopy.

Mydriasis and air sac perfusion technique

Induction of mydriasis is indispensable for the examination of the posterior eye segment (ophthalmoscopy). A major difference between the mammalian and the avian eye however is that the commonly used mydriatics of atropine and tropicamide have little effect in the avian patient due to a striated rather than smooth intraocular musculature. Therefore the iris is partly unter voluntary control. It is essential to have a dilated pupil (mydriasis) to perform an ophthalmoscopy (i.e., examination of the posterior eye segment including the vitreous, the fundus and the pecten oculi).

Neuromuscular blocking agents such as d-tubocurarine may be used.[1] As the drug penetrates the cornea insufficiently it has to be administered directly into the anterior chamber by paracentesis using a 27- to 30-gauge needle. This technique includes substantial risk for injuries of intraocular structures causing hyphema, increasing intraocular pressure (IOP), transmission of conjunctival flora with consecutive uveitis and systemic side effects if larger doses than recommended are used. It is recommended to use this technique just for therapeutic reasons (prevention of posterior of anterior synechia resulting from uveitis and consecutive miosis) and situations where long-lasting mydriasis (12-18 h) is necessary. The method of choice, conjunctival application of rocuronium (12-35 µl/eye), resulting in a mydriasis with species-specific differences related to lag of action after administration (up to 90 minutes in common buzzards and African grey parrots), completeness of mydriasis, and duration of action (20-360 minutes) must be considered. There is a need for further scientific studies.

An alternative for routine induction of a mydriasis, as well as for intraocular surgery and surgery in the head area, is air sac perfusion anesthesia (APA).[2,3] In principle, APA consists of a retrograde perfusion of the lung-airsac-system through a perfusion catheter via the left caudal thoracic airsac. Carrier gas of O_2 (0.3 L/min/kg BW) is used. Effect of nitrous oxide application is a low potentiation of isoflurane (approximately 11%) and thus improvement of the circulatory situation and reduction of isoflurane waste gases affecting the surgeon. Higher perfusion rates than recommended result in respiratory alcalosis due to a CO_2-wash-out-effect causing severe cardiac arrhythmias. Isoflurane maintenance concentrations vary (dependent on bird species) between 1.0-2.4 vol% (*Columba livia* Gmel, 1789). Pulse oximetry is indispensable as APA causes a reversible apnea due to reduced CO_2 partial pressure causing a missing stimulation of the respiratory centre. Advantages of APA, a long period anesthesia, which is used for routine ophthalmoscopy, electroretinography and head surgery in birds, are free surgical access to the head for intraocular surgery, stable or decreasing intraocular pressure and reversible apnoea with an absolute immobilisation of the patient. Achievement of mydriasis for ophthalmoscopy may be optimized by systemic administration of 0.2 mg/kg BW of the muscle relaxant vecuronium which allows a complete mydriasis and areflexia with a lag period of approximately 26 seconds and a duration of 25.6 minutes in pigeons (*Columba livia* Gmel, 1789) and a reduction of isoflurane consumption of approximately 25% at the same time. This technique allows examination including the lens periphery with the annular pad and the extreme fundus periphery.

APA principles:

- Anesthesia-induced mydriasis and free surgical access to the head area.
- Perfusion of lung-airsac-system with oxygen-(nitrous oxide)-isoflurane or sevoflurane mixture.
- Retrograde perfusion via left caudal thoracic air sac using a specific air sac catheter.
- Induction of a reversible apnoea due to lowering of the CO_2-partial pressure with subphysiologic values (below 48 mmHg; no stimulation of respiratory center).
- Modified anesthetic machine with low flow flowmeter (0.01-0.5 L/min).

APA performance:

- Carrier gas O_2.
- Induction: Head chamber.
- Maintenance: Air sac catheter.
- Isoflurane concentration: 1.0-2.7 vol%.
- Monitoring: Pulse oximetry (no respiratory movements), reflex score, blood gases, body temperature.

APA advantages:

- Routine method for ophthalmology and head surgery.
- Benefits of inhalation anesthesia and long period anesthesia.
- Free surgical access to the head.
- Stable or decreasing intraocular pressure (IOP).
- Reversible apnoea, absolute immobilisation of the patient.
- Achievement of mydriasis for ophthalmology.

Ophthalmoscopy, a technique to examine the fundus oculi (a clinical term, describing ocular structures, which are situated behind the lens) using a focused light beam reflected from the fundus, can be carried out by both monocular and binocular and direct or indirect ophthalmoscopy in combination with double aspherical ophthalmoscopic lenses (at 30, 40, 60, 78 and 90 diopters (D, Volk Bio II) refractive power (Figs 1 and 2). In all cases indirect binocular ophthalmoscopy using a head ophthalmoscope is advisable. A 30 D lens is used in birds with larger pupil diameters (nocturnal and diurnal raptors), ophthalmoscopy of pigeons and larger psittacine birds requires a 78 D lens, those in small birds (canaries, budgerigars) the use of a 90 D lens. Alternatively, monocular indirect ophthalmoscopy may easily performed in all birds, especially in smaller species. It should be pointed out that within a long term survey including more than 32 % of traumatized birds showed hemorrhages in the vitreous body originating from lesions of the pecten oculi. Thus ophthalmoscopy is obligatory in traumatized birds.

Part IV. Ocular Disorders in Birds

Fully functional vision is indispensable in (falconry kept) raptors and wild birds, however of great importance even in captive pet birds. Most often avian ocular diseases reflect ocular manifestations of systemic disorders, frequently representing zoonosises such as salmonellosis, tuberculosis, avian influenca and thus being of importance for both the patient and the patient owner. In addition, most emerging diseases such as borna virus infections are of high relevance. The presentation will give a broad overview on avian ocular disorders discussing (on an interactive basis) as well as diagnostic and therapeutic approches.

The ophthalmologist considers it logical to classify eye disorders according to the various ocular structures involved. A complete review on avian ophthalmology discords based on a morphologic basis will be given within the presentation.[1-4] Over a long period of investigation of ocular disorders in birds, an overall incidence of 7.6% of all the birds examined has been found. The highest incidence of eye disorders compared to other patient groups was generally found among wild birds (11.7%), among which most eye disorders were seen in Accipitriformes (26.1%), Strigiformes (20.0%), and Falconiformes (19.7%). In contrast, a lower incidence was found in falconry raptors (7.3%), but again Accipitriformes (6.3%) were most frequently affected by eye disorders, followed by falcons (5.5%) and owls (5.0%). Among disorders of the posterior eye segment (fundus oculi) most often trauma-related hemorrhages (most frequently arising from the pectin) may be diagnosed. In general chronic lesions caused by disorders located within the upper part of the fundus are resulting in a poor prognosis, as birds and especially raptors are orientating themselves primarily using the upper part of the fundus (with the central and temporal fovea included).

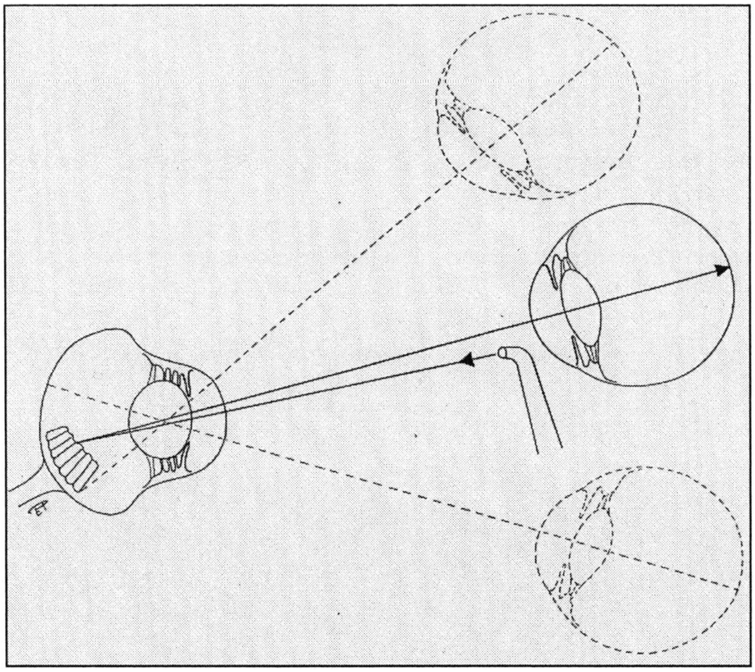

Figure 1. Optical principles of the (monocular direct) ophthalmoscopic examination of avian eyes. Ophthalmoscopy is the examination of ocular structures situated behind the lens, using a focused light beam directed through the pupil and performed using light, reflected from the fundus oculi. As a rule, a nearly coaxial direction of the light beam directed through the (dilated) pupil (mydriasis) of the patient's eye (left) on the one hand and the reflected light beam from the patients fundus to the examiners eye (right; viewing axis) on the other hand is indispensable in order to project an image on the examiner's retina.[4]

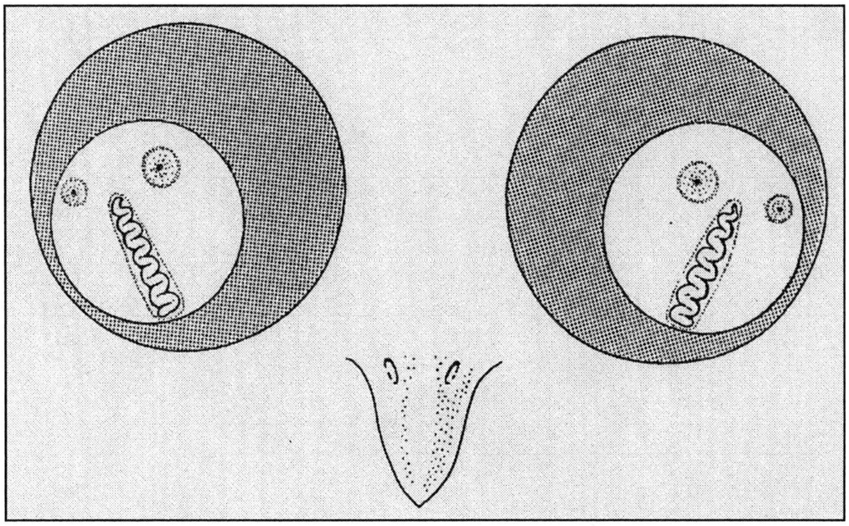

Figure 2. Topographic landmarks within the avian fundus, the pecten oculi and the central and temporal fovea. Accessible viewing field (bright area) using monocular direct ophthalmoscopy.[4]

References

1. Murphy CJ. Raptor ophthalmology. *Compend Contin Educ Pract Vet.* 1987;9(3):241-260.

2. Korbel R. Augenheilkunde bei Vögeln: Erkrankungen, Luftsack-Perfusions-Anästhesie und Ophthalmologische Photographie. *Vet. Habil*, München. 1996.

3. Korbel RT. Disorders of the posterior eye segment in raptors – examination procedures and findings. *Proc 3rd Int Conf Raptor Biomed.* 2000:179-194.

4. Kern TJ. Exotic animal ophthalmology. In: Gelatt KN, ed. *Veterinary Ophthalmology*. Vol 2. Ames, IA: Wiley-Blackwell;2007:1370-1405.

Section 8

AAV Master Classes Track 3

Avian Radiographic Anatomy

Michelle Fabiani, DVM, Dipl ACVR

Session #297

Affiliation: From Gulf Coast Veterinary Specialists, 1111 West Loop S #160, Houston, TX 77027, USA.

Gastrointestinal Tract

The crop is located to the right of the cervical air sacs cranial to the coelomic cavity. There are two parts to the stomach in birds: the proventriculus and ventriculus. As in mammals these are to the left of midline on the VD radiograph. The proventriculus is tubular shaped located dorsal to the heart and liver in the mid-coelomic cavity on the lateral view. The ventriculus usually contains mineralized grit, is circular in shape, and is located at the level of the coxofemoral joints on the lateral radiograph. The connection between the proventriculus and ventriculus is called the isthmus and this location is predisposed to neoplasia such as adenocarcinoma.

The duodenum us located in the right ceoleomic cavity but radiographically cannot be differentiated from other bowel loops in the caudal coelome. The large intestine is also usually indistinguishable from other small bowel segments, unless barium is in the lumen. The large intestine leads into the cloaca. The cloaca can contain either gas or fluid and is located in the caudal most aspect of the coelome.

Respiratory Anatomy

Respiratory disease is a common clinical problem in pet birds. The unique anatomy of the upper and lower respiratory tract creates diagnostic and therapeutic challenges.

Trachea and syrinx

The trachea is unique in that the tracheal rings are complete. The syrinx is located at junction of trachea and bronchi, and functions to produce voice/vocalizations.

Lungs

The lungs in birds are dorsally located in the cranial coelomic cavity. The trachea divides into primary and secondary bronchi, though only the primary bronchi are definitively radiographically evident. The lungs normally have a reticular pattern.

Air sacs

The air sacs are vital to respiration in birds. It may be clinically helpful to think of the air sacs as "cranial air sacs" (cervical, clavicular, and cranial thoracic) and "caudal air sacs" (caudal thoracic and abdominal). The first site of inspired air is the caudal air sacs, receiving air directly from the trachea during the first phase of inspiration. The cranial air sacs receive air from lung, and then air is expired.

- Clavicular - single
- Cervical - paired
- Cranial thoracic - paired
- Caudal thoracic - paired
- Abdominal – paired

Kidney Anatomy

Avian kidneys are anatomically located within the synsacrum, a fusion of the lumbar and sacral vertebrae. They are smoothly oval in shape and extend only slightly ventral to the bones of the spine on a lateral radiograph. The kidneys are not visible on the VD view.

Liver Anatomy

The liver is comprised of two lobes, a right and left lobe, which may be subdivided. Radiographically, the liver and heart should have an "hourglass" shape on a VD view. The liver margins should fall within the confines of an imaginary line drawn from the shoulder joint to the coxofemoral joint. Positioning is important; the legs should be pulled as far caudally as possible, and the patient should be straight. The position of the ventriculus can also be used to assess hepatic size. If there is material evident in the ventriculus (grit, particles), then the ventriculus on the VD view should be just left of midline, at the level of the coxofemoral joint. On the lateral view, it should be at the level of the coxofemoral joint, about 2/3 of the distance from the spine to the keel. Caudal displacement of the ventriculus indicates hepatic enlargement.

Spleen Anatomy

The spleen is round in most birds. It is located on the left of midline, immediately dorsal to the isthmus of proventriculus and ventriculus.

Cardiovascular Anatomy

The cardiac silhouette makes up the top half of the hourglass on a VD view, with the bottom half of the hourglass being comprised of the liver silhouette.

Reproductive Anatomy

The ovary is located cranial to the kidneys in the dorsal coelomic cavity to the left of midline. Enlarged oviducts and follicles can form a mid to caudal coelomic mass effect that is ventral to the kidneys and displaces the intestines ventrally and caudally. Once a shell is deposited these are easily radiographically identifiable as eggs.

Testicles are paired and also located cranial to the kidneys in the dorsal coelomic cavity.

Avian Orthopedics: A Game Plan for Success

Julia B. Ponder, DVM, MPH, and
Patrick T. Redig, DVM, PhD

Session #310

Affiliation: From The Raptor Center, University of Minnesota, College of Veterinary Medicine, St. Paul, MN 55108, USA.

Abstract: Managing avian fractures requires taking a holistic approach to the patient. The purpose of this presentation is to use case-based material to review the principles of orthopedics in birds, and to examine different fracture management modes based on considerations of the individual case. Pre-surgical considerations and post-operative care will also be covered, including considerations of economics, life conditions of the bird and desired outcomes.

Introduction

Successful fracture management in the avian patient requires an understanding of orthopedic principles, appreciation of unique characteristics of birds and clear objectives. Whether the patient is a wild or companion bird, the goal is to minimize risk of complications while maximizing patient outcome. In determining what an acceptable outcome might be, the long-term needs of the bird must be taken into consideration. For wild birds, the ultimate goal is full restoration of flight and function so that the bird can be returned to its natural existence. For a companion bird, full restoration is still an ideal goal, although it may be less critical. When full restoration of function is not possible for whatever reason, it may be appropriate to set a goal of returning a patient to a comfortable and pain-free life. Most companion birds have their food supplied daily (often in abundance) and may be able to thrive without perfect flight or function. While best medical practices will guide the veterinarian's approach to treating the patient, external factors often come into play. There are often limited available resources for treating and rehabilitating wild patients, while companion bird owners will make the final decisions in those birds. In addition, the knowledge and skill of the veterinary practitioner must be taken into account during the development of a treatment plan.

Orthopedic Principles

The goal of avian orthopedics is to capitalize on the body's healing potential while minimizing secondary morbidity. To this end, several specific objectives must be met for fracture management. The fracture segments must be stabilized early with complete rigidity while maintaining the length and rotational alignment of the bone. While it is critical to re-establish the normal longitudinal and axial alignment of the bone, exact anatomical reduction of bone fragments is not crucial.[1] Preservation and protection of soft tissues, which can be impaired by efforts to achieve exact anatomical reduction, will enhance and improve bone healing.[1] Where possible, load-sharing between the fixation device and the bone should be promoted. As an oblique fracture has no inherent ability for load-sharing, the fixation method must be able to bear all forces that would normally be applied to the bone.[1] A transverse fracture, however, has good load-sharing capacity, which can be taken into consideration as part of the fracture management plan. The decision whether to use a surgical or non-surgical fracture management technique in any given case is based on nature of the fracture, the ability to meet orthopedic objectives and goals, and the resources and expertise available. Overall outcomes will be improved if normal joint movement, range of motion and bird mobility can be maintained during the healing process, goals often best achieved with surgical fixation.

Relative to other species and taxons, bird anatomy presents some unique characteristics and challenges. Due to the thin and brittle nature of the cortices of avian bones, traumatic injury often results in the bones shattering when they break. Long bone fractures are frequently severely comminuted and are often open because there is minimal soft tissue protection. The lack of soft tissue covering over the long bones means that bone fragments are easily exteriorized and at risk of devitalization. Most birds have a strong ability to incorporate bone fragments into a healing fracture if soft tissue connections are maintained. When soft tissues are damaged either through initial trauma or during treatment, bone fragments may lose their vascular supply, creating a high risk of sequestrum formation.

In uncomplicated avian fractures with good fixation, bone healing is rapid. A simple humeral fracture with good fixation and vascular integrity will heal in as little as three weeks. Complicating conditions such as open fractures, severely comminuted fractures, infection, poor patient nutrition and distal limb fractures will result in delayed healing times at best.

Assessing the Situation

Patient assessment

The fixation method is only one half of the equation in fracture repair; the patient is the other. A holistic approach must be incorporated from initial assessment through final treatment. Patient assessment, which is often performed concurrently with emergency stabilization, may require a sequential approach if the bird's condition is too delicate to withstand the required handling. A minimum database should include physical exam, orthopedic exam, hematology, serum chemistries and radiology. The use of anesthesia during the physical exam and treatment should be considered as it is often beneficial. While a fracture may be the most obvious problem when initially presented it is important to recognize the potential that the trauma might be far more extensive than just one bone. Traumatic incidents often result in eye, head or intracoelomic trauma. An ophthalmological exam is particularly important in the traumatized wild patient as ocular impairment is common and may affect decisions regarding the bird's to be released back to the wild. It has been reported that 14-28% of raptors admitted to wildlife hospitals have ocular damage. A vision assessment must be combined with knowledge about the patient's ability to survive in the wild with compromised eyesight.[2] Although chronic ocular lesions in a well-conditioned wild bird suffering from acute trauma is often considered proof that the animal was hunting well despite eye lesions, it may be significant if the bird was hit by a vehicle on the same side as the eye lesions while feeding on carrion by the roadside.

In addition to trauma, the patient may have underlying health issues such as poor nutrition, poor body condition or chronic disease. As the strength of bone is also directly related to the forces applied to it (Wolff's Law), bone integrity is often poor in non-flighted or generally inactive companion birds. In addition, malnutrition and associate poor bone condition is common in companion birds. Deficits in nutrients such as calcium and vitamin D often result in weaker bones to start with and can impact healing times and overall outcomes.[3] Reproduction can also impact bone integrity as hens mobilize large amounts of calcium from medullary bone during egg-laying. In situations of excessive egg-laying, this condition can be a factor in predisposing the animal to fractures.[4] Underlying bone disease such as neoplasia or osteomyelitis are infrequent conditions that can predispose an avian patient to a fracture.

When examining the avian patient, it is helpful to begin with hands-off observations, providing opportunity for the patient to unmask clinical signs and for the clinician to evaluate mentation, posture, respiration and general appearance. Indications of respiratory distress may include a tail bob in rhythm with breathing pattern or excessive movement of the lateral abdominal/coelomic musculature. In addition to fracture imaging, radiographs may reveal evidence of internal trauma or sources of respiratory compromise. Clinical signs of fractures may include wing droop, local swelling, apparent loss of limb function, localized pain and/or altered limb positioning. Shoulder girdle fractures (coracoid, furcula or scapula) are often more subtle, clinically evident as a slight wing droop. One key observation of a bird with a coracoid fracture or luxation is the inability to elevate the wing above the horizontal plane.

Fracture Assessment

When evaluating a fracture patient's treatment options and prognosis, it is helpful to understand the factors that contributed to the breaking of the bone, as well as the characteristics of the fracture itself. Important considerations are the type of fracture, bone integrity, soft tissue compromise (open or closed fracture, additional soft tissue trauma), and degree of fracture displacement. Long bone fractures can be transverse, oblique, spiral, segmental or comminuted. [5]

The forces that cause fractures can be high or low energy. A high-energy force is more likely to shatter a bone, resulting in a comminuted fracture while a low energy force often results in a simple transverse or oblique fracture. High-energy fractures are also more likely to have significant soft tissue damage associated with them.[1] For these reasons, fractures resulting from high-energy impacts often are more difficult to treat, are slower to heal and have a poorer overall prognosis.

Since there is minimal soft tissue covering most avian long bones, fractures often present open with exposed bone. Exteriorized bone is often nonviable as it is typically separated from its blood supply.[1, 6]

The presence of significant joint involvement such as luxations or intra-articular fractures will influence prognosis and may be significant considerations in the decision to pursue treatment, especially for wild birds. While acute elbow luxations may be reduced and stabilized with no long-term sequela, chronic luxations and articular damage are frequently associated with degenerative joint disease, loss of range of motion, and chronic pain. The ethics of pursuing treatment in such a situation depends on the individual case presentation - will the injury impair the bird's survival in the wild? Will it condemn the bird to a lifetime in captivity and is the bird suited for such a life? Can the bird be medicated long-term for pain? Clearly, the answers to these questions are often very different for companion birds than for wild birds.

In situations of extreme trauma, amputation is often considered as a salvage procedure. Whether considering wing or leg amputation, this is a decision with many ramifications. Again, the bird's long-term welfare must be considered and its biology, management and behavior taken into account. Amputations on birds in wildlife rehabilitation situations may not be done without permission from authorities (United States Fish and Wildlife and/or state natural resource agencies); raptors in particular are poor candidates for amputation of anything more than a minor digit. While some states will permit amputations of the wing below the elbow, The Raptor Center does not support wing amputations in raptors at any site as experience shows that these patients routinely have long-term issues resulting from lack of balance, chronic trauma and continued efforts at flight. Likewise, leg amputations are the ultimate case of uneven weight bearing, which leads to chronic and intractable bumblefoot in raptors. While raptors may present the extreme in these situations and companion birds are known to be more tolerant of amputations, it is incumbent on the clinician to consider the legal and ethical considerations in all cases when contemplating such a surgery.

Treatment of a Fracture Patient

As most avian fracture presentations are emergency situations or acute traumas, these patients must first and foremost be considered critical care candidates. While fracture stabilization and soft tissue care are essential to minimize morbidity, careful consideration must be given to what the patient can withstand in emergency care. Preserving the patient's life is the principle consideration upon presentation. A holistic approach should be used, addressing immediate critical care concerns, assessing the overall health status of the patient and constantly evaluating the patient's condition and ability to withstand additional handling.

Fluid therapy to address dehydration, hypovolemia, blood loss or shock, can be administered subcutaneously (inguinal area or dorsally between scapulae), intravenously (medial metatarsal or basilic vein catheter), or intraosseously (ulna or tibiotarsal catheter). Fluid choice depends upon patient status. In non-critical cases, crystalloids given subcutaneously are readily absorbed and effective. Recommended dose is 60 ml/kg/day for maintenance plus replacement volumes; total dose can be up to 100 ml/kg/day. In hypovolemic cases, the addition of a colloid such as hetastarch (10-15 ml/kg administered intravenously or intraosseously) will help restore or maintain intravascular pressure.

While all open wounds should be cultured prior to institution of antimicrobial therapy, it is important to begin with a broad spectrum antibiotic pending culture and sensitivity results. Commonly used antibiotics include clindamycin (50 mg/kg q12h), enrofloxacin (10-20 mg/kg PO q6h) and amoxicillin/clavulanate potassium (50-100 mg/kg PO q8h).[7] Prophylactic antifungal therapy should be given on admission for any traumatized bird of a species predisposed to aspergillosis such as gyrfalcons, goshawks, northern owl species, African grey parrots, golden eagles or juvenile bald eagles. Commonly used antifungal medications include itraconazole (7 mg/kg q12h for 5 days, then q24h for 16 additional days) and voriconazole (12 mg/kg q12h)

Recommendations for pain management are changing rapidly in the field of avian medicine as new research increases our knowledge of pharmacokinetics and effectiveness. Multimodal therapy is recognized to have greater efficacy on pain management with less chance of negative side effects or toxicity. Opioids are indicated for moderate or severe pain in birds, especially in initial management and peri-operatively in surgical cases. Although fairly short-acting, butorphanol (1-4 mg/kg) [7] is commonly used in birds. Buprenorphine, which has a slower onset and longer duration, is being used increasingly. Non-steroidal anti-inflammatory drugs (NSAIDs) have both anti-inflammatory and analgesic effects with substantially longer duration than opioids. NSAIDS used routinely in avian medicine include meloxicam (0.2–1 mg/kg IM or PO q12h) and carporfen (2 mg/kg IM q12h).[7] Several studies have demonstrated extensive species variability in metabolism of all of these drugs. As pain management in aves is an area of active research with new information routinely being published, the reader is encouraged to refer to recent literature in this area.[8]

Attention to the preservation of soft tissue and bone during preliminary treatment of the avian patient is a key component of optimizing healing potential. Many avian long bone fractures are open due to the paucity of soft tissues. Debridement of wounds should be performed without excessive manipulation to preserve soft tissues and avoid compromising circulation to bone fragments. Protection of the bone ends protruding through the skin should be done as quickly as possible through coverage with a water-based, water-soluble lubricant (K-Y Jelly®, Johnson & Johnson) or saline-moistened gauze as the first bandage layer (no direct contact between dry gauze and bone). The treatment of the exposed bone ends will prevent this tissue from becoming devitalized and preserve bone integrity.

Approaches to Fracture Management (Fracture Management Options)

There are three basic options for management of fractures: confinement/cage rest, coaptation, and surgical fixation. Cage rest or confinement is an important initial step for the trauma patient, although only suitable as a definitive treatment in a very small number of situations. The most appropriate use of cage rest for fracture management is with injuries not amenable to bandaging or surgery such as pelvic fractures. It may also be used with extremely small, high stress species that do not tolerate coaptation well, recognizing that a return to full function may be compromised.

Coaptation may be used with or without surgical fixation. Although coaptation is most commonly used for initial or temporary pre-surgical stabilization, it may be the definitive management for some fractures such as shoulder girdle fractures, non-displaced metacarpal fractures, and tibiotarsal or metatarsal fractures of small birds.[9-11] Coaptation may also be used post-surgically to provide protection and/or additional support. Ideally, however, surgical fixation will be robust enough that prolonged coaptation is not required thereby allowing the bird to have normal use of the limb during bone healing.

Although coaptation is typically cheaper than surgical fixation and often requires less skill or experience, there are downsides to its use as primary stabilization for many long-bone fractures in birds. In general, functional alignment is poorer with coaptation, resulting in poorer expected outcomes. With forewing fractures, the use of coaptation will result in a much higher incidence of synostosis, especially if the fracture is in the distal third of the radius, ulna or both.[12] Prolonged use of coaptation may lead to decreased joint range of motion or loss of patagial elasticity.

Coaptation

Body wrap: A body wrap may be used alone or in conjunction with a Figure-of-8 bandage. When combined with physiotherapy, it is the recommended definitive treatment for shoulder girdle fractures (coracoid fractures and luxations, scapular fractures or furcular fractures), even when highly displaced. It is also indicated for temporary stabilization of wing fractures and may be combined with a Figure-of -8 bandage for fractures of the radius or ulna (recognizing risk of synostosis) or with a metacarpal splint for treatment of distal wing fractures.

Figure-of-8: The figure-of-8 bandage is used to stabilize fractures of the wing distal to the elbow. It is not indicated for humeral fractures. Short-term use (7-10 days) is recommended as the figure-of-8 bandage is often associated with complications such as reduced joint range of motion or patagial constriction.

Curved edge splint: Using moldable thermoplastic, a curved edge splint can be shaped to fit the carpo-metacarpal area to support and maintain alignment of metacarpal fractures. This may be a temporary technique prior to surgical stabilization or definitive repair in cases with minimal displacement. The splint is sandwiched in place with tape and a body wrap applied. It is important to monitor soft tissue swelling and adjust the fit as needed.

Splint: A tape splint is used for tibiotarsus and metatarsus fractures in small birds (pigeon size or smaller). A tape "sandwich" is applied with the leg in normal standing position and can be hardened by coating the completed splint in Vetbond®. In addition, a syringe case or other similar type of splint can be used in smaller birds.

Surgical fixation

Surgical fracture management often presents the best opportunity for complete return to normal function in the avian patient. A well-constructed surgical fixator will meet all fracture management objectives and stabilize against the forces of tension, torsion, shearing and bending. A variety of configurations using intramedullary (IM) pins, k-wires, external fixator pins, bars and clamps provides the clinician with needed flexibility given the broad diversity in the size and shape of avian patients. As healing progress, these fixators can often be sequentially dismantled to gradually increase load-sharing with the bone to promote healing.[13]

Intramedullary pin: A traditional Steinmann IM pin provides excellent opposition to the bending forces of a long bone and is relatively easily applied. It is most commonly used as the sole fixation in radius fractures and small bird femoral fractures. Due to the IM pin's inability to resist torsional, compressive or tensile forces, it is typically used with coaptation or as part of a hybrid fixator. IM pins can often be inserted in such a way as to avoid joint involvement by taking advantage of the curved anatomy of most avian long bones. In the humerus, radius, and femur, the IM pin is inserted retrograde, using surgical approach to the fracture site. In the ulna and tibiotarsus (and some humeral fractures), a closed approach and normograde insertion can be used, minimizing soft tissue damage and reducing overall morbidity. Due to its normal hyperflexed position, the stifle tolerates the presence of an IM pin exiting through the joint short-term. The elbow, however, does not respond well to any joint insult.

External skeletal fixator: Type I, II and III external skeletal fixators (ESF) are very good at resisting rotational movement, as well as compressive or tensile forces and have moderate capacity to oppose bending if sufficiently robust. Construction of the ESF requires careful pin selection and placement. The success of an ESF is enhanced by the use of partially threaded, positive-profile ESF pins, designed for birds, which provide good cortical anchoring and do not loosen when properly applied. Connecting bars can be stainless steel, acrylic, or epoxy although the weight of the former must be taken into consideration.

External skeletal fixators are applied in situations where a hybrid fixator (see below) cannot be used do to one's inability to insert an IM pin i without joint impingement or where fractures are highly comminuted and unstable. Metacarpal fractures are very amenable to a Type I ESF, as are some ulna fractures. A Type II fixator can be used for metatarsal fractures in raptors and other large birds.

The hybrid fixator: The most versatile system for long-bone fixation in birds is a hybrid system combining an external skeletal fixator with an IM pin tie-in (ESF-IM tie-in).[1,14] This system is inexpensive, easily learned, lightweight, adaptable and very effective in most cases. The fixator construct consists of an IM pin, two external skeletal fixator (ESF) pins and an external connecting bar that joins these components. When applying this system, every attempt should be made to reduce soft tissue disruption at the fracture site and preserve joint integrity. IM pins are inserted normograde where possible and ESF pins placed in the proximal and distal bone segments. Bending the IM pin 90 degrees, aligning it with the ESF pins and then using a penrose drain as a mold to create an acrylic connecting bar across all pins completes the tie-in. This construct resists all forces applied to the bone and its integrity can be increased with the addition of extra ESF pins in larger birds or more comminuted fractures.[15] As healing progresses, sequential dismantling of the fixation can be done over two weeks or more to transfer load-sharing to the bone. The hybrid fixator works for most diaphyseal long-bone fractures in birds and can be adapted for more proximal or distal fractures.

Post-operative care

There are several phases to post-operative care including immediate recovery, convalescence, and physical rehabilitation post-healing. Avian patients, both captive and wild, present unique considerations and challenges in each of these areas. Failure to prevent or address complications in the post-surgical phases can result in a failed outcome for the patient despite successful surgery.

Prevention of new trauma is a key concern immediately post-op. Patients should be adequately restrained during anesthetic recovery and not returned to a cage until fully able to balance and control their activity. The recovery cage should be prepared by removing or lowering any perching and providing a padded cage bottom. Many birds are less anxious when able to perch, yet are too unsteady to safely climb and perch immediately following the surgical procedure. A rolled up towel or wooden dowel covered in vetwrap on the bottom of the cage floor can be used to allow safe perching. Climbing should be prevented by using a solid sided cage, which should be placed in a low stress environment away from dogs, cats, humans, and any other predators. The use of low light and limited visibility will also encourage quiet recovery. Many patients will benefit from supplemental heat. A quiet, warm (85-90°F) environment with easy access for observation is ideal. In raptors, attention should be paid to preservation of flight feathers; it is routine to use a tail protector to prevent breakage of the rectrices during convalescence in most species (exceptions include great horned and small owls).

As in any patient, nutritional support is critical. Many avian patients are nutritionally compromised when presented for treatment; wild birds may have been down and unable to eat for several days or weeks prior to presentation and captive birds, while well fed, are likely to be on nutritionally-inadequate diets. Calcium and vitamins A, D_3, and K may be marginal or deficient in birds that eat only seeds and may need enhanced nutritional supplementation.

Diets designed for critical care are important initially, but getting the patient eating appropriate food items on its own reduces stress. The immediate post-operative phase is not the time to do long-term diet correction in a companion bird. Some wild species such as osprey are notorious for not eating in captivity; the clinician should work closely with a knowledgeable rehabilitator on nutritional support of wild birds. In addition to the type of food being offered, it is important to note presentation of the food. Consideration must be given to the patient's ability to access food, grasp and manipulate it, and eat.

As appropriate, both antibiosis and pain medications should be continued in the convalescing patient. Antibiotic choice should be adjusted based on culture and sensitivity results. The use of antibiotics in stressed birds, however, is often associated with oral/pharyngeal candida infections. The development of anorexia in a post-operative avian patient on antibiotics may be a sign of candidiasis.

Surgical patients should be examined 24 hours post-operatively. Suture lines and pin holes (fracture repair) should be cleaned and any bandage materials changed. Prior to discharging the patient into the care of the owner, all details of the post-operative plan (feeding, medication, wound care, housing, physical limitations and rechecks) should be reviewed with the owner. Attention should be paid to the owner's ability or inclination to provide appropriate care at home.

To maximize return to normal function, orthopedic patients require additional post-operative care. Although the need to treat soft tissue wounds and other concurrent issues may require a more frequent treatment schedule, the orthopedic patient should receive gentle passive physical therapy to maintain or increase range of motion of limb at least twice a week. Physiotherapy is typically performed under inhalant gas anesthesia. The limb is extended and held in a gentle stretch several times over a 5-minute period. Each joint is evaluated for range of motion relative to the normal limb. The fixation apparatus should be monitored for loosening of pins or exudate around pin tracts/surgical sites, as well as any signs of causing iatrogenic trauma. As soon as possible, coaptation such as body and wing wraps should be removed, allowing the limb to return to normal function (assuming adequate fixation!). These can normally be removed a few days after surgery except in the case of metacarpal fractures, where additional support is required due to the weight of fixation at the end of the wing. If fixation allows for normal use of the limb with no coaptation, the requirement for physiotherapy is greatly reduced.

Radiographic images of the surgical/fracture site should be taken at 10-14 days and again at 20-24 days post-operatively. These images should be evaluated for normal fracture healing, as well as potential complications such as reaction to the fixation device, ESF pin loosening or development of osteomyelitis or sequestrum. Many avian fractures have significant healing occur by day 21, although the process will be slower in open, highly comminuted, and/or contaminated fractures. As healing proceeds, load sharing can be gradually transferred to the bone through the process of dynamic destabilization of the fixator device.

An integral part of the follow-up care is careful prevention of co-morbidity situations. One of the most common sequela to fracture repair in a leg is the development of contra lateral bumblefoot (infectious pododermatitis) in raptors. The Raptor Center routinely uses an interdigitating bandage or other prophylactic bandaging on the primary weight bearing foot in orthopedic cases. Close monitoring of the feet for lesions during the convalescent phase is critical.

Summary

Understanding the guiding principles of orthopedics and the unique challenges presented by birds is key to enabling the practitioner to successfully address the broad diversity in patients and injuries seen in avian medicine. There are ethical as well as medical implications to decisions on how to treat fractures while potential complications and increased morbidity can be pro-actively mitigated during the treatment phase. Clinicians who may not have the expertise to provide orthopedic care and choose to refer may still have significant impact on the case outcome if they familiar with the unique concerns of the avian patient.

References

1. Redig PT, Ponder, JB. Orthopedic surgery. In: Samour J, ed. *Avian Medicine.* 3rd ed. London: Elsevier;2016:215-248.

2. Pauli A, Klauss G, Diehl K, Redig P. Clinical techniques: considerations for release of raptors with ocular disease. *J Exotic Pet Med.* 2007;16(2):101-103.

3. de Matos R. Calcium metabolism in birds. *Vet Clin North Am Exot Anim Pract.* 2008;11(1):59-82.

4. Tully TN. Basic avian bone growth and healing. *Vet Clin North Am Exot Anim Pract.* 2002;5:23-30.

5. Piermattei DL, Flo GL, DeCamp CE. Brinker, Piermattei, and Flo's Handbook of Small Animal Orthopedics and Fracture Repair, 4th ed. St. Louis; Saunders/Elsevier; 2006.

6. Redig PT, Cruz L. The avian skeleton and fracture management, a sub-section in trauma-related medical conditions. In: Samour J, ed. *Avian Medicine.* 2nd ed. London: Elsevier;2008:215-248.

7. Hawkins MG, Barron HW, Speer BL, et al. Birds. In: Carpenter JW, Marion CJ, eds. *Exotic Animal Formulary.* 4th ed. St. Louis, MO: Elsevier Saunders;2013:183-437.

8. Paul-Murphy J. Pain management. In: Harrison GJ, Lightfoot TL, eds. *Clinical Avian Medicine.* Palm Beach, FL: Spix Publishing Inc.; 2006:233-240.

9. Bueno-Padilla I, Arent LR, Ponder JB. Tips for raptor bandaging. *Exotic DVM.* 2011;12(3):29-47.

10. Redig PT, Nicolas Francisco O, Froembling M, et al. Coracoid fracture management in raptors: assessment of conservative approach. *Proc Annu Conf Assoc Avian Vet.* 2009;351-352.

11. Murray M. Management of metacarpal fractures in free-living raptors. *Proc Annu Conf Assoc Avian Vet.* 2012;283-284.

12. Beaufrere H, Ammersbach M, Nevarez J, et al. Successful treatment of a radioulnar synostosis in a Mississippi kite *(Ictinia mississippiensis)*. *J Avian Med Surg.* 2012;26(2):94-100.

13. Egger EL. External skeletal fixation – general principles. In: Slatter DH, ed. *Textbook of Small Animal Surgery.* 2nd ed. Philadelphia, PA: WB Saunders;1993:1641-1656.

14. Redig PT. The use of an external skeletal fixator-intramedullary pin tie-in fixator for treatment of long bone fractures in raptors. In: Lumeij JT, Poffers J, eds. *Raptor Biomedicine* III. Lake Worth, FL: Zoological Education Network;2000:239-254.

15. Van Wettere AJ, Redig PT, Wallace LJ, et al. Mechanical evaluation of external skeletal fixator-intramedullary pin tie-in configurations applied to cadaveral humeri from red-tailed hawks *(Buteo jamaicensis)*. *J Avian Med Surg.* 2009;23(4):277-85.

Tips and Tricks for Compounding Medications for Avian Patients

Lauren V. Powers, DVM, Dipl ABVP (Avian), Dipl ABVP (Exotic Companion Mammal)

Session #292

Affiliation: From the Avian and Exotic Pet Service, Carolina Veterinary Specialists, Huntersville, NC 28078, USA.

Abstract: The compounding of drugs into formulations suitable for clinical use is an essential part of avian medicine. Few medications are available in FDA-approved formulations practical for use for the myriad of avian species treated in daily practice. While extralabel drug use and extemporaneous compounding are generally permitted by federal law, there are important exceptions such as extralabel restrictions and prohibitions in food-producing animal species such as chickens and turkeys. The avian practitioner should be familiar with published standards and guidelines for drug compounding as well as pertinent federal and state laws and regulations. The compounding pharmacist is an invaluable resource both for filling compounded prescription requests and for help and guidance on in-house extemporaneous compounding.

Introduction

In order for a drug to be successfully delivered to an avian patient, it must be provided in a vehicle suitable for delivery and well-tolerated by the bird. Ease of administration and palatability are critical factors when formulating medications for avian patients due to the stress and potential trauma associated with handling and restraint. Few medications are approved by the Federal Drug Administration (FDA) for use in avian species or are available in approved formulations suitable for use.

Extralabel Drug Use

Extralabel drug use means the use of an FDA-approved drug in a manner different from what is indicated on the approved label. The Animal Medicinal Drug Use Clarification Act of 1994 (AMDUCA), an amendment to the Federal Food, Drug, and Cosmetic (FD&C) Act of 1938, allows veterinarians to legally use and prescribe FDA-approved veterinary and human drugs in an extralabel fashion with certain important exceptions, some of which are discussed below.

What is Drug Compounding?

Any manipulation of a drug not stipulated on the approved label is considered compounding and is considered as extralabel use. This includes (but is not limited to) changes to the concentration of a liquid drug (e.g., dilution), flavoring, mixing of two or more drugs together, and the creation of a liquid formulation for oral administration from tablets, capsules, injectable solutions, or bulk substances (active pharmaceutical ingredients, or APIs).

Standards for Drug Compounding

Some important factors to consider when compounding medications include potency (strength), stability over time, purity, absorption from the delivery site, and potential toxicity. Some drugs are not suitable for compounding into certain formulation due to rapid inactivation and loss of potency or lack of systemic absorption. For example, aqueous formulations of enrofloxacin compounded from commercially available 68-mg tablets and 22.7 mg/ml injectable solution into a commercial liquid sweetener or into corn or cherry syrup maintained drug stability in amber-colored vials at room temperature for 56 days.[1] However, the stability of doxycycline hyclate compounded from 100-mg tablets into aqueous solution using syrup and commercial liquid sweetener cannot be assured beyond 7 days.[2] Itraconazole compounded from API to 10 mg/ml in hydrochloric acid and diethylene glycol monoethyl with 500 mg/ml Pluronic L64 surfactant (Professional Compounding Centers of America, Houston, Texas 77099, USA) administered at 7 mg/kg PO to black-footed penguins (*Spheniscus demersus*) resulted in significantly lower oral absorption when compared to a commercially available formulation (Sporanox Liquid 10 mg/ml, Janssen Pharmaceuticals, Titusville, New Jersey 08560, USA) at the same dosage, likely due to the presence of hydroxypropyl-b-cyclodextrin, a carrier molecule shown to improve the absorption of itraconazole in the commercially available liquid product.[3]

Other stability studies for drugs commonly used in avian practice have recently been published, such as for meloxicam and carprofen,[4] voriconazole,[5] and enalapril.[6] Several other excellent resources are available that provide monographs for compounded drug formulations,[7,8] including monographs published by the United States Pharmacopeia (USP).[8]

United States Pharmacopeia (USP) guidelines

The USP publishes official drug quality standards in chapters addressing quality assurance practices and in chapters that provide standardized monographs for drug quality, purity, and uniformity.[7,8] Chapters in the USP guidelines numbered less than 1000 are considered federally enforceable by FDA. The USP Compounding Compendium is available for purchase from the USP and is an excellent resource.[8] Although USP is a scientific nonprofit organization, it was granted federal authority for writing US drug quality standards by the FD&C Act, Section 501(b).[9] While the USP has always written monographs for compounded drugs in addition to manufactured drugs, USP has only recently written enforceable general chapters for compounding pharmacy practice.[8,9] Most state boards of pharmacy have also adopted USP's pharmacy practice standards and enforce them as state law.[9]

Extemporaneous Compounding in the Veterinary Practice

Extralabel drug use and extemporaneous drug compounding for in-hospital use and dispensing for veterinary patients is permitted by federal law under AMDUCA provided that a valid veterinarian-client-patient relationship exists. Practitioners must also comply with state pharmacy and veterinary medicine laws and regulations.

The most common drug compounding practice performed in the avian practice is the preparation of liquid formulations for oral use from tablets, capsules, or injectable solutions. Minimum supplies for extemporaneous drug compounding include mortars and pestles, compounding liquids (e.g., Ora-Plus and Ora-Sweet, Perrigo, Dublin, Ireland; Syrpalta Syrup Vehicle, Humco, Austin, Texas, USA), and liquid dispensing bottles. Flavoring agents (e.g., LorAnn Oils, Lansing, MI, USA; FLAVORx, Columbia, MD, USA), rubber spatulas, and a digital gram scale (ideally a microgram balance with a draftshield) can also be useful. In the event that commercially formulated compounding liquids are unavailable or are unsuitable, thickening and sweetening agents such as corn syrup, cherry syrup, and lactulose can also be used for many compounded formulations. Commercially available kits with supplies and recipes for flavoring and compounding drugs are available (e.g., FLAVORx).

Working with Compounding Pharmacies

Compounding pharmacies are an invaluable resource for avian practitioners. Compounding pharmacies have access to essential equipment, supplies, and resources to help prepare high quality compounded drug formulations that meet USP standards and comply with federal and state laws and regulations. Avian practitioners can benefit greatly from establishing close relationships with local and national compounding pharmacies. Visiting local compounding pharmacies is encouraged. It is the veterinarian's responsibility to ensure that the compounding pharmacy is familiar with veterinary prescriptions, strives to meet the highest standards, and complies with federal and state regulations.[9] The compounding pharmacy should be:

- licensed to practice pharmacy in the state in which they reside as well as the state of dispensing, if not the same (Check with the state Board of Pharmacy [National Association of Boards of Pharmacy, www.nabp.net] to determine if the pharmacy is licensed and in good standing[9]), and
- fully compliant with USP Chapters 795 and 797.

Ideally, the pharmacy should be:

- accredited by the Pharmacy Compounding Accreditation Board (PCAB),
- a member of the International Academy of Compounding Pharmacists (IACP) and/or American Society of Health-System Pharmacists (ASHP), and
- obtaining APIs from FDA- and DEA-registered suppliers (All APIs should be substances of USP or USP-NF [USP National Formulary] grade and be accompanied by a valid certificate of analysis).

Compounding from Bulk Substances (Active Pharmaceutical Ingredients, or APIs)

Bulk substances (bulk drugs) are the active pharmaceutical ingredients (APIs) used to create finished compounded formulations. Bulk drugs lack excipients often found in manufactured drugs such as preservatives, stabilizers, fillers, and buffers. Excipients may detrimentally affect the final compounded drug formulation. Many compounding pharmacies prefer to work with APIs in part to avoid excipients.

The FDA considers APIs as unapproved new drugs and only considers compounding from FDA-approved drugs as legal based on its interpretation of AMDUCA. FDA compliance policies and regulatory action guidance for FDA staff regarding veterinary drug compounding based on AMDUCA (Compliance Policy Guide (CPG) 608.400, Compounding of Drugs for Use in Animals) were drafted in 1996 and revised in 2003 but were rescinded in May of 2015 when the Draft Guidance for Industry (GFI) #230 (Compounding Animal Drugs from Bulk Drug Substances) was issued by the FDA Center for Veterinary Medicine (CVM) in response to changes in human compounding regulation resulting from the Drug Quality and Security Act of 2013 in order to bring regulatory enforcement for animal compounding more in line with that for human medicine. GFI #230 represents FDA's current thinking on the use of bulk drug substances when compounded for veterinary use. Public input to

GFI #230, including a "List of Bulk Drug Substances that may be Used by an Outsourcing Facility to Compound Drugs for Use in Animals" without regulatory action, was closed in November of 2015. According to GFI #230, drugs compounded from bulk substances for veterinary use must be prepared for individual patients after the receipt of a valid prescription; must not be used in food-producing animal species; the bulk drug must be accompanied by a valid certificate of analysis; the drug must be compounded in accordance with USP standards; any adverse effects must be reported to the FDA; among other requirements.

Drug Compounding for Food-Animal Species

The FDA has much stricter rules for drug compounding for food-producing animal species, regardless of their intended use. For example, chickens and turkeys are considered food-producing animal species even if the owner keeps the animals strictly as pets and does not intend to eat, sell, or give away their meat or eggs. Certain drugs are prohibited from extralabel use in any food animal species, such as fluoroquinolones (e.g., enrofloxacin) and nitroimidazoles (e.g., metronidazole). Extralabel use of cephalosporins is prohibited in major-use food animal species such as chickens and turkeys. The Food Animal Residue Avoidance Databank (FARAD, www.farad.org) is a risk-management program supported by the US Department of Agriculture. FARAD is an excellent resource for prohibited and restricted extralabel drug use lists and for extralabel drug use meat and egg withdrawal times for individual animals (upon request).

Compounding Controlled Substances

While compounding of controlled substances is generally permitted under federal law, there are additional steps that must be followed. In addition to FDA regulations, prescriptions must comply with Drug Enforcement Administration (DEA) rules and regulations and comply with state laws. If working with a compounding pharmacy, the pharmacy must dispense the product directly to the client. If the compounded medication enters a practice or is prepared in a practice, it must be logged into appropriate controlled substance records.

When Not to Compound

FDA's requirements limit extralabel drug use in animals to situations where an animal's health is threatened or where the animal may suffer or die without treatment. A drug should not be compounded when a FDA-approved product is available for the intended use in a suitable formulation. A drug should never be compounded for the sole purpose of reducing cost. As stated previously, extralabel use of certain medications is prohibited or restricted in food-producing animal species.

Resale of Compounded Medications

State laws vary in regard to the administration and dispensing of compounded veterinary drugs. Practitioners should be familiar with pertinent state laws and regulations. Many states do not permit the resale of compounded medications but permit their use in the hospital setting within the scope of professional practice. A helpful State Summary Report provided by the AVMA can be found at https://www.avma.org/Advocacy/StateAndLocal/Pages/compoundinglaws.aspx.

What's in a Label?

The prescription label should meet USP Chapter 17 (Prescription Container Labeling) guidelines and comply with federal and state pharmacy and veterinary medicine laws and regulations. Labels should feature only the most important patient information needed for safe and effective understanding and use and should include:

- Patient's full name.
- Drug name. Compounded medications are no longer branded products and therefore only the generic name should be listed (e.g., "enrofloxacin" instead of "Baytril").
- Drug strength and quantity. For compounded liquid formulations, the concentration (e.g., mg/ml) and volume (e.g., total ml in bottle) should be listed.
- The label should indicate that the formulation is compounded.
- Fill date.
- Beyond-use-date (BOD) (see below).
- Explicit clear instructions on use (e.g., amount, frequency, duration).
- Prescriber name, address, and phone number.
- Pharmacy name, address, and phone number (if using a compounding pharmacy).

Beyond-use-date (BUD)

Manufactured drugs are given expiration dates. Compounded drugs are given beyond-use

-dates (BUDs). Beyond-use-dates are provided in the USP-NF General Chapters for Compounding (USP-NF 795).[8] In the absence of stability data for a specific drug and preparation, general USP guidelines should be used. For example, the general guideline for maximum BUD for aqueous (water-containing) oral formulations stored at controlled cold temperatures is 14 days.[8]

References

1. Petritz OA, Guzman DS, Wiebe VJ, et al. Stability of three commonly compounded extemporaneous enrofloxacin suspensions for oral administration to exotic animals. *J Am Vet Med Assoc.* 2013;243(1):85-90.

2. Papich MG, Davidson GS, Fortier LA. Doxycycline concentration over time after storage in a compounded veterinary preparation. *J Am Vet Med Assoc*, 2013;242(12):1674-1678.

3. Smith JA, Papich MG, Russell G, et al. Effects of compounding on pharmacokinetics of itraconazole in black-footed penguins (*Spheniscus demersus*). *J Zoo Wildl Med.* 2010;41(3):487-495.

4. Hawkins MG, Karriker MJ, Wiebe V, et al. Drug distribution and stability in extemporaneous preparations of meloxicam and carprofen after dilution and suspension at two storage temperatures. *J Am Vet Med Assoc*. 2006;229(6):968-974.

5. Nguyen KQ, Hawkins MG, Taylor IT, et al. Stability and uniformity of extemporaneous preparations of voriconazole in two liquid suspension vehicles at two storage temperatures. *Am J Vet Res*. 2009;70(7): 908-914.

6. Nahata MC, Morosco RS, Hipple TF. Stability of enalapril maleate in three extemporaneously prepared oral liquids. *Am J Health Syst Pharm*. 1998;55(11):1155-1157.

7. Trissel LA. *Trissel's Stability of Compounded Formulations*, 5th ed. Washington, DC: American Pharmacists Association; 2012.

8. US Pharmacopeial Convention. *USP Compounding Compendium*. Rockville, MD: US Pharmacopeial Convention; 2016.

9. Davidson G, Powers LV. Compounds for avian and exotic patients: regulation and science. *Proc Annu Conf Assoc Avian Vet*. 2011;171-184.

Clinical Cardiovascular Anatomy and Disease States

Brenna Fitzgerald, DVM, Dipl ABVP (Avian)

Session #326

Affiliation: From The Medical Center for Birds, 3805 Main St, Oakley, CA 94561, USA.

Manuscript adapted from: Fitzgerald BC, Beaufrère H. Cardiology. In: Speer B, ed. *Current Veterinary Therapy in Avian Medicine and Surgery*. St. Louis, MO: Elsevier Saunders;2015:252-328.

Abstract: Cardiovascular disease is common in companion birds. Recognition requires the practitioner to have a thorough knowledge of avian cardiovascular anatomy and physiology, and of clinical disease states and their risk factors. Of clinical significance are distinct anatomic and physiologic differences between avian and mammalian cardiovascular systems and adjacent respiratory structures. Atherosclerosis is likely an underlying factor in the majority of non-infectious cardiovascular diseases diagnosed in pet birds. Lesions are most frequently recognized in the ascending aorta, brachiocephalic trunks, and pulmonary arteries. Unlike humans, recognizable clinical disease is primarily the product of progressive, flow-limiting arterial stenosis rather than thromboembolism. Patients with clinically significant atherosclerosis most often present for falling or collapse, transient or persistent weakness and dysfunction of one or more limbs, with or without altered mentation. In addition to age, female sex, and species, risk factors for development of atherosclerosis also include high-calorie and fat diets, dyslipidemia (e.g. hypercholesterolemia), and limited physical activity. Congestive heart failure is a frequent clinical end-point encountered in companion birds. Heart failure can result from primary myocardial failure, ventricular pressure overload, volume overload, conduction disturbances, or diastolic dysfunction. Dilative and hypertrophic cardiomyopathy have been reported in psittacine birds, as well as myocardial disease of infectious, metabolic, nutritional, neoplastic, or toxic origin. Pericardial effusion can develop due to inflammatory changes of the pericardium (pericarditis), trauma or aneurismal rupture, cardiac or pericardial neoplasia, or as a manifestation of metabolic derangements, toxic insult, or, commonly, right-sided congestive heart failure. Infectious pericarditis and associated effusion can occur as an extension of epicarditis, myocarditis, or inflammatory or infectious conditions affecting other adjacent tissues.

Introduction

Cardiovascular disease has traditionally been thought to be a rare occurrence in companion birds, but a growing body of evidence collected over the last few decades indicates otherwise. It is frequently encountered in practice and poses a serious threat to the quality of life and longevity of many avian species. Successful intervention requires a foundational understanding of relevant anatomy and physiology, heightened awareness of risk factors, and recognition of clinical disease states. These proceedings review avian cardiovascular anatomy, as well as recognized disease states and their pathophysiology, clinical and pathologic features, and prevalence.

Anatomy and Physiology

The high-performance cardiovascular system of birds has evolved to meet the high aerobic demands of flight and other intense activity by providing adequate oxygen transport with maximal efficiency. Although heart rate relative to body mass is lower in birds than in mammals, the avian heart is relatively larger and its stroke volume and cardiac output greater. The budgerigar in flight has a cardiac output approximately 7 times that of mammals during exercise. Mean arterial pressure is higher in birds than in mammals, whereas total peripheral resistance is lower.[1,2]

The avian heart is 4-chambered, with left and right atria and left and right ventricles. The left ventricle perfuses the systemic circulation, and the right ventricle perfuses the pulmonary circulation. As resistance to blood flow is lower in the pulmonary circulation than in the systemic circulation, the right ventricle must produce less systolic pressure; it is smaller and the myocardium is thinner than in the left ventricle. The left ventricular wall is 2 to 3 times thicker than the right ventricular wall and can generate 4-5 times the systolic pressure.[1,2] The cone-shaped left ventricle extends to the cardiac apex, and its right wall forms the interventricular septum. The right ventricle is a crescent-shaped cavity that wraps around at least one half of the left ventricle. Its free wall is continuous with the outer wall of the left ventricle and it does not extend to the cardiac apex.[1-3] Paired cranial venae cavae and the single caudal vena cava empty into the right atrium, while the paired pulmonary veins enter the left atrium.[1,2,4] The left atrioventricular (AV) (mitral) valve is membranous and tricuspid (although the cusps are poorly defined) and is attached to papillary muscles via chordae tendineae. These are lacking in the right AV valve, which consists of only a spiral muscular flap.[1,2,5] The pulmonary and aortic valves are each composed of three semilunar cusps.[1-3] The ascending aorta extends from the left ventricle and unlike in mammals, curves to the right. It gives rise to the coronary arteries and paired brachiocephalic trunks, each dividing into common carotid and subclavian arteries, then continues caudally as the descending aorta, giving off the coeliac and cranial mesenteric arteries as its first major branches. The pulmonary trunk arises from the right ventricle and splits into right and left pulmonary arteries.[1,2,4,5]

The heart is enclosed within a fibrous, relatively non-compliant pericardial sac, which normally contains a very small volume of serous fluid that serves to lubricate the movement of the heart within the pericardial sac. The outer fibrous layer of the pericardium is continuous with the surface of the great vessels and attachments exist between the pericardium and the sternal plate, liver, adjacent air sacs and, indirectly, to the vertebral column. The heart is thereby well anchored within the cranioventral coelom, its apex partially enclosed within the lobes of the liver.[1-3]

Adjacent respiratory structures include the paired, non-expansile lungs, resting dorsal to the heart, and the large, complex clavicular air sac which lies within the thoracic inlet and surrounds the heart base and great vessels.[5,6] The position of these structures relative to the heart has great clinical relevance given the potential for cardiovascular involvement in cases of primary respiratory disease.

The cardiac conduction system of birds bears a close resemblance to that of mammals, consisting of the sinoatrial (SA) node, atrioventricular (AV) node, the bundle of His, and its bundle branches. However, there are 3 bundle branches (left, right, and middle) as well as an AV ring of Purkinje fibers surrounding the right AV opening that is unique to birds. A figure 8 is formed as the middle bundle branch passes around the aorta and joins the AV ring. The SA node lies near the openings of the cranial and caudal venae cavae into the right atrium and its pacemaker cells set the heart rate. The wave of excitation initiated in the SA node propagates through the atrial muscle and is transmitted to the ventricles via the AV node, which gives rise to the Purkinje fibers of the AV ring, bundle of His, and the bundle branches. The location of the AV node is a subject of some debate, but is most likely located at the right base of the interatrial septum. Left and right bundle branches emerge from the interventricular septum to supply the left and right ventricles, following the tracts of the coronary arteries into the myocardium (periarterial Purkinje fibers).[1,2]

Vascular microanatomy

Structurally the arteries can be classified into the elastic arteries, consisting of the aortic arch, thoracic aorta up to the coeliac artery, brachiocephalic trunks, and extrapulmonary portions of the pulmonary arteries, and the muscular arteries comprising the remainder of the arterial system. The walls of elastic arteries are composed of 3 layers, the *tunica intima* which is a single-cell layer of endothelial cells adjacent to the arterial lumen, the *tunica media* which is composed of concentric layers of collagen and elastic fibers composed of smooth muscle cells and elastin, and the *tunica adventitia* which is an outer layer of collagen fibers and connective tissue. Muscular arteries have the same 3 tunicae, but the media is only composed of smooth muscle cells and elastin and the adventitia is much thicker and has less compliance.[2]

Systemic arterial blood pressure is a function of cardiac output and the resistance of the arterial system; the latter is partly determined by afterload. Afterload is determined by 1) impedance of the proximal aorta itself, 2) compliance (distensibility) of the elastic arteries, and 3) peripheral resistance to blood flow afforded primarily by small arteries and arterioles. The characteristic impedance of the proximal aorta increases with increased aortic stiffness and decreased radius, resulting in greater left ventricular afterload. Decreased compliance of elastic arteries, as well as increases in peripheral resistance (occurring with decreases in vessel radius), further contribute to afterload.[2,7]

Vasoconstriction (and venoconstriction) are modulated by sympathetic tone and circulating catecholamines (EPI and NE acting via α-adrenergic receptors). Vasoconstriction is also mediated by activation of the renin-angiotensin-aldosterone system (RAAS), arginine vasotocin (AVT, the homolog of antidiuretic hormone in mammals), and endothelin-1.[2]

Recognized Cardiovascular Disease States

Atherosclerosis and stroke

Atherosclerosis is a chronic inflammatory and degenerative disease of the arterial wall wherein the lumen narrows by progressive accumulation of inflammatory cells, fat, cholesterol, calcium, and cellular debris that forms fibrofatty atheromatous plaques within the intima. Potential complications include stenosis, ischemia, thrombosis, hemorrhage, and aneurysm.[8-10] Atherosclerosis is likely an underlying factor in the majority of non-infectious cardiovascular diseases diagnosed in pet birds and is undoubtedly the most common lesion of the cardiovascular system identified post-mortem in companion psittacine birds.

The etiology and development of atherosclerotic lesions can be broadly explained by the response-to-injury hypothesis which postulates that damage to the endothelium lining the artery sets the stage for atherogenesis. Lesion formation involves endothelial dysfunction, damage, and oxidative stress that promote increased endothelial permeability to lipoproteins and intimal adherence and migration of inflammatory cells (particularly monocytes). Subendothelial accumulation of lipoproteins, particularly oxidized low-density lipoproteins (LDL), further attracts monocytes, which once within the subintimal space, differentiate into macrophages and internalize lipoproteins and store cholesterol, thereby becoming foam cells. These cells eventually die by necrosis or apoptosis, resulting in further accumulation of extracellular lipid and necrotic debris. In this way, the process of intimal lipid accumulation, chemoattraction of macrophages, and formation of foam cells drives the progression of the atherosclerotic lesion and eventual development of a necrotic core. Smooth muscle cells play a significant role in lesion development by proliferating and migrating into the intima and subintimal space and producing extracellular matrix components (including collagen and elastic fibers). This accounts for the formation of fibrous tissue which in advanced lesions further enlarges the lesion and creates a fibrous cap overlying the lipid and necrotic core. Smooth muscle cells also accumulate cholesterol and become foam cells.[8-10]

In psittacine birds, lesions are characterized by accumulation of the components discussed above in the intima and luminal side of the vascular media. Lesions can be classified into 7 types (I-VII); types I and II are considered early lesions, type III intermediate, and types IV-VII advanced. Lesion type VI is distinguished from IV and V by the additional complication of fissure, hematoma, or thrombosis, and type VII are calcific lesions characterized by large calcium plaques and osseous metaplasia. Clinical signs are attributed to advanced lesions, while early and intermediate lesions are generally silent and subclinical.[8-10]

Atherosclerotic lesions are most frequently recognized in the ascending aorta, brachiocephalic trunks, and pulmonary arteries. Although not as often appreciated, lesions also appear in peripheral vasculature, most notably the coronary arteries, carotid arteries, and descending aorta, as well as the subclavian arteries, coeliac artery, and ischiatic arteries. Coronary and aortic aneurysms have been known to arise secondary to atherosclerosis owing to decreased compliance of the affected vessels compared to normal elastic arteries.[9]

Unlike humans, recognizable clinical disease in birds is primarily the product of progressive, flow-limiting arterial stenosis rather than thrombosis and hemorrhage of disrupted plaques that results in thromboembolism and acute arterial obstruction. The difference is likely accounted for by the inability of the avian thrombocyte to form shear-resistant arterial thrombi and 3-dimensional aggregates. Acute myocardial infarction is rendered less likely by the differences in avian coronary vasculature and greater collateral circulation compared to humans. In one study, advanced lesions were commonly characterized by severe arterial stenosis and occlusion, but atherothrombotic lesions, aortic dissection and rupture, and lesional hemorrhage were identified rarely (in 1.9% of lesions).[8-10]

Patients with clinically significant atherosclerosis most often present for falling or collapse, frequently accompanied by transient or persistent weakness and dysfunction of one or more limbs, and with or without altered mentation. Extension and rigidity of one leg (and sometimes the ipsilateral wing), clenching of the toes, ataxia or difficulty perching, and tremors or convulsions may be reported. These signs are considered most consistent with stroke, but rarely is this confirmed diagnostically. These events can be difficult to distinguish from seizure activity, and indeed seizures can occur subsequent to ischemic stroke. Signs may be acute and transient, or there may be persistent neurologic abnormities, including reduced mentation, blindness, seizures, vestibular signs, paresis of one or both pelvic limbs, and ataxia. Alternatively, sudden death may occur as a result of a cerebral infarct, aneurysmal rupture, or a lethal cardiac arrhythmia secondary to myocardial ischemia.

Reported prevalence of atherosclerosis in psittacine birds varies widely (1.9 to 91.8%), with the highest recently reported among Amazons and African grey parrots (*Amazona* spp. and *Psittacus erithacus*, respectively). Most recently, prevalence and epidemiology of clinically relevant atherosclerotic lesions (types IV-VI) were investigated by review of over 7600 psittacine cases representing 5 genera. An overall prevalence of 6.8% was reported, but a significantly higher prevalence of advanced lesions was found for grey parrots, Amazon parrots, and cockatiels (*Nymphicus hollandicus*); cockatoos (*Cacatua* spp.) and macaws (*Ara* spp.) appeared less susceptible. There was a positive association between advanced atherosclerosis and increasing age, female sex, reproductive disease (predominately female), hepatic disease, and concurrent myocardial fibrosis.[9] These findings are subjectively consistent with the author's experience in practice, where severe atherosclerosis is most often encountered in grey parrots and Amazon parrots over 20 years of age, followed by cockatiels, lovebirds (*Agapornis* spp.), Eclectus parrots (*Eclectus roratus*), cockatoos, and macaws.

In addition to age, female sex, and species, proposed (but unconfirmed) risk factors for development of atherosclerosis also include high-calorie and fat diets, dyslipidemia (e.g., hypercholesterolemia), and limited physical activity. The predisposition of female psittacine birds to development of atherosclerosis likely relates to the profound effects of estrogens on lipid, protein, and calcium metabolism in reproductively active hens. Estrogen promotes increased plasma total cholesterol, triglyceride, calcium, and protein levels, as well as production of specific lipoproteins used to transport lipid from the liver (where lipogenesis predominately occurs in birds) to the developing yolk. These changes promote atherogenesis and may at least partially explain the association between reproductive tract disease and increased prevalence of advanced atherosclerotic lesions.[8,9] Species differences in plasma total cholesterol have been found to correlate with relative prevalence of severe atherosclerosis, with more susceptible species having significantly higher total cholesterol levels. One possible explanation is that some species are poorly adapted to the dietary changes related to captivity and are less equipped to metabolize dietary lipids. Diet-induced atherosclerosis has been demonstrated experimentally in budgerigars (*Melopsittacus undulatus*) and Quaker parrots (*Myiopsitta monachus*). Budgerigars fed a 2% cholesterol diet developed advanced atherosclerotic lesions by 3 months, which further progressed by 6 months; hepatic lipidosis developed concurrently. In a recent study, Quaker parrots fed a 1% cholesterol diet developed severe dyslipidemia and advanced atherosclerotic lesions in the aorta, brachiocephalic trunks, and coronary arteries within 4 months.[8-10] In the author's experience, companion birds ultimately diagnosed with advanced atherosclerosis often have a long-term history of a high-calorie and fat diet, regular provision of animal-based products (such as chicken bones, meat, and cheese), clinical obesity, and/or chronic reproductive activity (female birds).

Heart failure

Heart failure occurs when cardiac output becomes inadequate to maintain arterial blood pressure (manifested clinically by low output signs of lethargy, depression, and weakness) and when the heart is unable to empty the venous reservoirs, manifested by vascular congestion and transudation of fluid within tissues and body cavities (congestive signs). In the case of right-sided congestive heart failure, peripheral venous congestion, hepatic congestion, ascites, and pericardial effusion are often present. Pulmonary edema and congestion of the pulmonary veins occur with left-sided congestive heart failure, and a combination of signs may be seen with biventricular failure.[7] Compared to mammals, birds are thought to have a greater propensity for developing pulmonary hypertension and right-sided, rather than left-sided congestive heart failure owing to the morphology of the right AV valve, less deformable nucleated erythrocytes, and the rigid, non-distensible lungs which limit the ability of the blood capillaries to expand and accommodate greater blood flow.

Congestive heart failure is not a primary disease in itself, but an ultimate consequence of structural or functional abnormalities of the cardiovascular or pulmonary systems, compounded by the chronic effects of compensatory mechanisms.[7] Not all cardiovascular disease necessarily leads to congestive heart failure, but it is a frequent clinical end-point encountered in companion birds. The pathophysiology of congestive heart failure has been studied primarily in mammals, but is likely similar across vertebrate taxons owing to shared neuroendocrine regulatory pathways of circulation and hemodynamic constraints.

Heart failure can result from primary myocardial failure, ventricular pressure overload, volume overload, conduction disturbances, or diastolic dysfunction. Myocardial failure can occur with either primary disease of the myocardium or secondary to chronic pressure or volume overload. Pressure overload results from outflow obstruction (e.g., from congenital or acquired valvular stenosis, narrowing of the aorta). The affected ventricle(s) ultimately undergoes concentric hypertrophy, in which the ventricular wall thickens with a corresponding decrease in chamber volume. With time, myocardial fibrosis develops and impairs both systolic and diastolic function. As myocardial failure progresses, the ventricle ultimately dilates. Volume overload results from valvular insufficiency (e.g., endocardiosis, vegetative endocarditis) and abnormal communications between chambers (e.g., ventricular septal defect). The result is eccentric hypertrophy of the affected ventricle(s), characterized by chamber enlargement and increased myocardial mass with little increase in wall thickness. With progression of hypertrophy, eventual myocardial failure ensues, with further chamber dilatation. Diastolic failure results from decreased ventricular compliance, abnormal myocardial relaxation, and cardiac tamponade or constrictive pericardial disease.[7]

Dilative and hypertrophic cardiomyopathy have been reported in psittacines, with dilative cardiomyopathy generally considered idiopathic.[2] Hypertrophic cardiomyopathy most often develops secondary to the arterial luminal stenosis and decreased compliance characteristic of atherosclerosis, or to pulmonary hypertension related to atherosclerosis, chronic pulmonary disease, or left-sided congestive heart failure.

Myocardial changes consistent with ischemic injury, including myocardial necrosis, inflammation, degeneration, and fibrosis, have been associated with atherosclerosis of the coronary arteries in psittacines. Hypertrophic cardiomyopathy develops secondary to chronic increases in afterload, as occur with the arterial luminal stenosis and decreased compliance characteristic of atherosclerosis. Myocardial ischemia, combined with increased workload and myocardial oxygen demand, can result in reduced contractility, arrhythmias, and reduced cardiac output that may ultimately culminate in congestive heart failure.[7,9]

Myocardial disease may also be of infectious, metabolic, nutritional, neoplastic, or toxic origin. Myocarditis can be caused by a variety of infectious agents, including bacterial, fungal, and viral pathogens, as well as protozoal and helminth parasites.[3] Bacterial infection may involve the epicardium, myocardium, or endocardium and occur by hematogenous spread or by extension from infected adjacent tissues. Fungal infection, most commonly with *Aspergillus* spp., can spread to the heart from the adjacent lungs and air sacs and disseminated infection with *Aspergillus* spp., *Candida* spp., and Zygomycetes can have cardiac involvement by angioinvasion and hematogenous spread. Proventricular dilatation disease (PDD), for which avian *Bornavirus* (ABV) has been recently identified as the etiologic agent, is characterized histologically as lymphoplasmacytic infiltration of nerve ganglia throughout the body. Inflammatory infiltrates are frequently found in the epicardium and myocardium particularly in proximity to Purkinje fibers, potentially precipitating arrhythmias and sudden death. Other pathogens that can produce myocarditis are avian polyomavirus (APV), West Nile virus, *Sarcocystis falcatula*, and *Toxoplasma gondii*.

There are few published studies examining the prevalence and etiology of cardiac disease in psittacines and a relatively small number of case reports. Cardiac disease was recognized in 26 (9.7%) of 269 psittacine birds submitted for necropsy to Ohio State University, College of Veterinary Medicine between 1991 and 1995. In 15 (58%) of these, congestive heart failure was ruled the cause of death, while in the remaining 11 (42%) cardiac lesions were considered incidental and secondary to systemic infectious diseases including pulmonary aspergillosis, bacterial septicemia, APV infection, and PDD. In cases of right-sided congestive heart failure (10 birds), hepatic congestion, capsular fibrin, and ascites were consistently found. Pericardial effusion was identified in 3 of these cases. Pulmonary congestion and edema were observed in cases of left-sided congestive heart failure (5 birds), and birds with biventricular failure exhibited both hepatic and pulmonary lesions.[11]

In another study, 36% of 107 psittacine birds submitted for necropsy exhibited grossly visible abnormalities of the heart, great vessels, or both, while histologic changes in these tissues were found in 99% of birds; 15% had either myocardial hypertrophy or dilatation and 60% had inflammatory infiltrates into the myocardium. Myocardial hemorrhage (19%), aortic chondroid metaplasia (21%), and myocardial fibrosis (10%) were also identified. Atherosclerosis of the great vessels was found in 13% of cases and half of these also had calcification of the coronary arteries. Atherosclerotic disease and congestive heart failure accounted for the majority of non-infectious cardiovascular disease, while a similar spectrum of infectious disease states were identified as in the Ohio State study.[12]

In the author's practice, cardiac disease most consistently takes the form of congestive heart failure subsequent to advanced atherosclerotic disease, and as such, the represented age and species groups mirror those with higher prevalence of atherosclerosis. Cases of valvular insufficiency (including vegetative endocarditis), chronic pulmonary disease with probable hypertension, neoplasms or fungal granulomas encroaching on the heart base and great vessels, infectious myocarditis, cardiac arrhythmias, and congenital defects are encountered less frequently.

Pericardial disease and effusion

Pericardial effusion is characterized by an inappropriate accumulation of fluid within the pericardial sac. It can develop due to inflammatory changes of the pericardium (pericarditis), trauma or aneurismal rupture, cardiac or pericardial neoplasia, or as a manifestation of metabolic derangements, toxic insult, or, commonly, right-sided congestive heart failure. In some cases, an underlying cause cannot be identified and the condition is ruled idiopathic. Depending on the etiology, the effusion may be a pure transudate, modified transudate, or an exudate. Pure transudates accumulating within the pericardial sac (also termed hydropericardium) typically are the result of increased vascular permeability or low vascular oncotic pressure such as occurs with vasculitis or hypoproteinemia, respectively. Modified transudates accumulate with irritation or inflammation of the pericardium (pericarditis), hemorrhage into the pericardial sac (hemopericardium), or increased hydrostatic pressure, as occurs with central venous congestion subsequent to right-sided congestive heart failure.[3]

Infectious pericarditis and associated effusion can be a manifestation of systemic disease or occur as an extension of epicarditis, myocarditis, or inflammatory or infectious conditions affecting other adjacent tissues, including the air sacs, lungs, and liver. Bacterial, fungal, viral, and parasitic agents have been implicated. Visceral gout is a common non-infectious cause of pericarditis and pericardial effusion.[3]

Cardiac tamponade is defined as compression of the heart by a restrictive pericardium or accumulation of pericardial contents, resulting in impairment of ventricular filling (diastolic dysfunction) and subsequent decrease in stroke volume and cardiac output. Chamber filling requires positive intramural pressure (calculated as the difference of intracardiac pressure and intrapericardial pressure), but as intrapericardial pressure increases, intramural (filling) pressure progressively declines. Furthermore, given the Frank-Starling relationship between myocardial stretching and force of contraction, the systolic function of the underfilled cardiac chambers is diminished. The intramural pressure of the thinner-walled right ventricle is overcome more rapidly than that of the left. Therefore, cardiac tamponade results in right-sided congestive heart failure in advance of left-sided failure.

Cardiac tamponade may arise by constriction with a fibrotic pericardium subsequent to chronic, fibrinous pericarditis and adhesion to the epicardium. When pericardial effusion is the cause of compression, the clinical severity of cardiac tamponade depends in large part on the rate at which the effusion develops. As the pericardium is relatively non-compliant, acute accumulations of even small fluid volumes can have profound hemodynamic effects. Conversely, a more chronic onset of pericardial effusion allows time for expansion of the pericardial sac and compensatory physiologic mechanisms, including increased blood volume and heart rate via activation of the renin-angiotensin-aldosterone system and sympathetic stimulation.

Pericardial disease in psittacine birds has been examined in a handful of studies and case reports. In one study, pericardial effusion, accompanied by moderate to marked hepatomegaly, was identified in 6% of 107 psittacine birds submitted for necropsy. Among these birds, pericardial effusion occurred with infectious disease, visceral gout, or was accompanied by right ventricular dilatation and ascites consistent with right-sided congestive heart failure. In 15% of birds with pericardial effusion, fibrinous pericarditis was identified in association with chlamydiosis, psittacid herpesvirus infection, respiratory tract mycosis, PDD, or bacterial septicemia.[12]

In a survey of 60 psittacine and birds of other taxonomic groups presented with varying degrees of illness over a year period, 5 (8.3 %) were diagnosed with pericardial effusion via echocardiography. Four of these birds, 3 grey parrots and 1 blue-fronted Amazon parrot, were presented with lethargy, weakness, and pectoral muscle wasting. In the grey parrots, right ventricular dilatation, hepatic congestion, and ascites were also evident, as consistent with right-sided congestive heart failure. Severe lower respiratory mycosis was diagnosed in all 3 birds. Low-grade pericardial effusion in the Amazon had arisen secondary to right-sided congestive heart failure owing to incompetency of the right AV valve.[13]

In the author's practice, 43 cases of pericardial disease and/or pericardial effusion were reviewed. Infectious pericarditis was identified in 14 cases (32%), including 3 cases of APV infection, 5 cases of aspergillosis with extension from the respiratory tract to the pericardium, 1 case of suspected chlamydiosis, 3 cases of bacterial pneumonia with extension to the pericardium, and 2 cases of bacterial sepsis. Visceral gout occurred in 12 cases (28%) and hypoproteinemia with transudation into the pericardial sac in 2 cases of hepatic failure. Pericardial effusion related to congestive heart failure (either right-sided or biventricular) occurred in 9 cases (20%). One cockatiel developed acute hemopericardium and died as a result of cardiac tamponade. Pericardial thickening was noted in a grey parrot with multicentric lymphoma. In 5 birds (11%), pericardial effusion was ruled idiopathic.

References

1. King AS, McLelland J. Cardiovascular system. In: King AS, McLelland J. *Birds - Their Structure and Function.* 2nd ed. London: BaillièreTindall; 1984:214-228.

2. Smith FM, West NH, Jones DR. The cardiovascular system. In: Whittow GC, ed. *Sturkie's Avian Physiology.* 5th ed. San Diego, CA: Academic Press;2000:141-231.

3. Schmidt RE, Reavill DR, Phalen DN. Cardiovascular system. In: Schmidt RE, Reavill DR, Phalen DN, eds. *Pathology of Pet and Aviary Birds.* Ames, IA: Iowa State Press; 2003:3-16.

4. Baumel J. Systema cardiovasculare. In: Baumel J, King A, Breazile J, Evans H, Vanden Berge J, eds. *Handbook of avian anatomy: Nomina Anatomica Avium.* 2nd ed. Cambridge, MA: Nuttall Ornithological Club;1993:407-476.

5. Nickel R, Schummer A, Seiferle E, Siller W, Wight P. Circulatory system. In: Nickel R, Schummer A, Seiferle E, Siller W, Wight P, eds. *Anatomy of the Domestic Birds.* Berlin: Verlag Paul Parey-Springer Verlag; 1977:85-107.

6. King AS, McLelland J. Respiratory system. In: King AS, McLelland J. *Birds - Their Structure and Function.* 2nd ed. London: BaillièreTindall; 1984:110-144.

7. de Morais HA, Schwartz DS. Pathophysiology of heart failure. In: Ettinger SJ, Feldman EC, eds. *Textbook of veterinary internal medicine.* 6th ed. St. Louis, MO: Elsevier Saunders; 2005:914-940.

8. Beaufrere H: Atherosclerosis: comparative pathogenesis, lipoprotein metabolism, and avian and exotic companion mammal models. *J Exotic Pet Med.* 2013;22(4):320-335.

9. Beaufrere H: Avian atherosclerosis: parrots and beyond. *J Exotic Pet Med.* 2013;22(4):336-347.

10. Beaufrere H, Nevarez JG, Holder K, et al.: Characterization and classification of psittacine atherosclerotic lesions by histopathology, digital image analysis, transmission and scanning electron microscopy. *Avian Pathol.* 2011;40(5):531-544.

11. Oglesbee BL, Oglesbee MJ: Results of postmortem examination of psittacine birds with cardiac disease: 26 cases (1991-1995). *J Am Vet Med Assoc.* 1998;212:1737-1742.

12. Krautwald-Junghanns ME, Braun S, Pees M, et al.: Research on the anatomy and pathology of the psittacine heart. *J Avian Med Surg.* 2004;18:2-11.

13. Straub J, Pees M, Krautwald-Junghanns ME: Diagnosis of pericardial effusion in birds by ultrasound. *Vet Rec.* 2001;149:86-88.

Diagnosis and Management of Cardiovascular Diseases in Birds

Brenna Fitzgerald, DVM, Dipl ABVP (Avian Practice)

Session #327

Affiliation: From The Medical Center for Birds, 3805 Main St, Oakley, CA 94561, USA.

Manuscript adapted from: Fitzgerald BC, Beaufrère H. Cardiology. In: Speer B, ed. Current Veterinary Therapy in Avian Medicine and Surgery. St. Louis, MO: Elsevier Saunders;2015:252-328.

Abstract: Cardiovascular disease is common in companion birds, but poses a particular diagnostic challenge, in part due to the limited sensitivity of available investigative methods. Successful intervention requires accurate and timely diagnosis, and the practitioner must be familiar with diagnostic options and their strengths and weaknesses. Key elements of a baseline diagnostic work-up include comprehensive serum biochemistry panel, complete blood count (CBC), radiographic imaging (potentially enhanced with contrast), ultrasonography, electrocardiography (ECG), and in some clinical settings, endoscopy and advanced imaging [computed tomography (CT) and magnetic resonance imaging (MRI)]. Diagnosis begins with careful history collection and thorough physical examination. Clinical pathology, overall, has a low sensitivity for specific cardiovascular disease. ECG is an appropriate diagnostic step for characterization of arrhythmias, but must be used in conjunction with imaging techniques. Radiographic imaging allows assessment of the size, contour, and radiodensity of the cardiac silhouette and great vessels, while coelomic ultrasound allows identification of specific cardiac abnormalities, pericardial effusion, ascites, and hepatic parenchymal changes that cannot be distinguished radiographically. Angiography can supplement echocardiographic findings and identify vascular abnormalities. Endoscopy allows both direct visual inspection of cardiovascular structures and targeted collection of diagnostic samples. CT can be used to image the vasculature and to diagnose cardiomegaly, ventricular dilatation, pericardial effusion, ascites, pulmonary edema, and venous congestion. Successful intervention in avian cardiovascular disease requires innovative treatment approaches, incorporating both available information specific to birds and components of therapy in human and small animal cardiovascular medicine. Pharmacologic data in avian species is limited, but treatment approaches that have been reported, utilized by the author, or show promise for future application are discussed.

Introduction: Diagnosis

Cardiovascular disease has traditionally been thought to be a rare occurrence in companion birds, but a growing body of evidence collected over the last few decades indicates otherwise. It is frequently encountered in practice and poses a serious threat to the quality of life and longevity of many avian species. Successful intervention requires accurate and timely diagnosis, but cardiovascular disease poses a particular diagnostic challenge given the potential variability of its clinical manifestations (to include asymptomatic, slowly progressive disease), and the limited sensitivity of available investigative methods. These proceedings review current diagnostic options that can facilitate detection of pathologic changes. A baseline diagnostic work-up is appropriate in any patient presenting with clinical signs consistent with cardiovascular disease, in order to begin to assess overall systemic health status. Key elements may include comprehensive serum biochemistry panel, complete blood count (CBC), and radiographic imaging (which may include imaging enhanced with contrast), ultrasonography, electrocardiography (ECG) and blood pressure measurement to further characterize the nature and extent of cardiovascular disease. Endoscopy and advanced imaging, including computed tomography (CT) and magnetic resonance imaging (MRI), also have application in diagnosis and characterization of cardiovascular disease in some settings.

History, Clinical Signs, and Physical Examination Findings

Diagnosis of cardiovascular disease begins with careful history collection and thorough physical examination, to be performed in a manner in keeping with patient stability. Examination findings, as well as relative predisposition to cardiovascular disease given the species, age, gender, lifestyle, and diet should be taken under consideration when formulating initial differential diagnoses and prioritizing diagnostic maneuvers. Clinical presentations of cardiovascular disease share many commonalities, particularly as more than one condition can exist concurrently, but there are distinguishing features of some disease states that warrant mention. Many cases are likely subclinical and go unrecognized for many years until well advanced. Clinical signs can have a subtle, insidious onset, with owners reporting progressively declining activity level, reduced appetite, and waning interest in household activities, toys, and vocalizations that were historically of interest to the bird. In some cases, no premonitory signs are appreciated prior to sudden death. Presenting complaints common to most forms of cardiovascular disease are lethargy, weakness, reduced appetite, respiratory distress, and exercise intolerance. Physical examination findings may include cachexia, tachypnea, dyspnea, harsh lung sounds, cyanosis, tachycardia, arrhythmia, systolic murmur, coelomic distension and ascites, peripheral edema, and altered mentation.

Atherosclerosis

Patients with clinically significant atherosclerosis most often present for falling or collapse, frequently accompanied by transient or persistent weakness and dysfunction of one or more limbs, and with or without altered mentation. Extension and rigidity of one leg (and sometimes the ipsilateral wing), clenching of the toes, ataxia or difficulty perching, and tremors or convulsions may be reported. These signs are considered most consistent with stroke, but rarely is this confirmed diagnostically. These events can be difficult to distinguish from seizure activity, and indeed seizures can occur subsequent to ischemic stroke.[1] Signs may be acute and transient, or there may be persistent neurologic abnormities, including reduced mentation, blindness, seizures, vestibular signs, paresis of one or both pelvic limbs, and ataxia. In cases with concurrent congestive heart failure, dyspnea and exercise intolerance may be part of the history and appreciable upon physical examination along with other congestive signs.

Congestive heart failure

Birds with congestive heart failure present most consistently with respiratory embarrassment, owing to either intracoelomic air sac compression caused by ascites and organomegaly, pulmonary edema, or a combination of the two. Coelomic distention is among the congestive signs appreciable upon physical examination in cases of right-sided or biventricular failure. Peripheral venous congestion and edema may also be noted. In the case of cardiogenic pulmonary edema, increased respiratory noise is typically heard upon auscultation of the lung fields in a calm bird, but characteristic "crackles and wheezes" are generally absent except in severe cases, and cough is exceptionally rare.

Clinical Pathology

Clinical pathology, overall, has a low sensitivity for specific cardiovascular disease. Abnormal findings are more often attributable to secondary effects on other organ systems, underlying primary disease processes, or to concurrent, unrelated disease. In principle, disease affecting the myocardium may cause elevations in aspartate aminotransferase (AST), creatine kinase (CK), and lactate dehydrogenase (LDH).[2] Other biochemical abnormalities may be seen depending on the systems secondarily or concurrently involved, particularly elevations in uric acid and bile acids reflecting renal and hepatic functional impairment, respectively. These may occur as a consequence of renal hypoperfusion or hepatic congestion secondary to heart failure, or as concurrent manifestations of an underlying disease process affecting multiple systems. Alternatively, hyperuricemia can be causally associated with cardiac disease (i.e., visceral gout), as can hypoproteinemia (which may result in pericardial effusion) and electrolyte disturbances which can also produce cardiac arrhythmia. Hematologic changes, including elevated total white blood cell (WBC) count and shifts in the leukocyte differential are seen with many forms of inflammatory or infectious disease, as well as with stress, and can be seen in patients with cardiovascular disease.

Electrocardiography (ECG) and Blood Pressure

ECG is an appropriate diagnostic step for characterization of arrhythmias, and can provide information to suggest certain cardiac abnormalities, including chamber enlargement. However, it must be used in conjunction with other diagnostic modalities, namely imaging techniques, to diagnose specific cardiac disease. Severe pathology and mechanical dysfunction can exist in the absence of ECG changes.[2]

Accurate measurement of arterial blood pressure requires the direct method, but given that this technique is challenging and often impractical in the clinical setting, indirect methods are most commonly used in veterinary practice. However, it has been consistently demonstrated that values obtained with this method do not agree with direct systolic blood pressure measurements and may therefore be of low clinical value as a diagnostic tool.[3] In general, hypotension is defined as a systolic blood pressure lower than 90 mmHg and a mean below 60 mmHg. Values for hypertension have been poorly defined in birds, but are expected to be higher than in mammals owing to their greater blood pressure. Systolic values over 200 mmHg have been proposed as hypertensive.[4]

Radiography

Radiographic imaging is generally an insensitive method to identify cardiovascular disease, but allows assessment of the size, contour, and radiodensity of the cardiac silhouette and great vessels, including the aorta, brachiocephalic trunks, and pulmonary arteries and veins.[5] In large psittacines, using high-resolution digital imaging, the common carotid and axillary arteries may be visible on the ventrodorsal view, while some major branches of the descending aorta, namely the coeliac and cranial mesenteric arteries, may be seen on the lateral view.

Enlargement of the cardiac silhouette can be seen with pericardial effusion, myocardial hypertrophy or inflammation, chamber dilatation, cardiac or pericardial masses, and aneurismal dilatation. A rounded, globoid cardiac silhouette is suggestive of pericardial effusion, but accurate confirmation and differentiation between possible etiologies is impossible without echocardiography. The size of the cardiac silhouette on the ventrodorsal view can be quantified by taking standardized measurements and calculating ratios relative to skeletal landmarks. In addition to enlargement of the cardiac silhouette, radiographic abnormalities that may be seen in cases of congestive heart failure include an enlarged hepatic silhouette and caudodorsal displacement of the adjacent gastrointestinal tract, increased radiodensity of the pulmonary parenchyma, intracoelomic air sac compression, or loss of coelomic detail due to ascites.[5] The hepatic silhouette may appear enlarged due to hepatomegaly, perihepatic effusion (fluid accumulation in the hepato-peritoneal cavities), or cardiomegaly or pericardial effusion, considering that the cardiac apex is positioned between the liver lobes. Increased pulmonary density, particularly of the reticular pattern, may be seen with cardiogenic pulmonary edema or with primary disease of the pulmonary parenchyma (inflammatory or neoplastic infiltration, fibrosis).[5] Varying degrees of compression of the cranial and caudal thoracic and abdominal air sacs are typically seen with ascites and/or organomegaly. The clavicular air sac normally provides the contrast necessary to visualize the heart base and great vessels. When this space is lost or diminished, as occurs with space-occupying masses or accumulation of exudate within the air sac, visibility of the great vessels is obscured.

Increased radiodensity and tortuosity of the great vessels is suggestive, but not conclusive, of atherosclerosis.[5] Assessment of increased opacity is highly subjective and subject to variability in radiographic technique, though focal and linear mineralization associated with one or more vessels is highly suggestive of advanced, calcified atherosclerotic lesions. In psittacines, these appear most often along the ascending and descending aorta, brachiocephalic trunks, and occasionally along smaller arteries, including the coeliac artery.

Ultrasonography

Coelomic ultrasound allows identification of cardiac abnormalities, pericardial effusion, ascites, and hepatic parenchymal changes that cannot be distinguished radiographically beyond enlargements of the cardiac and hepatic silhouettes.[5] It can most often be performed in the awake or sedated patient, and represents a good option for those too compromised to withstand extended restraint or general anesthesia. Birds can be allowed to stand or be held in an upright position to reduce positional respiratory or circulatory compromise[5] and should ideally by fasted prior to examination to reduce interference from the gastrointestinal tract. Even so, the presence of echogenic material (e.g. grit) within the ventriculus can hamper visibility.

Instrumentation and technique, anatomic limitations and approaches

Two-dimensional (B-mode) echocardiography is a well-established technique to assess morphologic and functional status in birds, but M-mode is not useful because only longitudinal and semitransverse views are attainable in most birds.[5-7] M-mode echocardiography is possible in pigeons and chickens. Minimal technical requirements of the ultrasound system include a frame rate of 100 frames per second, high-frequency probe (≥ 7.5 MHz) with a small coupling surface (microconvex array, phased array, or one end of a linear array), Doppler function, and capability to record still images and video clips for later review.[5,7] Simultaneous ECG facilitates interpretation of images in relation to the cardiac cycle, but is not essential as end-diastolic and end-systolic frames can be visually selected.

Ultrasonographic examination is complicated by unique anatomic features of birds, most notably the presence of a large sternal plate and air sacs, that substantially limit available acoustic windows.[6,7] Cardiac views in particular are constrained by the situation of the heart in an indentation of the sternum and surrounded by the clavicular and cranial thoracic air sacs. Given these limitations, two approaches are possible for transcoelomic ultrasound: ventromedian and parasternal. The ventromedian approach is the most commonly used for psittacine and raptorial birds. The probe is placed just caudal to the sternum where a normally featherless area (apterium) lies along the ventral midline between adjacent feather tracts in most birds (excluding waterfowl). The beam plane is directed craniodorsally and the liver serves as an acoustic window, avoiding the air sacs laterally and the sternum ventrally.[5-7]

The heart can be visualized in two longitudinal views, horizontal (four-chamber view) and vertical (two-chamber view), the latter obtained by rotating the probe 90°. The horizontal view allows visualization of the left and right ventricles, interventricular septum, left and right atria, left atrioventricular (AV) valve, muscular right AV valve, and aortic valve, though it is difficult to distinguish the borders of the atria from the surrounding tissue. The right ventricle appears triangular in shape, tapering toward the cardiac apex. The vertical view permits visualization of the left atrium, left ventricle, and left AV valve; a marginal sliver of the right ventricle may also be visible adjacent to the sternum. Reflection of the ultrasound beam off the sternum may produce a mirror artifact in which a duplicated partial image of the heart appears on the opposite side of the sternum; this can create a false impression of chamber dilatation or aneurysm.[5-7]

The hepatic parenchyma should appear homogenous and of moderate echogenicity, with a fine, granular texture. Hepatic vessels appear as anechoic channels passing through the parenchyma. The parasternal approach can be utilized in birds whose ribs do not extend as far caudally, namely pigeons and gallinaceous birds, allowing sufficient space to accommodate the probe on the lateral flank between the last rib and the pelvis. The right lateral approach is generally chosen to avoid the ventriculus which often contains echogenic material that obstructs visibility of the heart. The leg is extended either cranially or caudally and the probe placed caudal to the last rib and dorsal to the lateral margin of the sternum, with the beam plane directed craniomedially. This approach provides longitudinal and semitransverse views of the heart.[5-7] Although views are limited for transcoelomic echocardiography, acoustic windows are greatly improved and visibility facilitated by cardiomegaly, hepatomegaly, ascitic fluid, and pericardial effusion, all frequently present in cases of congestive heart failure.

Subjective assessments

Subjective assessments of chamber size, wall thickness, ventricular contractility, diastolic function, and valvular morphology and function can be made.[7] Chamber dilatation, myocardial hypertrophy, valvular insufficiency and anomalies (to include vegetations), cardiac masses, septal defects, some aneurysms, pericardial effusion, ascites, and hepatic venous congestion can be identified by ultrasound assisted by color Doppler.[5-7] Ascitic fluid can be collected via ultrasound-guided coelomocentesis and pericardial fluid by pericardiocentesis.[5]

Hyperechogenicity of the walls of the ascending aorta is suggestive of mineralization that can accompany advanced atherosclerosis. In cases of right-sided congestive heart failure, the right ventricle is often demonstrably dilated such that it is of nearly equal size as the left. Thickening of the myocardium, to include the muscular right AV valve, is often appreciated. The right atrium may be dilated such that it is readily visible and pericardial effusion, when present, appears as an anechoic area between the heart and pericardium. Dilatation of the hepatic veins and uniform hyperechogenicity of the hepatic parenchyma are consistent with passive congestion.[7] Similarly, ventricular dilatation, myocardial thickening, and reduced myocardial contractility can be appreciated in cases of left-sided congestive heart failure and a combination of these findings is seen with biventricular congestive heart failure.

Cardiac measurements and calculation of functional parameters

An adequate morphologic and functional assessment can often be made qualitatively in cases where disease is advanced and pathologic changes readily apparent. However, quantitative evaluation may allow the practitioner greater sensitivity in identifying more subtle changes and monitoring trends in a given patient over time. Ventricular and atrial dimensions, aortic diameter, and interventricular septum width can be measured in both systole and diastole, and some reference values are available for pigeons as well as for a few psittacine and raptorial species.[5-7] The probe is positioned such that the cardiac chambers appear maximally expanded and measurements are taken at their widest points using the inner edge method. It should be noted that accurate measurement of the atria (particularly the right atrium) can be hindered by indistinct borders and that of the right ventricle by its complex three-dimensional configuration and motion of the right AV valve. Measurements cannot be successfully taken in all individuals due to variation in image quality from one bird to the next. Ventricular widths can be used to calculate fractional shortening

$$\left[(diastolic\ diameter - systolic\ diameter)\times\frac{100}{diastolic}diameter\right]$$

for evaluation of ventricular contractility (reference values are available for some psittacine species).[5,6] Accuracy and precision of echocardiographic measurements have been questioned by some authors considering the small size of the structures imaged, rapid heart rates of avian species, current equipment resolution, and the fact that different observers can add up to 30% variability. Measurements should be taken from several cardiac cycles and averaged to obtain representative values. Follow-up measurements should be obtained by the same operator using the same equipment and technique, and only changes in measurements larger than 20% considered genuine. Transcoelomic echocardiography is also thought to underestimate the true fractional shortening of the highly efficient avian heart, as values obtained through transesophageal echocardiography were much higher.[8]

Angiography

Angiography, which can be performed using either fluoroscopy or computed tomography (CT), can provide an additional diagnostic modality to supplement echocardiographic findings, and to identify vascular abnormalities.[5,9] Fluoroscopic angiography allows visualization of the heart and vascular tree in real time. The brachiocephalic trunks, aorta, pulmonary arteries, pulmonary veins, and caudal vena cava can be visualized on the ventrodorsal view. Ventricular hypertrophy, chamber dilatation, valvular or vascular stenosis, and aneurysms can be identified using fluoroscopic angiography.[5] CT angiography (CTA) shows great promise for identification of vascular disease, including arterial calcification and luminal stenosis related to atherosclerosis, aneurysms, and congenital vascular anomalies.[9]

Endoscopy and Advanced Imaging

Endoscopy allows both direct visual inspection of cardiovascular structures and targeted collection of diagnostic samples as appropriate. Endoscopy can only detect gross abnormalities, including cardiomegaly, pericardial effusion, pericardial thickening or exudate, arterial discoloration (to suggest atherosclerotic disease) or gross structural changes, and neighboring masses or granulomas. Standard lateral approaches permit visualization of the heart, great vessels, lungs, and liver via the cranial and caudal thoracic air sacs and of the descending aorta and its major branches, caudal vena cava, and ischiatic veins via the abdominal air sacs. The interclavicular approach (via the clavicular air sac) is the best means of visualizing the heart base, ascending aorta, brachiocephalic trunks, carotid and subclavian arteries, pulmonary arteries, and jugular veins. The ventral midline approach, with entry into the ventral hepato-peritoneal cavities, permits evaluation of the liver, heart, and pericardium.[2,9]

CT and magnetic resonance imaging (MRI) are seldom used to image the avian heart as scans cannot be gated to the fast cardiac cycle in birds to reduce motion artifacts and improve diagnostic value. However, CT can be used to image the vasculature and can readily diagnose arterial calcification associated with advanced atherosclerosis, as well as cardiomegaly, ventricular dilatation, pericardial effusion, ascites, pulmonary edema, and venous congestion. Cerebral complications such as ischemic and hemorrhagic strokes can be diagnosed using CT or MRI, though concurrent atherosclerosis cannot be detected unless lesion calcification is adequately severe.[1,9]

Ancillary Diagnostics

Fluid analysis, cytology, culture, histopathology, and pathogen-specific testing of antemortem diagnostic samples can assist in determining the etiopathogenesis of cardiovascular disease, particularly when it involves infectious disease or neoplasia. Certainly myocardial samples cannot realistically be obtained in the living bird, but pericardial and ascitic fluid, caseous exudates, and biopsies of the pericardium, other affected organs (such as liver, lung, and air sac), or tumors can be collected for analysis.

Introduction: Management

Successful intervention in avian cardiovascular disease requires innovative treatment approaches, which should incorporate both available information specific to birds and components of therapy in human and small animal cardiovascular medicine. Although limited pharmacologic data in avian species complicates effective medical management of cardiovascular disease, treatment approaches that have been reported, utilized by the author, or show promise for future application are discussed. Medications that have been used empirically, and those for which pharmacokinetic and pharmacodynamic data are available are presented in Table 1.

Atherosclerotic Disease

Treatment of atherosclerosis involves both controlling risk factors and managing sequellae, including peripheral hypoperfusion, ischemic stroke, and congestive heart failure. Atherosclerotic lesions cannot be resolved, but diet, husbandry, and lifestyle changes may help to prevent, slow progression, or decrease the size of lesions. Peripheral vasodilation may have symptomatic benefit by promoting improved peripheral perfusion and reducing afterload by decreasing vascular resistance. Isoxsuprine is a peripheral vasodilator that causes vascular smooth muscle relaxation predominately through α-adrenoceptor blockade. Its empirical use in birds with clinical atherosclerotic disease has resulted in apparent symptomatic improvement, including reduced frequency and severity of stroke-like events and neurologic signs.[10] Angiotensin-converting enzyme (ACE) inhibitors, including enalapril and benazepril, effect vasodilation by blocking the formation of angiotensin II (AII). AII promotes vasoconstriction and venoconstriction by mediating release of catecholamines, which act on vascular smooth muscle via α-adrenergic receptors. Although the relative vasodilatory effect of an ACE inhibitor compared to isoxsuprine in birds is not known, it is conceivable that the two used in combination would have synergistic effects: an ACE inhibitor by limiting α-adrenoreceptor stimulation and isoxsuprine by α-adrenoreceptor antagonism. In mammals, pentoxifylline promotes passage of erythrocytes through damaged microvasculature by increasing their flexibility. It may have value in improving peripheral perfusion in birds with atherosclerotic disease. It has been used in birds as an adjunct therapy for frostbite, at a dose of 15 mg/kg PO q 8-12h for 2-6 weeks with no apparent adverse effects.

Patients with signs of stroke may have marked neurologic deficits to include reduced mentation, limb paresis, and ataxia that prevent normal eating and drinking and impair mobility. They may experience seizures or suffer injuries from falls. Supportive care measures include fluid and nutritional support, analgesia, anticonvulsant therapy when needed, and management of secondary conditions such as trauma and aspiration pneumonia. Exercise restriction should be part of the longer-term treatment plan, as well as appropriate housing modifications to accommodate and protect birds with persistent deficits. Operant learning methods should be employed to facilitate low-stress medication administration. Managements that may also have preventative value, particularly for at-risk species, include moderation of dietary calories and fat to prevent and resolve obesity, avoidance of dietary sources of cholesterol, and control of female reproductive activity. It is possible that increasing opportunities for exercise beginning early in life, generally lacking for captive birds, may have preventative value.

Table 1. Selected agents for treatment of cardiovascular disease in birds. *Broad-breasted white turkey poults. †Some birds receiving carvedilol at 20 mg/kg had significantly decreased heart rate and blood pressure for up to 8 hours.

Drug	Species	Dose / Frequency	Basis
Diuretics			
Furosemide	Parrots, raptors, mynah birds	0.1-2.2 mg/kg IM, PO q	EU[a]
	Parrots	6-24h	
		1-5 mg/kg IM q 2-12h	EU (author)
		(acute treatment)	EU (author)
		1-10 mg/kg PO q 8-12h	PD[b]
	Chickens (*Gallus gallus*	(maintenance)	PD
	domesticus)	2.5 mg/kg IM	
		5 mg/kg PO	
Spironolactone	Chickens	1 mg/kg PO	PD
	Grey parrot (*Psittacus*	1 mg/kg PO q12h	EU
	erithacus)		
Vasodilators			
Isoxsuprine	Amazon parrot (*Amazona*	10 mg/kg PO q 24h	EU
	auropalliata)		
	Parrots	10 mg/kg PO q 12h	EU (author)
Enalapril	Pigeons (*Columba livia*)	1.25 mg/kg PO q 8-12h	PK[c]
	Amazon parrots		
	Parrots	1.25 mg/kg PO q 8-12h	PK
		2.5-5 mg/kg PO q12h	EU (author)
Benazepril	Grey parrot	0.5 mg/kg PO q24h	EU
Positive inotropes			
Digoxin	Budgerigars (*Melopsittacus*	0.02 mg/kg PO q24h	PK
	undulatus)		
	Quaker parrots (*Myiopsitta*		
	monachus)	0.05 mg/kg PO q24h	PK
	Sparrows		
	Indian ringneck parakeet		
	(*Psittacula krameri*)	0.02 mg/kg PO q24h	PK
	Indian hill mynah (*Gracula*	0.025 mg/kg PO q 24h	EU
	religiosa)		
	Chickens		
		0.01 mg/kg PO q 24h	EU
		0.01 mg/kg PO q 24h	EU
Pimobendan	Hispaniolan Amazon parrots	10 mg/kg PO q12h	PK
	(*Amazona ventralis*)		
	Parrots		
		6-10 mg/kg PO q 12h	EU (author)
	Harris' hawk (*Parabuteo*		PK,EU
	unicinctus)	0.25 mg/kg PO q12h	
	Parrots		
			EU
		0.25 mg/kg PO q12h	

Drug	Species	Dose / Frequency	Basis
Dobutamine	Hispaniolan Amazon parrots (*Amazona ventralis*)	5-15 µg/kg/min (CRI)	PD
Dopamine	Hispaniolan Amazon parrots	5-10 µg/kg/min (CRI)	PD
Negative inotropes			
Carvedilol	Turkeys (*Meleagris gallopavo*)*	1-20† mg/kg PO q 24h	PD
Propranolol	Turkeys*	10 mg/kg PO q 8h	PD
	Turkeys*	10-30 mg/kg PO q 8h	PD
Atenolol	Most species	0.2 mg/kg IM, 0.04 mg/kg	EU
	Turkeys*	IV 10-30 mg/kg PO q 24h	PD
Carteolol	Turkeys*	0.01-10 mg/kg PO q 12h	PD
Nifedipine	Turkeys*	10-50 mg/kg PO q 8h	PD
Diltiazem	Most species	1-2 mg/kg PO q8-12h	EU
Parasympatholytics			
Atropine	Most species	0.01-0.1 mg/kg IM,IV	EU
	Moluccan cockatoo (*Cacatua moluccensis*)	0.05 mg/kg IM	EU
Glycopyrrolate	Most species	0.01-0.02 mg/kg IM	EU
Other Antiarrhythmics			
Lidocaine (preservative-free)	Chickens	6 mg/kg IV	PD
	Chickens	2.5 mg/kg IV	PK,PD
Propantheline	Moluccan cockatoo	0.3 mg/kg PO q8h	EU

[a]EU = empirical use. [b]PD = pharmacodynamic study. [c]PK = pharmacokinetic study.

Congestive Heart Failure

Appropriate treatment requires that known or hypothesized underlying cause(s) be addressed, if possible. The mainstays of treatment of congestive heart failure in small animal medicine, namely diuretics, ACE inhibitors, and positive inotropes, can be applied to treatment of the condition in birds. Beta-blockers, considered part of standard treatment in humans, may also have potential application. The goal of initial treatment is to stabilize the patient presenting with acute or decompensated failure, followed by design of a longer-term management strategy. In the author's experience, this has met with highly variable success, but some patients have been maintained in stable condition for up to 3 years.

Diuretics

Marked reduction of excessive circulating plasma volume (hypervolemia), edema, and effusion is an immediate treatment priority accomplished through the use of diuretics, principally furosemide. In the author's experience, a dose range of 1-5 mg/kg IM has been most efficacious when stabilizing psittacine patients with severely decompensated disease. An initial frequency of 2 hours is usually followed by a shift to 6-12 h intervals. Once the patient has been stabilized, furosemide can be administered orally, with the dose increased at least 2-fold (presuming oral bioavailability is 60-75% as in humans) at a frequency of 8-12 hours. The ultimate goal for long-term management is to identify the lowest dose that controls congestive signs. Spironolactone may be used concurrently with furosemide when congestive signs cannot be controlled with furosemide and an ACE inhibitor alone.

ACE inhibitors

ACE inhibitors are an essential component of long-term medical management of congestive heart failure by blunting the effects of the renin-angiotensin-aldosterone system (RAAS). Enalapril has been the most commonly used ACE inhibitor in birds and empirical evidence suggests it is both safe and efficacious.[5] With inclusion of enalapril in the treatment regime, increased quality of life and longevity have been reported in birds with severe cardiac pathology.

Positive inotropes

Positive inotropes are used to enhance myocardial contractility and are appropriate for treatment of heart failure due to systolic dysfunction. They are contraindicated in cases of hypertrophic cardiomyopathy (where diastolic dysfunction is the primary problem), and of outflow obstruction (e.g. aortic stenosis).[5] Likewise, it should be questioned whether their use is appropriate in birds with heart failure secondary to atherosclerotic disease and luminal stenosis of major arteries. In those cases with severe systolic dysfunction, however, it is fair to consider their inclusion in the treatment regime, at least in the short term, in an effort to stabilize the patient. Digoxin and pimobendan have been used or proposed for treatment of heart failure in avian species, but pharmacodynamic data is lacking and pharmacokinetic data and information as to their efficacy and margin of safety is extremely limited.[5] Digoxin, a digitalis glycoside, may have value in certain heart failure cases when used in combination with diuretics and ACE inhibitors, particularly when supraventricular tachyarrhythmia is a feature.[2] However, the narrow therapeutic index is a concern and chronic, maintenance administration may prove problematic.[5] Pimobendan enhances myocardial contractility without increasing myocardial oxygen consumption. Pharmacokinetic data suggests higher doses may be needed in psittacine species, relative to humans and dogs, to achieve therapeutic levels. Dopamine and doputamine, rapid-acting, intravenous sympathomimetics, increased heart rate and arterial blood pressure when administered by CRI to Hispaniolan Amazon parrots, but their effect on systolic function and cardiac output is unknown.

Negative Inotropes

Beta-blockers are part of core therapy for heart failure in humans, both in early and advanced stages. Treatment is generally initiated at a very low dose and gradually titrated upwards. The beneficial effects of β-blockers have also been documented in furazolidone-induced model of dilated cardiomyopathy (DCM) of turkey poults. Cardioprotective efficacy of β-blockers is dose-dependent and greatest with non-selective β-blockers; efficacy is ranked in descending order as: carteolol > propranolol > atenolol ≥ metoprolol. β-blockers were found to improve systolic performance and decrease mortality in turkeys with advanced furazolidone-induced heart failure. The cardioprotective effects of calcium channel blockers (nifedipine > verapamil) were found to rival those of β-blockers in furazolidone-induced DCM of turkey poults.[11] The author suggests that β-blockers may have application as adjunctive treatment of congestive heart failure, provided the condition is first stabilized using conventional management strategies. Hypothetically, β-blockers may also have merit for treatment of HCM in birds.

Antiarrhythmic agents

Prior to antiarrhythmic therapy, factors underlying or contributing to the arrhythmia (e.g. hypoxia, electrolyte distrubances, intoxications) must be identified and addressed. It is only appropriate to treat arrhythmias that are symptomatic, causing hemodynamic instability, and complicating or precipitating heart failure. Aside from digoxin, β-blockers and CCBs are utilized in small animal cardiology for treatment of supraventricular tachyarrhythmias, including atrial fibrillation, in order to slow the ventricular response rate. Lidocaine is a parenteral antiarrhythmic agent used to control life-threatening ventricular tachyarrhythmias. Antimuscarinic agents atropine and glycopyrrolate are used to treat bradycardias, including SA arrest and first- and second-degree AV block, and to determine whether these arrhythmias are related to increased vagal tone or other cause.[5] Another indication is as an antidote for organophosphate and carbamate intoxication. Propantheline is an oral antimuscarinic agent used in the treatment of anticholinergic-responsive bradycardias.

Supportive care and husbandry considerations

General supportive principles apply to the avian heart failure patient, namely rest, minimizing stress with judicious, limited handling and restraint, and provision of supplemental oxygen when appropriate. Oxygen supplementation is indicated for patients with pulmonary edema, while physical fluid removal is more efficacious to stabilize those with air sac compression and cardiac tamponade. Patients may be presented in a severely debilitated state. In-patient supportive care measures may include fluid and nutritional support, but parenteral fluid administration must be carefully questioned. Maintaining the delicate balance between management of hypervolemia and concurrent dehydration and renal dysfunction is a profound clinical challenge, requiring close monitoring for changes in clinical status and adjustment of the treatment plan accordingly. Longer-term husbandry considerations, including dietary changes (which should also entail sodium restriction), and lifestyle and housing modifications, mirror those discussed for atherosclerosis. Given that congestive heart failure often arises secondary to atherosclerotic disease in psittacine birds, preventative considerations may be applicable to both conditions.

Pericardial Effusion and Cardiac Tamponade

Treatment is based first on removal of the fluid, and second on treatment of the underlying etiology of fluid accumulation. Fluid removal can be accomplished either by ultrasound-guided or endoscopic pericardiocentesis, or by endoscopic or surgical fenestration of the pericardium. Diuretics are contraindicated in cardiac tamponade because they decrease the preload necessary to maintain ventricular filling pressure and cardiac output. However, furosemide with or without an ACE inhibitor may be useful to reduce pericardial fluid of low volume related to congestive heart failure.

References

1. Beaufrere H, Nevarez J, Gaschen L, et al.: Diagnosis of presumed acute ischemic stroke and associated seizure management in a Congo African grey parrot. J Am Vet Med Assoc. 2011;239(1):122-128.

2. Lumeij J, Ritchie B. Cardiology. In: Ritchie BW, Harrison GJ, Harrison LR, eds. Avian Medicine: Principles and Applications. Lake Worth, FL: Wingers Publishing;1994:695-722.

3. Acierno MJ, de Cunha A, Smith J, et al. Agreement between direct and indirect blood pressure measurements obtained from anesthetized Hispaniolan Amazon parrots. J Am Vet Med Assoc. 2008;233:1587-1590.

4. Lichtenberger M, Ko J. Critical care monitoring. Vet Clin North Am Exot Anim Pract. 2007;10(2):317-344.

5. Pees M, Krautwald-Junghanns ME, Straub J. Evaluating and treating the cardiovascular system. In: Harrison GJ, Lightfoot TL, eds. Clinical Avian Medicine. Vol 1. Palm Beach, FL: Spix Publishing, Inc.; 2006;379-394.

6. Krautwald-Junghanns ME, Schulz M, Hagner D, et al. Transcoelomic two-dimensional echocardiography in the avian patient. J Avian Med Surg. 1995;9:19-31.

7. Pees M, Krautwald-Junghanns ME. Avian echocardiography. Semin Avian Exotic Pet Med. 2005;14:14-21.

8. Beaufrere H, Pariaut R, Rodriguez D, et al.: Comparison of transcoelomic, contrast transcoelomic, and transesophageal echocardiography in anesthetized red-tailed hawks (Buteo jamaicensis), Am J Vet Res. 2012;73(10):1560-1568.

9. Beaufrère H, Pariaut R, Rodriguez D, et al. Avian vascular imaging: a review. J Avian Med Surg. 2010;24:174-184.

10. Simone-Freilicher E. Use of isoxsuprine for treatment of clinical signs associated with presumptive atherosclerosis in a yellow-naped Amazon parrot (Amazona ochrocephala auropalliata). J Avian Med Surg. 2007;21(3):215-219.

11. Glass MG, Fuleihan F, Liao R, et al.: Differences in cardioprotective efficacy of adrenergic receptor antagonists and Ca^{2+} channel antagonists in an animal model of dilated cardiomyopathy, Circ Res. 1993;73(6):1077-1089.

Section 9

AAV Master Classes
Track 4

Avian Metabolic Bone Diseases

M. Scott Echols, DVM, Dipl ABVP (Avian)

Session #339

Affiliation: Echols Veterinary Services, Salt Lake City, UT 84054, USA and Medical Center for Birds, Oakley, CA 94561, USA.

Abstract: Metabolic bone diseases relate to disorders of the bone and especially include osteoporosis. Until recently osteoporosis has been largely ignored in captive birds- especially adults. This discussion covers osteoporosis, diagnosis and management/treatment options in birds.

Metabolic Bone Disease

Metabolic bone disease (MBD) is a non-specific phrase pertaining to diseases of the bone that have many causes. These diseases of the bone are related to loss of trabecular and cortical bone mass.[1] Conditions such as osteoporosis (most commonly described), rickets and osteomalacia are considered MBDs. Most often MBD relates to degree of mineralization, bone formation and ultimately bone strength.

Rickets is generally due to vitamin D, calcium and/or phosphorus deficiency (usually from improper diet and sunlight [vitamin D]) or impaired metabolism in young animals.[2] As the animal grows, the bone may fracture or deform. In birds it is most often seen as a curved keel or misshapen long bones. Some of these young birds go on to develop improved bone mineralization and strength but retain their deformed bones. Because this is more of a developmental disorder, rickets is not further discussed in this article.

Osteomalacia is softening of the bones in adult animals. Osteomalacia is simply the adult form of rickets. However, bone deformation is far more pronounced in young growing animals with rickets than in adults with osteomalacia (that started with normal bones). Certain diseases (hyperparathyroidism), drugs, inadequate sunlight and more are known causes of osteomalacia in humans. The same correlations have been proposed but inadequately proven in birds.

Osteoporosis is simply decreased bone strength and is associated with an increased risk of fractures. Osteoporosis is more accurately described as lower than normal peak bone mass for the animal. At least in humans, bone density may be classified as normal, osteopenic, osteoporotic and severely osteoporotic based on a T-score generated from dual energy X-ray absorptiometry ('DXA' scan). While DXA scans have been performed in research birds, such data is limited.[3]

Of the definitions above, osteoporosis best encompasses the nature of discussion presented herein.

Osteoporosis

While osteoporosis conjures up images of old people hunched over and at risk of breaking their back or hip, the disease has far wider implications. In humans, osteoporosis has been associated with loose teeth (similar parallels have been made in non-human mammals), musculoskeletal pain, nerve pain, joint disorders and more. The typical age related bone changes in humans with osteoporosis including increased bone marrow space, endosteal resorption of cortical bone and increasing bone tissue porosity, are also recognized in birds.[1]

Cage layer paralysis was first recognized in the literature in the 1950s shortly after the introduction of battery cages.[4] The disorder is essentially due to osteoporosis and weakened bones.[4] Riddell, et al. even stated the disease should better be termed cage layer osteoporosis.[5] Osteoporosis is simply the loss of structural bone that leads to bone fragility and susceptibility to fractures. Spinal degeneration and/or transient irregularities in intracellular calcium metabolism may lead to abnormal muscle activity and paralysis seen with this disease.

Common boney abnormalities of chickens with cage layer fatigue include an S-shaped keel, rib deformities (swollen vertebral-sternal junctions), thin and easily broken bones including those of the cranium.[5] Additionally, vertebral fractures are common in poultry and narrowing of the 4th and 5th rib space adjacent to the spinal column due to collapse and fusion of the individual vertebrae have been described.[5] Overlying vertebral compression fractures have also been correlated with pulmonary hemorrhage and spinal cord degeneration and/or compression in poultry.[5] One study showed that by the time battery raised chickens had been fully processed, 98% of the carcasses had broken bones with an average of 6 fractured bones in each bird.[6] The authors noted that the bones most frequently broken are recognized to be relatively liable during periods of skeletal depletion.[6] Other studies have reported that up to 70% of laying hens have old healing fractures and the incidence varies based on husbandry and strain.[7] Osteoporosis is considered widespread in the poultry industry and is considered responsible for 30% of the fractures that occur during production and at depopulation.[4] Recognition of these problems has led to the serious welfare concerns of cage/battery raised poultry.

In chickens with cage layer fatigue, most soft tissue are histologically normal except the parathyroid glands which are often hyperplastic.[5] Of course, secondary soft tissue pathology (e.g., inflammation, spinal cord trauma) can be seen with osteoporosis.

Histologic changes in the bones with cage layer fatigue are far more pervasive. As expected the cortex of many long bones tends to be thin and absorption cavities are prominent. Also common is a layer of subperiosteal bone overlying the cortex.[5] In many areas this new bone can also be seen communicating with the medullary bone through the absorption spaces underlying the thin cortex. Medullary bone tends to be fragmented and osteoid is commonly greater than calcified medullary bone. Some of the weight bearing bones, femur in particular, may not display these characteristic changes in affected birds. Vertebral fractures and damage directly to the spinal cord are likely what results in the paralysis of cage layer paralysis.[5]

Based on micro computed tomography, review of radiographs through a consultation service and recent advanced X-ray technology taken of birds from across the United States, the author feels that osteoporosis is common in the captive avian population representing numerous orders of birds.

Formation of Bone in Birds

Understanding basic bone biology helps researchers better understand bone pathology. In the appendicular skeleton, longitudinal growth of the long bones occurs via endochondrial ossification. Resting chondrocytes within the germinal layer of the epiphyseal growth plate become committed to differentiate into proliferative chondrocytes. Nourished by epiphyseal capillaries, a zone of proliferation forms as columns of packed chondrocytes within an extracellular matrix that contains high amounts of type II collagen.[4]

As more matrix is secreted the chondrocytes within the columns separate, hypertrophy and start secreting type X collagen. This hypertrophic zone receives its blood supply from metaphyseal blood vessels. The development also creates a narrow avascular zone of pre-hypertrophic chondrocytes between the hypertrophic and proliferative zones.

In addition to collagen, the chondrocytes secrete proteoglycans, growth factors and other matrix components which all regulate further development of the cells.

In the lower hypertrophic zone, chondroclasts absorb matrix while fully hypertrophied chondrocytes secrete alkaline phosphatase (ALP). ALP initiates formation of hydroxyapatite crystals (aka, the bone mineral).

Next, osteoblasts formed from precursor cells in the marrow resorb chondrocytes undergoing apoptosis. Osteoblasts produce bone matrix (out of type I collagen fibrils and increased Ca^{2+} and PO_4^{3-}) that forms hydroxyapatite crystals in and outside of the matrix. Ongoing osteoclastic resorption combined with osteoblastic formation results in the development of woven trabecular bone which is based on the irregular structure of collagen fibrils.

Bone elongation occurs by continued chondrocyte proliferation at the head of the growth plate and mineralization and hypertrophy at the rear. The trabecular network is mostly resorbed forming bone marrow.

This above process is different from that in mammals in that birds don't form secondary ossification centers and do tend to have much better vascularized growth plates.

Long bone widening involves intramembranous ossification. Within the perichondrium, osteoblasts develop and produce bone spicules. These bone spicules merge to form bone with cavities lined by osteoblasts. The osteoblasts fill in the cavities by secreting concentric layers of lamellar cortical bone.

Meanwhile osteoclasts absorb bone on the endosteal surface so that bone widens on the outer surface with an expanding ring (of bone) and inner resorption (of bone). During early growth the expanding bone rings grow too rapidly to fill the resulting cavities with bone. As growth slows, bone remodeling increases and bone cavity filling rises.

Towards the end of the growth period, secondary osteons are formed by osteoblasts cutting a tunnel through the bone followed by osteoblasts forming concentric layers of lamellar bone. During this period of osteon formation, osteoblasts become entombed within the bone and differentiate into osteocytes. Finally, osteocytes develop a network of interconnections and regulate bone remodeling based on biomechanical forces.[4]

Bone changes during reproduction

Bone biology changes dramatically at the onset of sexual maturity in hens. Osteoblasts stop forming lamellar bone and start producing woven or medullary bone. This medullary bone is laid down on the surfaces of structural bone and in spicules within the medullary cavity and is designed to provide a source of calcium for egg shell production. This is most dramatic in the leg bones but can also be found in wing and pneumatic bones. Medullary bone deposition is fastest during the early stages of lay and then tends to accumulate slowly throughout the laying period.[4]

In the mature hen, rising estrogen stimulate osteoblasts causing them to lay down medullary bone rather than structural bone. The reverse occurs when birds are out of lay and estrogens decline. In human post-menopausal osteoporosis, declining estrogen increases bone resorption, trabecular bone loss and suppresses structural bone formation which is notably different from that in birds.[1,4,8]

During lay, structural bone formation may cease entirely to give way to medullary bone production. However, osteoclastic resorption continues resulting in declining structural bone content in the hen. During the early laying period, the trabecular bone loss in the spine and epiphyses of long bones can be marked. Cortical bone loss tends to be more gradual. Overall structural bone loss is noted throughout the skeleton especially in the long bones of the wing, leg, keel, vertebrae and others even though loss may be replaced by medullary bone. Through the lay period and with chronicity, the end result is a progressive weakening of the skeleton.[7]

Even though potentially more medullary bone is produced than structural bone is lost, accounting for stable or increased total bone content, the bones become weaker. Because medullary bone is a type of woven bone based on irregularly arranged collagen fibrils and is present in isolated spicules it is fundamentally not as strong as structural bone. The overall strength of the skeleton is weakened resulting in increased risk of fractures.

Diagnosis of Osteoporosis

Bloodwork

Unless the bird is suffering from advanced MBD, severe electrolyte disturbances (most commonly hypomagnesemia and hypermagnesemia resulting in hypocalcemia), is taking certain medications, has other disease states or competing elements in the diet, most labwork abnormalities will be non-specific in birds with osteoporosis and more likely related to secondary complications.

Radiographs

The classic radiograph of paper thin cortices and folding fractures related to MBD represents the extreme in disease. Radiographs truly are shades of grey and interpretation of bone loss is subjective. Additionally, the X-ray machine, settings, patient positioning, patient motion, developing process and interpretative software (if digital) all play a role in the image we view. Finally, post-imaging software (sharpening tools are commonly added to digital radiographs), viewing monitor and the viewer's anatomic understanding and interpretation also add to the subjective nature of reading radiographs.

Because medullary bone is as radiodense as structural bone, standard radiographs and bone ash measurements are not a good means to evaluate bone structure when medullary bone is present.[4]

When medullary bone is at a minimum, the cortex of long bones can be subjectively measured for relative thickness and density. It should also be noted that external bone diameter is not adequate when evaluating bone density. It is well described that in human aging osteoporotic bones, the external bone diameter increases along with the marrow space as the cortices thin. The same is believed to happen in avian bones.[7]

There are common areas on radiographs the author has recognized and correlated with CT and basic physical activity that may give veterinarians more insight into bone and animal health.

Birds that tend to not fly much have generally poor bone density along the thoracic girdle (coracoid, furcula and proximal humerus) and distal neck/proximal notarium vertebrae.

Birds that tend to not walk or climb much generally have poor bone density along the synsacrum, proximal femurs and sometimes extending down the legs.

Birds that tend to not walk or fly much have the aforementioned deficits plus a more generalized radiolucency along the entire spine.

Specific non-invasive bone mineral density diagnostics

The currently available non-invasive bone density diagnostics, other than the gold standard DXA, include pDXA (peripheral dual energy x-ray absorptiometry), QUS (quantitative ultrasound), QCT (quantitative computed tomography) and pQCT (peripheral quantitative computed tomography). Abnormal pDXA, QUS, QCT and pQCT are screening tests that would ultimately lead to an additional test- central DXA as the preferred technique to measure bone mineral density. There have been limited published works describing these modalities being used to evaluate bone density in birds.[3,9-11]

While DXA scans are typically discussed for use in adult humans as the gold standard in measuring bone mineral density, their use has been explored in birds.[3,9]

DXA scans use 2 X-ray beams with differing energy levels aimed at specific bones. The soft tissue is subtracted out and the bone mineral density is measured by the absorption of each beam by the bone studied. A T score is generated and these values are compared against normals to determine if the bone mineral density is low, normal or high for the bone studied.

DXA is not without fault as it cannot distinguish flat mineral dense versus thick mineral poor bone, shows ambiguity between elements of varying shape, depends on bone orientation during scanning and also depends on the animal's position and regions of interest.[3] Additionally, in the case of birds and some reptiles, reproductive activity can effect bone mineral density measurements via DXA (and really most density measurement tests). However, if the above factors are adjusted for, DXA can provide a precise T score for site specific measurements and can be used on animals as small as rats.[3]

While DXA scans have primarily been reserved for humans and research animals, there is the potential for use in avian clinical cases. However, this diagnostic tool needs additional study to better understand its role in avian medicine.

Computed tomography

Computed tomography or CT is considered the gold standard when measuring bone (or any tissue) density. As opposed to 2-dimensional radiographs, CT can give volumetric density in Houndsfield Units (HU).

The HU scale is a quantitative scale for describing radiodensity. Air is given a value of -1000 HU, water 0 HU and the number increases with increasing tissue density. Using CT, one can get a point HU value over any tissue scanned.

As described above, medullary bone can be as radiodense as structural bone and the overall bone content can be the same or higher when medullary bone is deposited at a greater rate than structural bone. The end result may be weaker bones; however, CT measures the density represented over a given volume. The HU reading may actually be higher in bones with increased medullary bone over those with normal structural bone (given the same size of bone).

With the above limitation, CT still has advantage over radiographs when evaluating bone density in a few situations. First CT should give representative HU values with typical structural bone (not filled in with medullary bone). Second, using computer sectioning, CT HU can be measured just over the cortex. In this situation, the 'volume' is just one cortex wall rather than the entire thickness of the bone. This is of course assuming the medullary bone has not invaded the cortex and the value is compared to cortex only readings from other comparable animals and bones. Third, CT can be used to visually see radiolucent holes in the bones. Once CT data is rendered the operator can 'peel away' densities. By digitally removing densities, soft tissues can be stripped away leaving only the denser tissues such as bone. If the density of the bone is the same or less than the stripped soft tissues, the bone will disappear. Visually the bone is either faintly visible or non-visible. It may also appear as if the bone has holes in it as density loss tends to be patchy rather than uniform.

Modification of Osteoporosis

Reproductive activity

Getting birds out of lay activity can be highly beneficial for those at risk of osteoporosis. Birds that are selected and genetically bred for high egg production tend to have shortened (or even absent) rest periods and are highly susceptible to osteoporosis.[4] Once lay activity has stopped, structural bone formation commences and medullary bone gradually disappears.

Interestingly, egg production is not directly correlated with osteoporosis. Studies in hens support that egg production has little or no effect on bone quality. After all most egg laying hens take in more calcium than put into shell production. Rather it is likely the duration of increased estrogen (related from prolonged egg laying periods) that correlates with poor bone quality.

Bone calcium metabolism is highly complex and not completely understood in the laying hen. At least in domestic laying hens, eggshell calcification occurs at night after eating ceases.[12] It is believed that the majority of eggshell calcium must come from bone stores and not the gut. Even though intestinal calcium absorption is more intense during eggshell formation there is simply less calcium available in the gut (which would be mostly empty during night time hours). However, in a normally functioning laying hen, the bone calcium reserves would be replenished from intestinal sources while the egg is plumping and eggshell is not being laid. Simply put intestinal (or dietary) availability of calcium is not sufficient to satisfy the high calcium required for eggshell formation.[12]

The remedy is to get pet birds to stop laying (or being broody). This can be accomplished by antecedent-behavior-consequence rearrangement (discussed elsewhere); behavior modification of bird-owner interactions (e.g., stop petting, stroking, cuddling the bird); environmental, dietary and general activity changes (also discussed elsewhere) and select medicinal (GnRH agonists) use. Rarely, surgical modification (removing cystic ovaries, neoplastic tissues and *questionably* ovariosalpingohysterectomies) can be used to manage excessive egg laying.

Genetics

As more breeds of various species of captive birds are created, genetics play an increasing role in the health of these animals. Some genetic variations, such as the English Trumpeter pigeon with feathered feet and altered leg bone formation, prevent normal movement and likely have an effect on weight loading and subsequent structural bone formation. Significant tibiotarsal breaking strength differences have been noted in certain lines of chickens.[4] The differences in some of these chicken strains have been primarily attributed to structural bone quantity.

Genetic selection has been shown to improve bone strength and resistance to osteoporosis in laying hens.[7] Genetic factors and appropriate breeding may also need to be considered for heavily inbred groups of captive birds.

Drugs

During lay, bisphosphonate (inhibits osteoclast activity) can alleviate structural bone loss.[4] Parathyroid hormones, other osteoclast inhibitors, estrogens and other drugs have been well described as means to alter the course of osteoporosis in humans. Even with the large array of drugs available, these medications are rarely effective as the only form of treatment of osteoporosis in humans.

It seems reasonable that by understanding bone biology, select drugs (some of which may already be available) could be beneficial in treating osteoporosis in birds. However, no such trials were found at the time of writing.

Diet

Deficiencies of phosphorus, calcium and cholecalciferol have been shown to result in osteomalacia (softened bones) bone loss.[4] Simply avoiding osteomalacia (with proper dietary phosphorus, calcium and cholecalciferol) has not been shown to prevent osteoporosis in egg laying hens. Particulate forms of calcium supplementation have been shown to improve shell and bone quality over powdered supplements. The belief is that the particulate calcium remains in the gut longer (than powders) and is more consistently available through the day and/or night for egg shell production.

While various forms of calcium supplementation (e.g., oystershell, limestone) may increase medullary bone and possibly reduce cancellous bone loss, calcium deficiency is currently not considered to be a primary cause of osteoporosis (at least in hens).[7] As is also true in postmenopausal women, dietary vitamin D and calcium supplementation are of limited benefit in osteoporotic chickens.[8]

Consumer reports show that 90% or greater of the captive pet parrot population is fed seed-based diets. These diets are generally considered to be nutritionally inadequate for long-term health. How exactly seed-based, pellet-based, fruit- and vegetable-based, combination and more diets relate to bone development and, more interestingly, long-term bone maintenance for captive birds is not exactly clear. Unfortunately, the general recommendation of feeding an appropriate diet (often consisting of a species-formulated pellet-based diet plus added vegetables and some fruits) for captive parrots (and other species) is currently the most common advice avian veterinarians give concerning captive parrots. The appropriate diet must be considered not only for the species in question, but also that animal's lifestyle, health status and more. There is no one magic diet.

One study used peripheral quantitative computed tomography to study bone mineral density and content of the left tibiotarsus in healthy male and female 2-year-old budgerigars (*Melopsittacus undulatus*) housed under identical conditions for 12 months.[10] One group received a commercial pelleted diet while the other a commercial seed diet. There was no difference in bone density and mineral content between the birds fed the different diets. However, the female birds of both groups had higher values than the males. The cause for the increased bone mineral content and density in the females was not proven. However, the authors postulated the increased values may represent hyperostosis and/or prophylactic storage of calcium during reproductively dormant periods.[10] This one study shows an interesting lack of effect of a pelleted diet compared to a seed diet fed for 1 year in budgerigars.

Omega-3 fatty acid (O-3 FA) supplementation has been shown to improve bone mineral density and/or reduce bone breakage in a number of poultry species. The benefits of O-3 FA supplementation have been widely published in the avian literature and should be considered when managing osteoporosis.

As most commercial bird foods are made from corn and soybean components, these diets are typically high in omega- 6 fatty acids (O-6 FA). Fish oils are highly unstable when manufactured in foods. Flax seeds provide a more stable source of O-3 FA in manufactured diets. Flax seed oil provides a more concentrated source (than flax seeds) of O-3 FA and can be added to the diet. Finally, fish oil (EPA and DHA) provides the most biologically potent form of O-3 FA and must be supplemented fresh due to its instability to heat, oxygen and physical mixing. Based on research in multiple animals, fish oil supplementation is likely the best means to rapidly increase body EPA and DHA levels- even for herbivorous species that would naturally consume more ALA.

Omega-3 FA contribute to several factors involving bone health. Studied 4-week-old quail fed identical diets for 7 months except for fat content (soybean oil [SBO], hydrogenated soybean oil [HSBO], chicken fat [CF] or menhaden fish oil [FO] all at 50g/kg of feed) had notable differences in bone parameters. [13] The ratio of O-6 FA:O-3 FA in the diets were as follows: SBO 12.55, HSBO 17.85, CF 18.47 and FO 0.66.

As O-3 FA's reduce the concentration arachidonic acid and subsequent production of PGE_2 (the opposite is true with O-6 FA), bone formation is increased. PGE_2's long term goal is to stimulate bone resorption. As expected, quail fed FO (and HSBO) had markedly improved cortical bone thickness and density compared to the other groups. Also, quail fed FO had higher percentages of tibial ash, Ca and P. Last, quail fed FO or HSBO had increased bone shear force over the other groups. All of the findings indicated bones from FO and HSBO groups were stronger. As a potential negative, HSBO quail had more trans-fatty acids in studied tissues.[13]

A similar study used 16-week-old chickens fed diets of varying O-6 FA:O-3 FA (ranging from 48.7:1 to 4.7:1) for 42 weeks. While the study did find increasing cortical bone thickness as the O-3 FA concentration increased (up to the highest O-3 FA level), other parameters such as bone strength and mineral content were similar between the different diets.[14] One of many differences is that this second study used a much lower O-3 FA content diet than in the quail study above. These findings support a role of supplemental O-3 FA in improving bone quality in young Galliformes and may similarly affect other species.

Free-range hens supplemented with alpha linolenic acid had a significant (40-60%) reduction in keel fractures, reduced breakage severity, increased bone mineral density and content, greater bone and trabecular volume and finally enhanced bone turnover compared with control groups that ate a typical diet high in O-6 FA. [9] Supplementation began between 23 and 30 weeks of age and birds were evaluated by numerous means (DXA score, histomorpometry, biomechanic studies and more) at 30, 50 and 70 weeks of age. Keel fractures were even lowered by 40% at the end of lay compared to control groups. [9]

In general, feeding diets mostly devoid of calcium, containing excessively high phosphorus and other deviations from diets 'normal' for the species puts the animal at risk of developing some form of MBD if fed too long. The exclusive red meat diet fed to raptors is a well-known risk factor for severe MBD including folding fractures.[15] MBD-inducing diets are rarely this obvious or cause such severe nutritional disease.

For granivores, seed-based diets are typically high in phosphorus and/or low in calcium and are likely the most common dietary cause of low grade MBD. Additionally, research supports that diets either high in O-6 fatty acids and/or low in O-3 fatty acids may be another risk factor for osteoporosis and other forms of MBD.

Exercise

While studies lack relating exercise as a treatment for avian osteoporosis, it is known that bone strength is related to the degree of biomechanical load placed on the bone.[4] Load bearing is considered essential for normal skeletal development and turnover, is well documented and often termed adaptive bone remodeling.[7] Long periods of inactivity are well known to be detrimental to bone mass in humans and chickens.[7] General recommendations for prevention of osteoporosis in humans include building bone mass prior to older age and/or estrogen decline (post-menopause) and maintaining load-bearing exercise.[8] Hens housed in battery cages have notably decreased bone strength compared to those housed in alternative systems that allow for exercise. Along the same lines, allowing hens (chickens) freedom of exercise improves bone strength.[7] All of the above comments are expected and common sense based on our understanding of bone development and maintenance across numerous classes of animals. Based on the author's work with radiographs and CT, inactivity does seem to play a role in avian bone density.

Tatara, et al. surgically denervated the right humerus (radialis and mediano-ulnaris neurectomy) in 22-day-old turkeys and then compared this bone to the left (untouched) humerus and from sham operated birds at 17 weeks of age.[1] Experimental group birds were not able to flap their right wing and experienced a moderately drooped wing. The denervated humerii had significantly decreased volumetric bone density, relative mean wall thickness, cross sectional area, overall strength and more that supported focal (one bone) cortical osteopenia resulting from disuse atrophy and possibly related to elimination of neuronal regulation of the bone.[1] This study supports that lack of physical exercise (in this case, due to neurectomy) can significantly and adversely affect bone mineral density in birds.

Lay activity in hens also has an important role in bone mineral density relative to exercise. In studied chickens, exercise does have a positive benefit on bone strength in the early lay period when osteoclasts and subsequently bone resorption are low.[7] Later in lay, continuous exercise does not continue to suppress osteoclast recruitment and structural bone resorption. As a result, even continuous exercise in late lay birds does not benefit bone quality.[7] This brings us back to earlier comments that it is important to cease lay activity in effort to improve bone mineral density.

The obvious remedy is to increase bone loading activities such as flying, climbing and walking for pet birds. This can be accomplished by providing larger space, creating an environment that encourages movement (e.g., foraging, outside play time, flight training) and even forced exercise (e.g., moving the food and water bowls away from the bird, training the bird to flap or perform various activities). What is not clear is just how much activity (and for how long) is needed to improve structural bone formation.

Sunlight

Numerous studies have evaluated the role of either natural unfiltered sunlight or supplemental UVA/B exposure to growing and adult birds. The effects of UV radiation on growing chicks of precocial species generally demonstrate benefits related to bone health. These effects are more difficult to measure in adults as mature bone is more slowly modified. Also the addition of UV light (natural and artificial) can affect behavior and has been related to increased activity. It is not always clear if the increased activity (and subsequent bone loading) or the actual UV light is improving bone quality. Either way, exposure to natural sunlight for diurnal species seems both humane and makes common sense for improving the quality of a captive bird's life.

Cousquer, et al. studied metabolic bone disease in wild collared doves.[16] The authors found that 51.2% of juvenile and 9.6% of adult birds admitted to a hospital in the United Kingdom had clinical signs of metabolic bone disease (e.g., bendy beaks, deformed bones). The histologic and blood work findings most supported vitamin D deficiency induced rickets in the birds. The incidence was highest between December and February and dropped to zero from June to August. Based on meteorologic data representing the time of the study, the average normal total sunlight hours from December to February (increased incidence of disease) were 195. For the period of May through August (lowest incidence of disease), the total sunlight hours were 721. While diet and other factors were not controlled, the authors postulated that inadequate UVB light exposure during the short winter months was the greatest risk factor for these birds developing rickets.[16] As a note, collared doves breed year round, are invasive to the United Kingdom and originate (at the end of the 19th century) from warm temperate and subtropical Asia from Turkey east to southern China and south through India to Sri Lanka.[17]

Additional potential therapies

It has been known that bone growth can be stimulated by heat. Additionally, heat treatment increases bone mineralization and heat shock protein expression in human cell cultures.[18] While a similar pathway has been suggested in avian species, heat therapy has not been proven to have similar effects in birds. However, considering this concept low level laser therapy has been used to improve the biomechanical properties and increase cortical bone volume in select osteoporotic bones of rats.[19,20] The author has been using low level laser therapy in select cases of μCT confirmed osteoporosis in captive birds. The author only brings this up as a possible treatment option to explore and further study.

References

1. Tatara MR, Krupski W, Majcher P, Studzinski T. Long-term denervation of the humerus in turkeys as an experimental model for osteopenia. *Poult Sci.* 2005;84:718-722.

2. Dinev I. Clinical and morphological investigations on the incidence of forms of rickets and their association with other pathological states in broiler chickens. *Res Vet Sci.* 2012;92:273-277.

3. Dirrigl FJ, Dalsky GP, Warner, SE. Dual-energy x-ray absorptiometry of birds: an examination of excised skeletal specimens. *J Vet Med Assoc.* 2004;51:313-319.

4. Whitehead CC. Overview of bone biology in the egg laying hen. *Poult Sci.* 2004;83:193-199.

5. Riddell C, Helmboldt CF, Singsen EP, Matterson LD. Bone pathology of birds affected with cage layer fatigue. *Avian Dis.* 1968;12:285-297.

6. Gregory NG, Wilkins LJ. Broken bones in domestic fowl: handling and processing damage of end-of-lay battery hens. *Br Poult Sci.* 1989;30:555-562.

7. Fleming RH, McCormack HA, McTeir L, Whitehead CC. Relationships between genetic, environmental and nutritional factors influencing osteoporosis in laying hens. *Br Poult Sci.* 2006;47:742-755.

8. Beck MM, Hansen KK. Role of estrogen in avian osteoporosis. *Poult Sci.* 2004;83:200-206.

9. Tarlton JF, Wilkins LJ, Toscano MJ, et al. Reduced bone breakage and increased bone strength in free range laying hens fed omega-3 polyunsaturated fatty acid supplemented diets. *Bone.* 2013;52:578-586.

10. Zulauf-Fischer I, Liesegang A, Haessig M, et al. Differences in bone mineral content and density between male and female budgerigars (*Melopsittacus undulatus*) during the non-reproductive season. *J Vet Med Assoc.* 2006;53:456-457.

11. Martinez-Cummer MA, Hurtig M, Leeson S. Use of apparent transverse quantitative ultrasonography to assess skeletal integrity in layers. *Poult Sci.* 2006;85:1648-1651.

12. Bar A. Calcium transport in strongly calcifying laying birds: mechanisms and regulation. *Comp Biochem Physiol* A. 2009;152:447-469.

13. Liu D, et al. Long-term supplementation of various dietary lipids alters bone mineral content, mechanical properties and histologic characteristics of Japanese quail. *Poult Sci.* 2003;82:831-839.

14. Baird HT, Eggett DL, Fullmer S. Varying ratios of omega-6:omega-3 fatty acids on the pre- and postmortem bone mineral density, bone ash, and bone breaking strength of laying chickens. *Poult Sci.* 2008;87:323-328.

15. Kubiak M, Forbes N. Effects of diet on total calcium, vitamin D and parathyroid hormone in falcons. *Vet Rec*. 2012;171:504.

16. Cousquer GO, Dankoski EJ, Patterson-Kane JC. Metabolic bone disease in wild collared doves (*Streptopelia decaocto*). *Vet Rec*. 2007;160:78-84.

17. Eurasian collared dove. Wikipedia. Available at: https://en.wikipedia.org/wiki/Eurasian_collared_dove. Accessed June 2, 2016.

18. Johnsson M, Jonsson KB, Andersson L, et al. Genetic regulation of bone metabolism in the chicken: similarities and differences to mammalian systems. *PLoS Genet*. 2015;11:e1005250.:1-19.

19, Mohsenifar Z, Fridoni M, Ghatrehsamani M, et al. An evaluation of the effect of pulsed wave low-level laser therapy on the biomechanical properties of the vertebral body in two experimental osteoporosis rat models. *Lasers Med Sci*. 2016;31:305-14.

20. Mohsenifar Z, Fridoni M, Ghatrehsamani M, et al. Evaluation of the effects of pulsed wave LLLT on tibial diaphysis in two rat models of experimentalosteoporosis, as examined by stereological and real-time PCR gene expression analyses. *Lasers Med Sci*. 2016;31:721-32.

Avian Bornaviral Ganglioneuritis in Clinical Practice

Robert D. Dahlhausen, DVM, MS,
Susan E Orosz, PhD, DVM, Dipl ABVP (Avian), Dipl ECZM (Avian)

Session #024

Affiliation: Avian & Exotic Animal Medical Center and Veterinary Molecular Diagnostics, Inc, 5989 Meijer Dr, Suite 5, Milford, OH 45150, USA (Dahlhausen) and Bird and Exotic Pet Wellness Center, 5166 Monroe St, Suite 350, Toledo, OH 43623, USA (Orosz).

Abstract: Avian *Bornavirus* (ABV) has been shown to be a causative agent of proventricular dilatation disease in birds. The avian *Bornaviradae* represent a genetically diverse group of viruses that are widely distributed in captive and wild populations around the world. They are widely distributed in tissues of affected birds and generally produce a non-suppurative ganglioneuritis in the gastro-intestinal tract and central nervous system. Diseased birds may or may not demonstrate ABV ribonucleic acid on ante-mortem tests or test positive on serological assays. Alternatively, clinically normal individuals may test positive on molecular and serological assays. The route of transmission in natural infections has not been thoroughly ascertained. Management of affected birds is beneficial and currently centered at reducing neurological inflammation, managing secondary complications, and providing nutritional support.

Introduction

Neuropathic gastric dilatation of psittacine birds was initially reported as a wasting disease in macaws imported into North America and Europe from Bolivia in the late 1970s.[1-8] Originally limited to macaw species (*Ara* spp.), the disease was later identified in other parrot species as well. One of the first reports was by Ridgeway and Gallerstein in 1983 followed by the report of impaction, dilatation, and degeneration of the proventriculus in 16 large psittacine birds by Clark in 1984.[1,6] Originally called macaw wasting disease, the disease has also been referred to as macaw fading syndrome, myenteric ganglioneuritis, infiltrative splachnic neuropathy, neuropathic gastric dilatation, and proventricular dilatation disease (PDD). The disease may be more appropriately termed avian myenteric ganglioneuritis, non-suppurative ganglioneuritis, or avian auto-immune ganglioneuritis.[9] These articulate the disease process better and remove the focus on the proventriculus.

PDD has been described in over 80 species of psittacine and non-psittacine birds worldwide, both captive and in the wild.[10,11] African grey parrots (*Psittacus erithacus*), macaws (*Ara* species), Amazon parrots (*Amazona* species), and cockatoos (*Cacatua* species) are the most common psittacine species affected.[10] The disease is minimally represented in Quaker parrots (*Myiopsitta monachus*) and lovebird species (*Agapornis*).[9] Lesions suggestive of PDD also have been described in canaries (*Serinus canaria)*, the greenfinch (*Carduelis chloris)*, long-wattled umbrella bird (*Cephalopterus penduliger),* a bearded barbet *(Lybius dubius),* Canada geese (*Branta canadensis*), toucans (*Rhamphastidae*), honey creepers (*Drepanidinae*)*,* roseate spoonbill (*Platalea ajaja*) and a peregrine falcon (*Falco peregrines*).[12-15] PDD is a progressive neurologic disease with a high case fatality rate once clinical signs are present.[16] The disease presents a serious threat to captive propagation and conservation efforts for endangered psittacines such as the Spix's macaw (*Cyanopsitta spixii*).

Proventricular Dilatation Disease

Clinical signs

Clinical signs are variable and depend upon the host species involved, severity of disease, distribution of lesions and affected organ system involved. They vary from case to case. They may present in the central, peripheral, and/or autonomic nervous system. Although neurogenic in nature, signs are generally classified as gastrointestinal (GI) or central nervous system in character. Birds may exhibit only gastrointestinal or neurologic signs or a combination of both. GI tract signs reflect pathology of the terminal ganglia of the vagus nerve (cranial nerve X). The vagus, also referred to as the pneumogastric nerve, is the autonomic nervous system's parasympathetic control of the heart and digestive tract. It regulates homeostatic function of the proximal GI tract, pancreatic endocrine and exocrine function, hepatic glucose production, and heart rate. Gastrointestinal signs reflect varying degrees of dysfunction and neurogenic atrophy including delayed crop emptying and impaired GI tract transit, regurgitation, anorexia, dilatation and sometimes impaction of the upper GI tract. Pyloric motility is disturbed with severe alteration of gastric emptying and stasis.

Affected birds have impaired ability to digest and absorb dietary nutrients leading to loss of body weight, passage of undigested food material in the feces and diarrhea. The vagus nerve is also a major constituent of the inflammatory reflex, a neural reflex that controls innate immune responses and inflammation during GI tract pathogen invasion and tissue injury.[17-19] Impairment of this reflex and reduced vagal stimulation of gastric acids impairs the natural resistance to bacterial overgrowth and leads to alterations in the intestinal microbiome. Overgrowth of pathogenic organisms such as clostridial species and fungal organisms often occur in affected birds. This classic form of the disease is most often observed in new world psittacines.

Central nervous system lesions commonly involve the cerebrum or cerebellum. Cerebral perivascular cuffing and glial cell injury can cause seizures in affected birds. Optic lobe lesions lead to a cortical blindness which may be reversed with effective treatment. Disruption of the Purkinje, glial, and granule cell layers of the cerebellum produce disorders in fine movement and equilibrium evidenced as ataxia, proprioceptive deficits, intention tremors, incoordination, dysarthria (vocalization abnormalities), motor deficits and reduced cognitive ability. Old World psittacines often exhibit these central nervous system signs although concurrent, non-clinical lesions in the GI tract are usually present.

Inflammation and myelin degeneration of the dorsal nerve roots, white matter, and associated ganglia have been identified at all levels of the spinal cord in PDD affected birds.[20] Thoracolumbar lesions were the most common and severe. The dorsal root ganglia contain cell bodies of sensory neurons that bring information from the periphery to the spinal cord. PDD associated peripheral neuritis has been implicated as a cause of feather picking and self-mutilation in affected birds.[20]

Myocarditis as a component of PDD has been previously reported.[21] Lesions are more frequent and severe in the right side of the heart which may reflect the higher density of nervous tissues in this area. Additionally, parasympathetic innervation of the heart is partially controlled by the right vagus branch which innervates the sinoatrial node. Dilatation of the right ventricle of the heart of affected birds has previously been described. Arrhythmias and alterations in blood pressure may also be observed in affected birds. Cardiac lesions can result in acute death in otherwise clinically normal birds.

A survey of pet birds with clinical signs of PDD found 66% of birds exhibited central nervous system signs, 22% GI tract signs, 9% feather picking and mutilation, and 9% acute death.[22]

Pathology

Characteristic histologic lesions of PDD involve lymphocytic, plasmacytic inflammatory infiltrates of nervous tissues, axonal swelling, myelin degeneration, and perivascular mononuclear cell infiltrates of blood vessels and connective tissue surrounding affected nerves.[20] Lesions often present in ganglia of the GI tract (ganglioneuritis), central nervous system (encephalitis, myelitis), peripheral nerves including the sciatic, brachial, and vagal nerves (neuritis), and retina (retinitis). Abnormalities may be observed at all levels of the spinal cord but are especially severe in the thoracolumbar portion. They include vacuolation and spongiosus of the white matter, axonal swelling with myelin degeneration, perivascular infiltrates in the grey and white matter and associated ganglia and dorsal nerve roots, and gliosis. Lesions may also be observed in the heart which demonstrates focal to diffuse areas of myocardial necrosis associated with infiltration of mononuclear cells. The intracardiac nerve plexus of the right side of the heart is often more severely affected.[21] Lymphocytes may be scattered diffusely throughout the adrenal medulla or localized in clusters adjacent to cortical tissue.

The disease process of PDD involves inflammation of the nerve ganglia which exposes normally sequestered ganglioside proteins to the host immune system. Studies by Rossi et al have shown that the release of G 50 proteins from nerve ganglia produce pathologic lesions similar to that noted in the autoimmune neuropathy of Guillian Barré syndrome. [23-27] Disease occurs when the host immune response is directed to these exposed proteins causing corresponding neurological dysfunction.

Diagnosis

PDD should be considered in the differential diagnosis of any bird exhibiting neurologic and/or gastrointestinal signs. A presumptive diagnosis can be made based upon the history, clinical signs, and radiographic evaluation of the digestive tract. Definitive diagnosis can be confirmed by demonstrating the characteristic histopathologic lesions in tissues from affected birds.

Plain and contrast radiography, and fluoroscopic evaluation of gastrointestinal motility, have been used for tentative ante mortem diagnosis. Radiology and ultrasonography often reveal various degrees of enlargement, thinning and/or impaction of the ingluvies, proventriculus, ventriculus, and proximal duodenum . The proventriculus is often severely dilated, filling the left side of the coelomic cavity. It often appears as a "J" shape causing the ventriculus to be displaced to the right and ventrally.[28] Dilatation and thinning of the wall of these organs may lead to impaction and rupture. Contrast studies often demonstrate prolonged transit times throughout the gastrointestinal tract.[29]

Clinical pathologic findings in PDD are inconsistent and generally reflect the state of malnourishment, dehydration, and secondary infections that may occur with this disease. Serum chemistry profiles and complete blood counts are often normal although a relative and absolute heterophilia, hypoproteinemia, anemia, gram-negative and clostridial bacterial enteritis have been reported.[28] Gastrointestinal stasis predisposes to overgrowth of the digestive tract with yeast and gram-negative bacteria.

Ante-mortem diagnosis of PDD can be confirmed by identification of the characteristic myenteric ganglioneuritis in biopsy samples of the crop, ventriculus, and adrenal gland.[10] Crop biopsy specimens should include a visible blood vessel/nerve complex. While reported to be an effective method of ante mortem diagnosis with an accuracy of 76%, it has been reported that only about 76% of PDD affected birds have crop lesions.[10, 30] In practicality, crop biopsies are reported to be indicative of PDD in only about 30 to 35% of cases.[10,31]

Post-mortem examination often reveals emaciation and the presence of a distended, impacted proventriculus and ventriculus. Thinning of the wall of these organs occurs and rupture of these organs may be present. Histological examination of a wide range of tissues should be conducted on birds suspected of succumbing to PDD. Tissues submitted should include the ingluvies, proventriculus, ventriculus, duodenum, adrenal gland, heart, spleen, and brain.

Differential diagnoses

Tumors or papillomas of the crop, proventriculus, ventriculus, and intestines, ingestion of foreign bodies, Macrorhabdosis, and parasitism can cause gastrointestinal signs identical to those seen in PDD.[10] Emptying of the proventriculus and ventriculus appears to be inhibited as long as the intestinal tract is distended. Inflammatory disease and neoplastic diseases of the ventriculus and proventriculus can also cause gastrointestinal stasis. Heavy metal poisoning is commonly associated with central nervous system signs but can cause gastrointestinal stasis as well. Internal papillomatosis may result in a chronic wasting disease that resembles PDD. PDD should be considered as a differential in any bird with CNS disease. Traumatic injuries, heavy metal poisoning, neoplasia, viral, bacterial and fungal infections of the CNS, nutritional deficiencies and hydrocephalus are additional diseases that can appear similar.[10]

Avian *Bornavirus* (ABV)

An infectious etiology for PDD has long been suspected based on its apparent spread through aviary collections. Transmission electron microscopy studies in the 1990s provided the first evidence of a viral etiology demonstrating inclusion bodies and enveloped virus-like particles in the myenteric plexus, celiac ganglia and fresh feces from affected birds.[32] Attempts to isolate an infectious agent were unsuccessful at the time.

In 2008, two independent research groups identified a novel virus in tissues from PDD-affected birds and named it avian *Bornavirus* (ABV).[33,34] It was proposed as the causative agent of PDD. Subsequent studies have confirmed the association between ABV infection and PDD.[35-42] Disease development was reproduced by parenteral inoculation of psittacine birds, such as cockatiels (*Nymphicus hollandicus*) or Patagonian conures (*Cyanoliseus patagonus*), with genotypes ABV-2 and ABV-4. However, ABV has also been found in healthy birds which may remain apparently disease free for years.[35-39]

Bornaviruses are enveloped, non-segmented single-stranded negative sense RNA viruses. They are members of the family Bornaviridae in the order Mononegavirales. Other families within the order include the Filoviridae (West Nile virus), Rhabdoviridae (rabies virus), and the Paramyxoviridae. Currently 15 ABV genotypes have been identified. The recent reclassification of avian bornaviruses is Psittaciform *Bornavirus* 1 (which includes parrot *Bornavirus* (PaBV) 1,2,3, 4,7), Passeriform *Bornavirus* 1 (CnBV-1,2,3, canary) and (MuBV-1, munia finch), Waterbird *Bornavirus* 1 (ABBV-1, aquatic bird *Bornavirus* 1), Passeriform *Bornavirus* 2 (EsBV-1,estrildid finch *Bornavirus* 1), and tentative, unclassified bornaviruses of avian *Bornavirus* MALL (ABV-MALL), parrot *Bornavirus* 5 PaBV-5, parrot *Bornavirus* 6 PaBV-6, and parrot *Bornavirus* 8 PaBV-8.[22,43] PaBV-2 and PaBV-4 are the predominate genotypes in psittacine birds. The classification of PaBV-5 and PaBV-6 is tentative and remains unclassified at the species level.

The genetic variability of ABV is much greater than that observed in *Borna* Disease Virus. There is a 91-100% shared nucleotide identity within a genotype and only 68-85% between genotypes. Different genotypes seem to cause different disease in different species and individuals but the relationship between genotype, species of bird, and observed clinical disease is obscure at this time. Infection with one genotype does not appear to be protective against another. Simultaneous infection with two genotypes can occur and may result in severely worse clinical disease.

ABV is widely distributed in the body of infected birds.[44] It reproduces in a non-cytopathic manner in the host nucleus and persists due to mechanisms that evade the host immune system.[45] As a result, ABV infections are considered chronic and life-long. It is therefor unlikely that anti-viral or vaccine therapy will effectively eliminate the viral infected state.

Infection rate

ABV is widely distributed in both captive and wild avian populations. Approximately 15-40% of normal healthy birds are positive for the presence of ABV. Lierz detected ABV RNA in 27 (45.8%) of 59 healthy appearing pet birds.[46] Thirty five of 77 (45%) healthy birds from an aviary with a past history of PDD were found positive for ABV-specific serum antibodies.[47] A survey of laboratory samples submitted for other testing revealed approximately 34% (271/791) avian samples tested positive from across the USA.[48]

Almost all collections of psittacine birds will contain individuals infected with ABV. A large-scale survey revealed that ABV infection is widespread among captive psittacines in Europe, with 23% of 1,442 birds considered infected.[36] Similarly, a high infection rate was observed in captive canaries in Germany as well as in certain populations of wild waterfowl in North America.[49,50]

While a significant percentage of the avian population is infected with ABV, the frequency of clinical borna viral disease is much lower. The majority of ABV positive birds do not exhibit clinical disease. When it does occur, disease presents on a continuum of intensity with many birds exhibiting only mild disease. The clinically "wasted" bird represents the more severe form of untreated, chronic disease.

Transmission

The epidemiology of BDV is currently not well understood. ABV is shed in the urine and feces of infected birds. The urofaecal-oral route is assumed to be important for horizontal transmission but other routes cannot be excluded. Transmission via the respiratory tract and vertically through the egg have also been discussed but supporting experimental evidence is lacking at this time. Piepenbring et al. did report successful ABV transmission to one cockatiel placed in contact with a group of cockatiels that were experimentally infected with PaBV-4.[38] Most other studies indicate that horizontal transmission of ABV by direct contact is inefficient in immunocompetent fully fledged birds.[51] Cockatiels were infected with PaBV-4 by oral and intranasal application. Clinical signs of disease were not observed in any of the birds during the 174 day observation period. At the end of the study, pathohistologic and immunohistochemical examination revealed neither lesions typical for PDD nor ABV specific antigen in any of the birds.[52] Kistler et. al. reported a dramatic PDD outbreak during which 13 unweaned chicks of various psittacine species died.[53] These observations suggest that ABV transmission may be much more efficient in unweaned nestlings with immaturely developed immune systems compared to older individuals.

Experimental studies have shown that ABV infections can be induced experimentally when the virus is injected intramuscularly, intravenously, and intracranially in birds. Gancz et al. (2009) were the first to demonstrate that PDD could be transmitted to healthy birds by the use of infected-brain tissue.[39] They inoculated cockatiels by multiple routes with a brain homogenate from either an PaBV-4 positive bird or from a PDD-/ABV-control bird. The birds inoculated with healthy control bird homogenate remained healthy, whereas all three birds inoculated with brain homogenate from ABV-infected birds developed both gross and microscopic lesions typical of PDD.[39] Gray et al. isolated ABV in cultured duck embryo fibroblasts (DEF).[40] After six passages, these infected cells were injected intramuscularly into two Patagonian conures. Clinical signs of PDD developed within 66 days post-infection in both challenged birds. The presence of typical PDD was demonstrated on necropsy and histopathology. RT-PCR demonstrated the presence of the inoculated strain in the brains of the challenged birds.[40] Piepenbring et al. inoculated 18 cockatiels by both the intracerebral and intravenous routes with an isolate of PaBV-4 cultured for six passages in a quail cell line (CEC-32).[38] All challenged birds became persistently infected but the clinical disease patterns that developed varied among individuals. Five birds developed clinical signs of PDD, while on necropsy 7 of the 18 had a dilated proventriculus. All infected birds did, however, show mononuclear cell infiltrates characteristic of PDD in a wide range of organs.[38] Strong suggestive evidence for vertical transmission of ABV from the hen to the egg has been reported in psittacines and canaries. Several studies found eggs from infected hens to be ABV-positive by PCR assays.[51,54-56] Final proof of vertical transmission provided by the detection of productive infection in embryos or chicks reared under isolated conditions is lacking at this time.

The incubation period of PDD is unknown. Clinical observations suggest that this could be as short as several weeks to as long as many years.[8,28] In general, transmission is believed to require long term, close contact among birds. Unlike other RNA viruses that are more stable (West Nile virus), endonuclease enzymes in the environment tend to rapidly degrade the avian *Bornavirus*. Soap and dilute bleach appear to be effective in disinfecting enclosures and items that come in contact with ABV-positive individuals. The sensitivity of ABV to desiccation suggests that virus could lose viability soon after being shed into the environment.[57]

Avian *Bornavirus* Diagnostics

Polymerase chain reaction

The diagnosis of PDD has traditionally been based on the identification of histologic lesions in biopsy or necropsy tissues. With the discovery of the association of ABV with PDD, diagnostic testing has focused on reverse transcription polymerase chain reaction (RT-PCR) for the detection of ABV ribonucleic acid (RNA) in affected birds. Ante mortem detection of viral infection in the naturally or experimentally infected bird is however challenging. Urine, feces, and cloacal swabs are samples reportedly most likely to contain virus. However a whole blood sample in combination with choanal/cloacal swabs may provide the greater sensitivity. Fecal swabs are less desirable as RNAases and other degrading or inhibiting agents may rapidly degrade the viral RNA in these samples. Cloacal sampling, however, likely underestimates the prevalence of ABV infection. Intermittent urofecal shedding of ABV is described in both experimentally and naturally-infected parrots which can lead to a false negative test result.[38,58,59] One research study recommended the use of feather calami for ABV RNA testing.[60] The consensus of opinion among most other researchers at the 2014 ABV Research Forum was that this was not an appropriate sample for accurate ABV testing.[9] It should be noted that there is no consistent or standardized method of testing among laboratories offering this service. Reported rates of ABV positive tests among several university and commercial laboratories range from 3-33%. With only about 68 to 85% sequence identity among the known ABV genotypes, some assays may not be able to detect all ABV genotypes. Primers have been designed to target the nucleocapsid (N) gene, matrix (M) gene, phosphoprotein (P) gene, and polymerase (L) gene. Assays designed for detection of the M and immunodominant N gene sequences appear to have a similar high sensitivity.[35] Those for the L and P genes are generally less accurate. Both gel-based RT-PCR and real-time RT-PCR have been successfully used to detect ABV RNA. Real-time PCR assays appear to be the more sensitive of the two techniques.

Serology

Serologic assays have been used to detect ABV exposed or infected birds. The indirect fluorescent antibody (IFA)assay is preferred by many investigators, particularly in Europe. Researchers in the USA tend to use the Western Blot assay.[61,62] Both appear to be appear to be sensitive and specific assays. They cannot however differentiate between diseased birds and healthy carriers. Not all birds shedding viral RNA are antibody positive.[47] The discrepancy between viral shedding, presence of antibodies, and clinical disease is now well recognized in parrots infected with ABV.[38,62,63] Viral shedding and presence of antibodies coincided in only one-fifth of the samples in a large study of captive psittacines.[36] In another study of free-ranging psittacines, 50% of ABV RNA positive birds did not show antibodies against ABV by IFA.[11] Many apparently healthy birds may however be seronegative while, at the same time, shedding ABV in their feces.

In a large-scale study conducted in captive psittacine birds in Europe, 17% of the birds had detectable antibodies.[36] Widespread persistent asymptomatic infection with less common clinical disease is a consistent feature of *Bornavirus* infections, and may be attributed to the lack of cytopathic effect of bornaviruses and their ability to escape recognition by the innate immune system.

Serologic testing for antiganglioside antibodies appears to more accurately detect clinically affected birds. Antiganglioside autoantibodies are utilized as markers of immune-mediated disease and are triggered by a variety of pathogens. High antiganglioside antibody levels were detected in 15.5% of 650 avian serum samples and 98% of symptomatic and histologically positive PDD birds.[24,25] Further studies to help determine the sensitivity and specificity of antiganglioside autoantibodies in the diagnosis of PDD are needed.

Proventricular Dilatation Disease Pathogenesis

Infection studies with ABV have satisfied Koch's postulates and shown that it is a cause of PDD in birds. Some researchers are of the opinion that it is the sole cause of PDD in birds.[64] While ABV is a causative agent of PDD, the relationship of ABV in the pathogenesis of this disease remains unclear. Although the virus is non-cytopathic it can induce inflammation and a selective loss of glial cells and neurons. This cell loss appears secondary to T cell cytotoxicity. The presence of CD8+ T cells parallels the onset of neurologic dysfunction. Rossi et.al. have demonstrated immunocomplex deposition around affected nerve ganglia associated with high positive expression of complement fractions by immunofluorescence in birds with clinical PDD.[24] They suggest that PDD is induced by an autoimmune mechanism similar to the pathogenesis of Guillain Barrè syndrome where gangliosides act as major antigens. To confirm this theory, they challenged 6 cockatiels intraperitoneally (IP) or orally with purified avian gangliosides. One month post infection, 100% of IP inoculated and 33% of orally challenged parrots developed neurological and GI tract signs compatible with PDD. Four of the birds demonstrated the classic lymphoplasmacytic ganglioneuritis in crop biopsies and all showed histopathological changes compatible with PDD.[23] These findings warrant further investigation into the mechanisms of autoimmune reactions against nerve ganglioside proteins and their possible involvement in the pathogensis of PDD.

Treatment

PDD has historically been a fatal disease of psittacine birds with mortality approaching near 100%. Even with appropriate supportive care, most birds succumb to progressive debilitation, starvation, secondary infections, or CNS disturbances. The ganglioneuritis and encephalomyelitis associated with PDD is inflammatory in nature, the chronicity of which, contributes to the progressive, debilitating nature of the disease. The rationale for a therapeutic approach was that diminishing this reaction was expected to lead to clinical improvement and possible resolution of clinical signs in affected birds. In 2002, we demonstrated the resolution of clinical PDD by cyclooxygenase 2 (COX-2) inhibition.[65] Since this time, clinicians have recognized the early signs of avian ganglioneuritis and the variable ways it can presents in a clinical setting. The use of preferential and selective COX-2 inhibitors has improved and extended the quality of life of affected birds.

Gastrointestinal prokinetic agents such as cisapride (Propulsid, Janssen Pharmaceutica Inc. Titusville, NJ,USA) and metoclopramide (Reglan, Schwarz Pharma, Seymour, IN,USA) are helpful in improving transit in birds with GI tract involvement, especially early in the course of therapy. Appropriate antibacterial and antifungal therapy should be instituted to control overgrowth of intestinal anaerobes, yeasts, and *Macrorhabdus ornithogaster* in these individuals. GI tract disease results in an altered intestinal microbiome. Probiotics and prebiotics (Sivoy, Rome, Italy) may prove helpful in restoring a normal intestinal environment. Omega fatty acids are also helpful in reducing inflammation and clinically beneficial in affected birds.[66] Semi-elemental diets such as Emeraid Omnivore and Carnivore (Lafeber Emeraid LLC, Cornell, IL, USA) require minimal digestion and provide a readily absorbable source of essential nutrients and omega fatty acids). The liver is often exposed to bacterial showers from the abnormal intestinal environment in affected birds. Herbal supplements like milk thistle (Milk Thistle, Low Alcohol, Gaia Herbs, Brevard, NC, USA) and ginger (Ginger Root, Certified Organic, Gaia herbs, Brevard, NC, USA) are helpful in reducing inflammation, preserving hepatic function, and improving GI tract transit.

Gabapentin (Neurontin, Pfizer, New York, NY, USA) is increasingly being used in birds to help treat self-mutilation and to control neurogenic pain. It is a useful adjunct in managing birds with avian ganglioneuritis. Clinical signs in affected birds tend to increase with the onset of breeding activity. Reduction of the stress of elevated hormonal activity through the use leuprolide acetate (Lupron Depot, AbbVie, Chicago, IL, USA) or deslorelin implants (Suprelorin, Virbac Animal Health, Fort Worth, TX, USA) is very beneficial. Offlabel use of Suprelorin-F deslorelin acetate implants is prohibited by federal law in the United States since the product is not FDA approved but legally marketed as an index drug.

Mammalian *Borna* disease virus shows a high degree of sensitivity to ribivirin (Virazole, Valeant Pharmaceuticals International,Bridgewater, NJ, USA), a ribonucleic analog that stops viral RNA synthesis. To date, the use of this and other antiviral agents in the treatment of avian bornaviral disease has not been rewarding or adequately researched.

Immunomodulation therapy through the use of *Mycobacterium bovis* extracts to redirected trafficking of activated antigen-specific CD4+T cells to local inflammatory sites has been shown to modulate the initiation and progression of a Th1-mediated peripheral and CNS autoimmune disease. This researched group found significant improvement in clinically diseased birds when this approach was combined with selective anti-COX-2 therapy.[67]

With patience, perseverance, prolonged therapy, reduced stress, and attention to correction of secondary problems, quality and longevity of life in many ABV clinically affected birds can be significantly improved. Implementation of appropriate therapeutic plans, early in the course of clinical disease, are most productive. Therapies may be minimally effective in severely affected individuals.

Formulary

Celecoxib (Celebrex, Pfizer Inc.,Mission,KS, USA), initially 30-40 mg/kg PO q12h, long term 15-30 mg/kg PO q12h. 60-80 mg/kg PO q12h for central nervous system involvement.

Robenacoxib (Onsior, Novartis Animal Health,North Ryde, NSW), 2-10 mg/kg IM weekly, and subsequently monthly.[67]

Meloxicam (Metacam, Boehringer Ingelheim Vetmedica Inc., St. Joseph, MO USA). a preferential COX-2 inhibitor, not selective.).0.5 mg/kg IM, PO q24h; Oral form is formulated for canine species and may have reduced bioavailability in avian species; once clinical trial has shown it use exacerbates signs of clinical disease in ABV infected cockatiels.[68]

Gabapentin (Neurontin, Pfizer, New York, NY, USA) 10-25mg/kg PO q12h up to 50 mg/kg PO q12h in self -mutilating birds (Cathy Johnson-Delaney, written communication, 2015)

References

1. Clark FD. Proventricular dilatation syndrome in large psittacine birds. *Avian Dis.* 1984;28(3):813-815.

2. Gerlach H. Uber das Sog. Macaw Wasting Syndrome. Vortrag, 4. *Arbeitstagung der Zootierarzte im deutschsprachigen Raum. Munster.* 24-25 Nov. 1984.

3. Graham DL. Infiltrative splachnic neuropathy, a component of the "wasting macaw" complex. *Proc Int Conf Avian Med.* 1984;275.

4. Hughes PE. The pathology of myenteric ganglioneuritis, psittacine encephalomyelitis, proventricular dilatation of psittacines, and macaw wasting syndrome. *Proc 33rd West Poult Dis Conf.* 1984;85-87.

5. Phalen DN. An outbreak of psittacine proventricular dilatation syndrome (PPDS) in a private collection of birds and an atypical form of PPDS in a nanday conure. *Proc Annu Conf Assoc Avian Vet.* 1986;27-34.

6. Ridgway RA, Gallerstein GA. Proventricular dilatation in psittacines. *Proc Annu Conf Assoc Avian Vet.* 1983;228-230.

7. Woerpel, RW, Rosskopf WJ. Clinical and pathological features of macaw wasting disease (proventricular dilatation syndrome). *Proc 33rd Western Poultry Dis Conf.* 1984;89-90.

8. Woerpel, RW, Rosskopf WJ. Proventricular dilatation and wasting syndrome: myenteric ganglioneurtitis and encephalomyelitis of psittacines; an update. *Proc Int Conf Avian Med.* 1984;25-28.

9. Avian *Bornavirus* Research Forum. *Annu Conf Assoc Avian Vet.* 2014.

10. Schmidt RE, Reavill DR, Phalen DN. *Pathology of Pet and Aviary Birds.* Ames, IA: Iowa State Press; 2003.

11. Encinas-Nagel N, Enderlein D, Piepenbring A, Herden C,et.al. Avian *Bornavirus* in free-ranging psittacine birds, Brazil. *Emerg Infect Dis.* 2014;20:12:2103-2106.

12. Daoust PY, Julian RJ, Yason CV, Artsob H. Proventricular impaction associated with nonsuppurative encephalomyelitis and ganglioneuritis in two Canada geese. *J Wildl Dis.* 1991;27(3):513-517.

13. Gregory CR, Ritchie BW, Latimer KS, et al. Progress in understanding proventricular dilatation disease. *Proc Annu Conf Assoc Avian Vet.* 2000;269-275.

14. Perpiñán D, Fernández-Bellon H, López C, Ramis A. Lymphoplasmacytic myenteric, subepicardial, and pulmonary ganglioneuritis in four nonpsittacine birds. *J Avian Med Surg.* 2007;21(3):210-214.

15. Shivaprasad HL. Proventricular dilatation disease in a peregrine falcon (*Falco peregrinus*). *Proc Annu Conf Assoc Avian Vet.* 2005;107-108.

16. Weissenock H, Bakonyi T, Sekulin K, Ehrensperger F, et.al. Avian bornaviruses in psittacine birds from Europe and Australia with proventricular dilatation disease. *Emerg Infect Dis.* 2009;15:1453-1459.

17. Pavlov VA, Tracey KJ. The vagus nerve and the inflammatory reflex linking immunity and metabolism. *Nat Rev Endocrinol.* 2012;8:743-754.

18. Andersson U, Tracey KJ. Reflex principles of immunological homeostasis. *Annu Rev Immunol.* 2011;30: 313-335.

19. Tracey KJ. The inflammatory reflex. *Nature.* 2002;420:853-859.

20. Berhane Y, Smith D, Newman S, et al. Peripheral neuritis in psittacine birds with proventricular dilatation disease. *Avian Pathol.* 2001;30:563–570.

21. Cazayoux Vice C.A. Myocarditis as a component of psittacine proventricular dilatation syndrome in a Patagonian conure. *Avian Dis.* 1992;36(4):1117-9.

22. Philadelpho NA, Rubbenstroth D, Guimara MB, et.al. Survey of bornaviruses in pet psittacines in Brazil reveals a novel parrot *Bornavirus*. *Vet Microbiol.* 2014;174:584-590.

23 Rossi G, Crosta L, Ceccherelli R, Pesaro S. New evidence in PDD pathogenesis: can ganglioside sensitization satisfy Koch's postulates? *Proc Annu Conf Eur Assoc Avian Vet.* 2009;155.

24 Pesaro S, Crosta L, Bertoni P, Ceccherelli R, et.al. Anti-ganglioside antibodies production as a theory of PDD pathogenesis. *Proc Annu Conf Eur Assoc Avian Vet.* 2009;89.

25. Rossi G, Ceccherelli R, Crosta L, et al. Anti-ganglioside specific auto-antibodies in ganglia of PDD-affected parrots. *Proc Annu Conf Eur Assoc Avian Vet, 1st Sci ECZM Meeting.* 2011;177-178.

26. Rossi G, Enderlein D, Herzog S, et al. Comparison of anti-ganglioside antibodies and anti-ABV antibodies in psittacines. *Proc Annu Conf Eur Assoc Avian Vet, 1st Sci ECZM Meeting.* 2011;187-189.

27. Rossi G, Crosta L, Pesaro S. Parrot proventricular dilation disease. *Vet Rec.* 2008;163:310.

28. Phalen DN. Implications of viruses in clinical disorders. In: Harrison GJ and Lightfoot TL, eds. *Clinical Avian Medicine.* Spix Publishing, Inc;2006:721-745.

29. Gerlach S. Macaw wasting disease-a 4 year study on clinical cases, history, epizootiology analysis of species, diagnosis and differential diagnosis, microbiological and virological results. *Proc Annu Conf Eur Assoc Avian Vet.* 1991:273-281.

30. Gregory CR, Latimer KS, Campagnoli RP, Ritchie BW. Histologic evaluation of the crop for diagnosis of proventricular dilatation syndrome in psittacine birds. *J Vet Diagn Invest.* 1996 Jan;8(1):76-80.

31. Bond MW, Downs D, Wolf S. Screening for psittacine proventricular dilatation syndrome. In: *Proc Annu Conf Assoc Avian Vet*, 1983;92-97.

32. Gough RE, Drury SE, Harcourt-Brown NH, Higgins RJ. Virus-like particles associated with macaw wasting disease. *Vet Rec.* 1996;139:24.

33. Kistler AL, Gancz A, Clubb S, Skewes-Cox P, et. al. Recovery of divergent avian bornaviruses from cases of proventricular dilatation disease: identification of a candidate etiologic agent. *Virol J.* 2008;5:88.

34. Honkavuori KS, Shivaprasad HL, Williams BL, , P. L. Quan PL, et.al. Novel borna virus in psittacine birds with proventricular dilatation disease. *Emerg Infect Dis.* 2008;14:1883-1886.

35. Lierz M, Hafez HM, Honkavuori KS, Gruber AD, Olias P, Abdelwhab EM, et al. Anatomical distribution of avian *Bornavirus* in parrots, its occurrence in clinically healthy birds and ABV antibody detection. *Avian Pathol.* 2009;38:491-496.

36. Heffels-Redmann U, Enderlein D, Herzog S, Herden C, Piepen-bring A, Neumann D, et al. Occurrence of avian *Bornavirus* infection in captive psittacines in various European countries and its association with proventricular dilatation disease. *Avian Pathol.* 2011;40:419-426.

37. Weissenböck H, Bakonyi T, Sekulin K, Ehrensperger F, Doneley RJ, Dürrwald R, et al. Avian bornaviruses in psittacine birds from Europe and Australia with proventricular dilatation disease. *Emerg Infect Dis.* 2009;15:1453-1459.

38. Piepenbring AK, Enderlein D, Herzog S, Kaleta EF, Heffels- Redmann U, Ressmeyer S, et al. Pathogenesis of avian *Bornavirus* in experimentally infected cockatiels. *Emerg Infect Dis*. 2012;18:234-241.

39. Gancz AY, Kistler AL, Greninger AL, et al. Experimental induction of proventricular dilatation disease in cockatiels (*Nymphicus hollandicus*) inoculated with brain homogenates containing avian *Bornavirus* 4. *Virol J.* 2009;6:11.

40 Gray P, Hoppes S, Suchodolski P, et al. Use of avian *Bornavirus* isolates to induce proventricular dilatation disease in conures. *Emerg Infec Dis*. 2010;16(3):473-479.

41 Mirhosseini N, Gray PL, Hoppes S, Tizard I, et.al.. Proventricular dilatation disease in cockatiels (*Nymphicus hollandicus*) after infection with a genotype 2 avian *Bornavirus*. *J Avian Med Surg*. 2011;25(3):199-204.

42 Payne S, Shivaprasad HL, Mirhosseinia N, Gray P, et.al. Unusual and severe lesions of proventricular dilatation disease in cockatiels (*Nymphicus hollandicus*) acting as healthy carriers of avian *Bornavirus* (ABV) and subsequently infected with a virulent strain of ABV. *Avian Pathol.* 2011;40(1):15-22.

43. Kuhn JH, Dürrwald R, Bào Y, Thomas Briese T, et.al. Taxonomic reorganization of the family Bornaviridae. *Arch Virol*. 2015;160:621-632.

44. Rinder M, Ackermann A, Kempf H, et al. Broad tissue and cell tropism of avian *Bornavirus* in parrots with proventricular dilatation disease. *J Virol*. 2009;83:5401-5407.

45. Reuter A, Ackermann A,Kothlow S, Rinder M, et.al. Avian bornaviruses escape recognition by the innate immune system. *Viruses* 2010;2:927-938.

46. Lierz M. Avian Borna virus research at the free University of Berlin. Presented at: *Research Forum, Annu Conf Eur Assoc Avian Vet*; 2009.

47. Herzog S, Enderlein D, Heffels-Redmann U, Piepenbring A, et.al. Indirect immunofluorescence assay for intra vitam diagnostic of avian Borna virus infection in psittacine birds. *J Clin Microbiol.* 2010;48(6):2282-2284

48. Dahlhausen RD, Orosz SE. Avian Borna virus infection rates in domestic psittacine birds. *Proc Annu Conf Assoc Avian Vet*, San Diego, CA.2010.

49. Rubbenstroth D, Rinder M, Stein M, Hoper D, Kaspers B, et.al.Avian bornaviruses are widely distributed in canary birds (*Serinus canaria* f. *domestica*). *Vet. Microbiol.* 2013;165:287–295.

50. Payne S, Covaleda L, Jianhua G, Swafford S, et.al. Detection and characterization of a distinct *Bornavirus* lineage from healthy Canada geese (*Branta canadensis*). *J Virol.* 2011;85(22):12053-12056.

51. Rubbenstroth D, Brosinski K, Rinder M, Olbert M, et.al. No contact transmission of avian *Bornavirus* in experimentally infected cockatiels (*Nymphicus hollandicus*) and domestic canaries (*Serinus canaria forma domestica*). *Vet Microbiol.* 2014;172:146-156.

52. Heckmann J, Enderlein D, Piepenbring A, Herzog , S et.al. Investigation on different routes of transmission of avian *Bornavirus* in cockatiels (*Nymphicus hollandicus*). *Proc Int Conf Avian Herp Exot Mammal Med*; 2015.

53. Kistler AL, Smith JM, Greninger AL, Derisi JL, et.al. Analysis of naturally occurring avian *Bornavirus* infection and transmission during an outbreak of proventricular dilatation disease among captive psittacine birds. *J. Virol.* 2010;84:2176-2179.

54. Kerski A, de Kloet AH, de Kloet SR.Vertical transmission of avian *Bornavirus* in Psittaciformes: avian *Bornavirus* RNA and anti-avian *Bornavirus* antibodies in eggs embryos, and hatchlings obtained from infected sun conures (*Aratinga solstitialis*). *Avian Dis.* 2012;56:471-478.

55. Lierz M, Piepenbring A, Herden C, Oberhäuser K, et.al. Vertical transmission of avian *Bornavirus* in psittacines. *Emerg Infect Dis.* 2011;17(12):2390-2391.

56. Monaco E, Hoppes S, Guo J, Tizard I. The detection of avian *Bornavirus* within psittacine eggs. *J Avian Med Surg.* 2012;26:144-148.

57. Kranz JB, Escandon P, Musser JMB. Environmental stability of avian *Bornavirus*: pH and drying. *Proc ExoticsCon.* 2015;89.

58. Raghav R, Taylor M, Delay J, Ojkic D, et.al. Avian *Bornavirus* is present in many tissues of psittacine birds with histopathologic evidence of proventricular dilatation disease. *J Vet Diagn Invest.* 2010;22:495-508.

59. Heatley JJ, Villalobos AR. Avian *Bornavirus* in the urine of infected birds. *Vet Med Res Rep* 2012;3:19-23.

60. de Kloet AH, Kerski A, de Kloet SR. Diagnosis of avian *Bornavirus* infection in Psittaciformes by serum antibody detection and reverse transcription polymerase chain reaction assay using feather calami. *J Vet Diagn Invest.* 2011;23(3):421-429.

61. Lierz M, Piepenbring A, Heffels-Redmann U, Sibylle Herzog S, et.al. Experimental infection of cockatiels with different avian *Bornavirus* genotype. *Proc Annu Conf Assoc Avian Vet.* 2012;62.

62. Villanueva I, Gray P, Mirhosseini N, Payne S. The diagnosis of proventricular dilatation disease: use of a Western blot assay to detect antibodies against avian *Borna* virus. *Vet Microbiol.* 2010;143:196-201.

63. Heffels-Redmann U, Enderlein D, Herzog S, Piepenbring A, et.al. Follow-up investigations on different courses of natural avian *Bornavirus* infections in psittacines. *Avian Dis.* 2012;56:153-159.

64. Payne SL, Delnatte P, Guo J, Heatley J, et.al. Birds and bornaviruses. *Anim Health Res Rev.* 2012;13(2): 145-156.

65. Dahlhausen B, Aldred S, Colaizzi E. Resolution of clinical proventricular dilatation disease by cyclooxygenase 2 inhibition. *Proc Annu Conf Assoc Avian Vet.* 2002;9-12.

66. Clubb SL: Clinical management of psittacine birds affected with proventricular dilatation disease. *Proc Annu Conf Assoc Avian Vet.* 2006;71.

67. Rossi G, Crosta L, Ceccherelli R. PDD: Our point of view after 7 years of research. *Proc Annu Conf Assoc Avian Vet.* 2012;79.

68. Hoppes S, Heatley J, Guo J, Turner D, et.al. Meloxicam treatment in cockatiels (*Nymphicus hollandicus*) infected with avian *Bornavirus*. *J Exot Pet Med.* 2013;22:275-279.

Sedation in Pet Birds

Christoph Mans, Dr. med. vet., Dipl ACZM

Session #306

Affiliation: From the School of Veterinary Medicine, University of Wisconsin-Madison, 2015 Linden Drive, Madison, WI 53706, USA.

Abstract: Sedation of pet birds has become an increasingly popular technique in order to facilitate common clinical procedures such as physical examination, blood collection or radiography. Sedation provides immobilization, reduces vocalization, and attenuates the stress-response caused by manual restraint. Midazolam and midazolam-butorphanol are the most commonly used drugs for sedation in pet birds, and provide dose-dependent sedation with no significant side effects, if used at the published dosages. Midazolam has amnestic and anxiolytic properties in birds, which provide additional benefits, besides it sedative properties. The intranasal route of administration is a non-invasive alternative to intramuscular administration, and has been shown to be safe and effective technique to rapidly and reliably induce sedation in pet birds. Reversal with flumazenil should be performed in most cases of midazolam or midazolam-butorphanol induced sedation, in order to achieve rapid and complete recovery.

Introduction

Sedation of dogs and cats in veterinary practice is daily routine for a variety of procedures, such as radiographs and ultrasonography, or other non-painful, but potentially stressful procedures. However, historically for avian patients, either manual restraint of conscious birds or general anesthesia is typically performed, in order to complete most clinical procedures. General anesthesia predisposes birds to cardiovascular and respiratory depression, and may cause aspiration of gastric or crop contents and hypothermia. In contrast, manual restraint in conscious birds is simple to perform, but can have negative consequences, including stress to the bird and/or handler, negative conditioning to the clinic environment (e.g., the person restraining or the towel used for restraint), hyperthermia, and the predisposition of trauma to the handler and/or bird. Several recent studies demonstrated that manual restraint of birds causes increased body temperature and respiratory rate.[1,2] In sick, old, or very stressed bird, acute collapse and death secondary to manually restraint have been reported. Therefore, sedation techniques provide a useful alternative for reducing physiologic stress in birds undergoing non-painful clinical procedures. Further sedation in birds, provides easier restraint and increases the safety of many clinical procedures (e.g., blood collection, radiography, ultrasonography) and allows for a more complete examination, which would otherwise only be achieved under general anesthesia. Using safe and effective sedative protocols in pet birds, provides substantial benefits to the patients as well as the veterinarian and staff, and should be considered for a variety of clinical procedures.

Route of Drug Administration

Historically sedative drugs have been most commonly administered by intramuscular injection in birds. Recently the intranasal administration of sedative drugs in birds has gained increased attention. Intranasal administration of midazolam in Hispaniolan Amazon parrots, ring-necked parakeets, budgerigars and canary finches resulted in rapid onset of dose-dependent degree sedation, which is completely reversible with intranasal flumazenil.[2-5]

Intranasal drug administration offers an alternative, non-invasive technique for drug administration in birds (Table 1). It is characterized by its ease of administration, high bioavailability, rapid onset of action, and reduced pain compared to intramuscular administration.

Table 1. Comparison of intramuscular and intranasal drug administration in pet birds.

	Intramuscular administration	Intranasal administration
Pro	Reliable drug delivery Faster administration	Non-invasive and not painful Higher client acceptance Reduced risk of needles stick injury No changes in the biochemical panel
Contra	Invasive Potentially painful Possible post-injection hemorrhage Changes in the biochemical panel	Risk of incomplete drug delivery Difficult to administer larger volumes May cause sneezing More difficult in birds with feathered nares

Elevation of muscle enzymes in biochemistry panels secondary to intramuscular drug administration is avoided, if intranasal administration is used instead. In addition, clients perceive the intranasal route as non-invasive, compared to intramuscular injection, which leads to better client compliance in cases in which sedation is recommended. The time of onset to sedation is rapid, typically within 3-5 minutes.[2-5] However, limitations of intranasal administration include incomplete drug delivery, due to sneezing during administration, physiologically narrowed nostrils (e.g., cockatoos) or upper respiratory disease (e.g., blocked or stenotic nostrils). In some cases in larger birds (e.g., macaws) the drug volume can also be limiting the effectiveness and produce excessive sneezing, therefore leading to incomplete drug delivery. Higher concentrated drugs (e.g., midazolam 50 mg/ml, Zoopharm, Windsor, CO) are available, but intramuscular administration might be more feasible in these cases.

Drugs used for sedation

Midazolam: Midazolam is currently the most commonly used drug for sedation of pet birds and has a wide margin of safety. Midazolam has sedative, muscle relaxing, anxiolytic, amnestic and appetite stimulating properties in birds.[2,6,7] The injectable form of midazolam (midazolam hydrochloride (5 mg/ml); Hospira Inc, Lake Forest, IL) or a more concentrated form (50 mg/ml, Zoopharm, Windsor, CO) can be administered intranasally and intramuscular, without side effects.[2-4] Dosages commonly used in pet birds range from 0.5-3 mg/kg.[8] At the University of Wisconsin, we routinely use 2 mg/kg of midazolam in pet birds, if administered intranasally and as the sole sedative agent. In smaller birds such as finches or budgerigars we routinely use 4-6 mg/kg of midazolam if administered alone.

Diazepam: Diazepam is of similar efficacy as midazolam in birds following intranasal administration, but has a longer onset time and duration of action.[4,5,9] While less commonly used for intranasal administration in birds, diazepam represents a suitable alternative, in cases in which midazolam might not be available. The intramuscular administration of diazepam should be avoided, due to delayed absorption and muscle irritation.[5] Dosages commonly used in pet birds range from 0.2-2 mg/kg, if used as a sole sedative agent. Dosages as high as 10-15 mg/kg have been administered to finches and budgerigars, without significant side effects.[4,5,9]

Butorphanol: Butorphanol is the currently the most commonly used opioid analgesic in birds. Besides its analgesic effects, butorphanol has sedative effects, which are potentiated by benzodiazepines (i.e., midazolam, diazepam). The combined administration of midazolam and butorphanol is recommended in birds for which midazolam alone provides only an insufficient level of sedation or which require deeper sedation for certain clinical procedures such as radiographic positioning. Butorphanol can be given in combination with midazolam, drawn into a single syringe and can be given parenterally as well as intranasally. No side effects of intranasal administration of butorphanol at a dose range of 1-3 mg/kg are seen in psittacine birds.[8] At the University of Wisconsin we routinely uses butorphanol 1-2 mg/kg combined with 2 mg/kg of midazolam in pet birds administered intranasally or by intramuscular injection.

Reversal of sedation

Reversal of sedation in pet birds will depend on the patient and the purpose of sedation. Sedation performed in order to facilitate physical examination and diagnostic sample collection, should always be reversed, in order to have the patient return to normal behavior and food intake as soon as possible. It is important not to discharge sedated patients, as owners do not tend to appreciate to have a partially sedated bird, which might be imbalanced, sleepy and refuses to eat.

Birds that underwent sedation for e-collar or bandage placement or that were sedated for control of seizures, should not be reversed. In these cases, birds should be carefully monitored and reversal considered, if the level of sedation is perceived too deep or the duration of sedation is prolonged, and might interfere with physiologic behavior, particular food intake.

Flumazenil: Flumazenil is a benzodiazepine antagonist, and is used to reverse the sedative effects of midazolam and diazepam in birds.[2,3,8,9] The injectable form of flumazenil (flumazenil hydrochloride (0.1 mg/ml), Abaxis Pharmaceutical Products, Schaumburg, IL) can be administered intranasally, intramuscularly or intravenously, without side effects. The recommended dosages range from 0.01-0.1 mg/kg. Alternatively a flumazenil to midazolam ratio of 13:1 has been recommended, but it has been shown that lower doses of flumazenil achieve complete recovery from midazolam induced sedation in birds.[2] The author prefers to administer 0.05 mg/kg initially. If reversal is deemed unsatisfactory, then the same amount of flumazenil can be administered repeatedly. Recovery from sedation is usually complete within 10-15 minutes. Reversal with flumazenil is recommended in all birds, which are intended to be discharged from the hospital soon after, in order to avoid injuries from falling of perches and to avoid negative client perception, of having a behavioral abnormal acting bird. Therefore it is important to ensure that complete reversal has occurred, prior to discharge from the hospital, by assessing complete recovery of the sensory and motor function (e.g., approaching the bird, rotating finger/perch).

Syringes used for intranasal drug administration

Syringes (0.5 ml insulin syringes or 1 ml tuberculin syringes) are recommended for use in most birds. These syringes allow for an accurate delivery of sedative drugs, even in small birds. The needles should be removed prior to intranasal delivery in order to avoid injury. The dead space in the hub of tuberculin syringes should be considered, then used for small volumes, and therefore insulin syringes without a hub are generally preferred.

Step-by-step procedure

1. Educate the bird owner about what to expect in regards to sedation of their bird. Sedation with benzodiazepines and butorphanol at the published dosages is safe in pet birds, but birds will appear sleepy, which owners are not used to. Inform them that in some cases regurgitation might occur, particularly in macaws.

2. Manually restrain the bird using a towel and immobilize the head.

3. Administer the sedative drugs into the nostrils over 5-10 seconds. Split the total dose between both nostrils. If intramuscular administration is performed, administer the sedative drugs by injection in the pectoral muscle using a 25-28 gauge needle.

4. Release the bird from manual restraint and allow 7-10 minutes for reaching maximum effect of sedation. Do not allow the bird to perch high, instead place it on the floor or in its carrier.

5. Manually restrain the sedated bird. Monitor respiration throughout the restraint period, as well as heart rate. Perform necessary clinical procedures (e.g., physical examination, blood collection, radiographs).

6. If reversal is indicated, administer flumazenil intranasally. Split total volume between nostrils. Alternatively flumazenil can be administered by intramuscular or intravenous injection.

7. Release the bird on the floor or in its carrier and monitor recovery. Do not allow perching high, in order to avoid trauma from falling, until the bird has completely recovered. Most birds will recover within 10-15 minutes. Assess recovery before discharging the bird. Administer a second dose of flumazenil is recovery is incomplete.

Useful tips

1. Each bird requires an individual assessment prior to choosing a suitable sedative drug protocol. Birds, which have no previous experiences with manual restraint, may require less sedation compared to birds, which had previous negative experiences, and become readily stressed or fearful in a veterinary clinic environment.

2. Macaws usually require a combination of midazolam and butorphanol in order to achieve a sufficiently deep level of sedation, suitable to perform a variety of clinical procedures.

3. Even though a bird might not appear sufficiently deep sedated while under manual restraint for a clinical procedure, is might become more sedated, once stimulation is discontinued, making it unsuitable to be discharged from the hospital. Therefore every bird, which is intended to be discharged shortly after completion of the procedures performed under sedation, should receive flumazenil.

4. Cockatoos often do not recover completely after the initial dose of flumazenil. A second dose of flumazenil is often necessary.

Conclusions

Sedation of pet birds offers a practical alternative to manual restraint of conscious birds and to general anesthesia for a large variety of clinical procedures. Sedation decreases the risks of complications by attenuating the side effects of manual restraints and avoids the potentially fatal side effects of general anesthesia (e.g., cardiovascular and respiratory depression, aspiration of crop content). The intranasal route of administration is an alternative non-invasive technique, which has been proven highly effective in birds and usually increases the client compliance agreeing on sedation for their bird. Intranasal or intramuscular sedation, using midazolam, with or without butorphanol, is a simple, safe, effective, reversible, and readily available option for sedation of avian patients in daily clinical practice.

References

1. Greenacre CB, Lusby AL. Physiologic responses of Amazon parrots (*Amazona* species) to manual restraint. *J Avian Med Surg.* 2004;18:19-22.

2. Mans C, Guzman DSM, Lahner LL, et al. Sedation and physiologic response to manual restraint after intranasal administration of midazolam in Hispaniolan Amazon parrots (*Amazona ventralis*). *J Avian Med Surg.* 2012;26:130-139.

3. Vesal N, Eskandari MH. Sedative effects of midazolam and xylazine with or without ketamine and detomidine alone following intranasal administration in ring-necked parakeets. *J Am Vet Med Assoc.* 2006;228:383-388.

4. Sadegh AB. Comparison of intranasal administration of xylazine, diazepam, and midazolam in budgerigars (*Melopsittacus undulatus*): clinical evaluation. *J Zoo Wildl Med.* 2013;44:241-244.

5. Vesal N, Zare P. Clinical evaluation of intranasal benzodiazepines, alpha-agonists and their antagonists in canaries. *Vet Anaesth Analg.* 2006;33:143-148.

6. Gilbert DB, Patterson TA, Rose SP. Midazolam induces amnesia in a simple, one-trial, maze-learning task in young chicks. *Pharmacol Biochem Behav.* 1989;34:439-442.

7. Farkas L, Crowe SF. The role of the benzodiazepine-GABA system in the memory processes of the day-old chick. *Pharmacol Biochem Behav.* 2000;65:223-231.

8. Mans C, Braun J, Feeny M. Intranasal sediation with midazolam or midazolam-butorphanol in psittacine birds: a retrospective evaluation of 114 cases (2010-2012). *Proc 1st Int Conf Avian, Herpetol Exot Mamm Med.* 2013;292.

9. Prather JF. Rapid and reliable sedation induced by diazepam and antagonized by flumazenil in zebra finches (*Taeniopygia guttata*). *J Avian Med Surg* 2012;26:76-84.

Introduction to Avian Endoscopy for the Practitioner

Stephen J. Divers, BVetMed, DZooMed, Dipl ACZM, Dipl ECZM (Herpetology), Dipl ECZM (Zoo Health Management), FRCVS

Session #183

Affiliation: From the Department of Small Animal Medicine and Surgery (Zoological Medicine), College of Veterinary Medicine, University of Georgia, 2200 College Station Road, Athens, GA 30602, USA.

Abstract: Probably one of the most expedient ways to make a definitive diagnosis is to collect a biopsy for histopathology and microbiology. Endoscopy provides a minimally invasive approach to achieve just that, and with greater accuracy than imaging-guided aspirates, faster than paired serologic titers, and less invasive than coeliotomy. Over the past few decades, endoscopy has become such an established part of avian medicine, and indeed exotic pet medicine in general. An avian/exotic service/practice without endoscopy capabilities runs the risk of losing credibility in today's veterinary profession. There are many publications in journals and books that advocate for the safety and improved diagnostics of endoscopy, so why are the numbers of exotic veterinarians using endoscopy still comparatively low? My answer to this dilemma is this masterclass! The plan is to cement the reasons why endoscopy is invaluable in practice and why you absolutely have to have it. My ultimate goal is to get you sufficiently interested at the end of this 1-hour session, that you will want to take that leap and attend a dedicated endoscopy training course.

Reasons Why Endoscopy is so Valuable

Much of what we do as veterinarians is governed by accurate diagnosis. It informs the client of the likely prognosis, which after all, is often their primary concern. An accurate diagnosis also tends to dictate appropriate therapeutic approaches, and removes the flawed guesswork associated with antibiotic choice, or whether surgery or medical treatment may be more appropriate.

It is still concerning that many avian diagnoses are made merely to the organ level with a presumed, generic cause (e.g., parrots with pneumonia, airsacculitis, kidney disease, liver disease) when what is often needed is a more definitive diagnosis to a histologic and etiologic level. A definitive diagnosis relies upon the demonstration of a patient's pathologic response, and identification of the cause. Essentially, the pathologic response can be demonstrated through histopathology, cytology, or paired rising serologic titers. There are relatively few serologic tests available for most exotic pets, and those that are available typically require several weeks between sample collection (the patient may well be better or dead by the time results are obtained). Consequently, they are typically used for retrospective confirmation. Cytology is most rapid, but suffers from a lack of tissue architecture, and therefore histopathology is often considered the gold standard. Demonstration of the causative agent relies on microbiologic culture or PCR for bacteria, fungi, viruses (and some parasites), sensitive assays to identify toxins, and a variety of other techniques for parasite identification.

It is well appreciated that the collection of samples from lesions/tissues for histopathology, microbiology, parasitology, and/or toxicology remains the best hope of reaching a specific diagnosis. Samples can be easily collected postmortem; however, diagnosis in the live animal requires an ante-mortem biopsy.

There are a variety of ways that ante-mortem biopsies can be collected, including surgical excision, image-guided (e.g., computed tomography [CT], ultrasonography), or endoscopic. Surgical access to a lesion or diseased tissue, unless part of the integument, typically requires an invasive approach that may involve extensive laparotomy/coeliotomy. CT and ultrasound-guided biopsies are certainly possible, and techniques have been reported in the exotic animal literature; however, iatrogenic trauma is certainly more likely than when compared to direct endoscopic visualization. There also multiple examples in the domestic and human literature to indicate the superior diagnostic capability of biopsy over ultrasound-guided aspirate cytology.[1,2] There also many examples of how endoscopy, being less invasive, is less traumatic and less painful than traditional surgical approaches. In human medicine, faster recoveries, reduced hospital stays, and decreased pain scores have been attributed to endoscopic procedures.[3]

The endoscopic approach is typically non- to minimally invasive. Endoscopic evaluation of the gastrointestinal (via the mouth, anus or cloaca) and respiratory tracts (via the glottis), require no surgical incision, and yet the endoscopist can access deep, internal structures. Coelioscopy and laparoscopy enable internal visceral evaluation and biopsy through a much smaller incision than would be required for traditional laparotomy and biopsy. The benefits of endoscopy and endoscopic biopsy have been well demonstrated in human and domestic animal medicine; however, an increasing number of studies and reports in the avian and exotic animal literature have also demonstrated the safety and effectiveness of these techniques in birds, reptiles, mammals and fish.[4] It should be appreciated that there are situations where endoscopy is more difficult or even contraindicated. For example, cases of ascites should not be approached from a lateral approach for fear of creating a connection between the hepatoperitoneal cavity and the airsac system (a ventral midline approach would facilitate that all important liver evaluation and biopsy). Severe obesity can hinder endoscopy, and where the air sac system has been severely compressed (e.g., massive reproductive disease or retained egg), there may be insufficient air sac space to insert and weld an endoscope. Pre-endoscopy radiography should always be performed prior to diagnostic coelioscopy.

Once competency has been achieved, endoscopic techniques tend to percolate into other areas of veterinary practice. For example, at the University of Georgia we are now recommending laparoscopic ovariectomies for exotic mammals and offering gender identification of juvenile turtles.

Instrumentation

Given the variation in size, species-specific anatomy, and variety of procedures that may be performed, a selection of different endoscopes and instruments may be required; however, when dealing with most companion birds, the 2.7-mm system offers flexibility and diagnostic power for an affordable investment (Table 1).[5]

Table 1. Endoscopic instrumentation for companion birds.

Telescopes and sheath	Light source, camera, and photodocumentation	Endoscopic instruments
2.7-mm x 18 cm telescope, 30° oblique	Endovideo camera and monitor	1.7-mm biopsy forceps
4.8-mm operating sheath	Xenon light source and light guide cable	1.7-mm single-action scissors
	Digital capture device (e.g., AIDA-VET)	1.7-mm remote injection needle
		1.7-mm grasping/retrieval forceps

The 2.7-mm system is most widespread in practice, offers greatest versatility, and can be expanded as individual practice caseload dictates.[6,7] This system offers several advantages including single-entry procedures, ports for gas or fluid infusion, and an operating channel for the introduction of 1.7-mm (5 Fr) instruments (Fig 1). In addition, the 1.9-mm telescope with integrated sheath, and the 1-mm semi-rigid miniscope are extremely useful for smaller species.

The telescope is connected via a fiber-optic light guide cable to a light source. While halogen light sources are far cheaper and very effective for small animals < 2 kg, xenon light sources provide better quality light and an intensity that can illuminate the body cavities of larger animals (Fig 2).

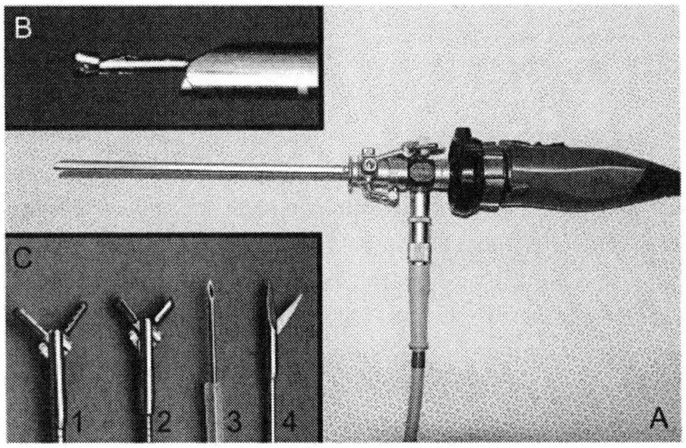

Figure 1. 2.7-mm telescope system. (A) 2.7-mm telescope housed within a 4.8-mm operating sheath, connected to a light cable and an endovideo camera. (B) 1.7-mm biopsy forceps inserted down the instrument channel and emerging directly in front of the telescope. (C) A variety of 1.7-mm instruments can be used through the operating channel, including retrieval forceps (1), biopsy forceps (2), remote injection/aspiration needle (3), and single-action scissors (4).

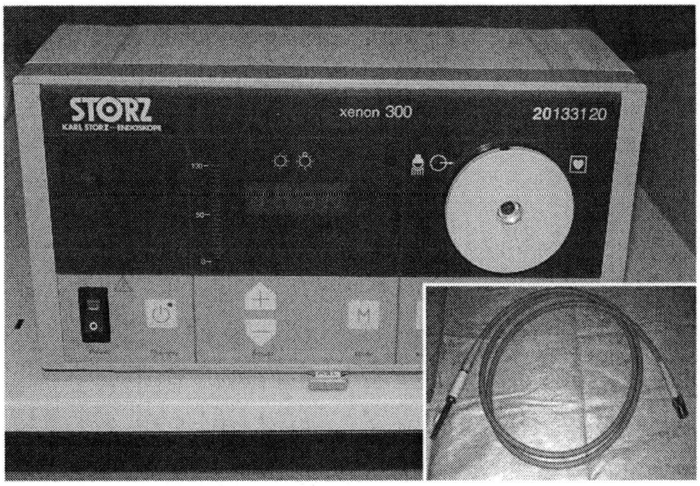

Figure 2. Xenon light source (300 watt) and fiber-optic light cable (insert).

An endovideo camera connected to the eye-piece of the telescope, although once considered a luxury addition, is an integral part of the endoscopy system and greatly facilitates the surgeon's performance. Cameras, available in both European PAL and American NTSC formats, can vary dramatically in cost from budget single-chip cameras to 3-chip digital high definition models; however, the fact is that any camera, however cheap, will greatly improve performance compared to using the eye-piece, and will facilitate photo-documentation. The Tele Pack system combines a hybrid xenon light source, endovideo camera, LCD monitor, and a PCMCIA card slot for digital image capture and storage into a portable unit that is well suited for veterinarians working in the field. Operating room set-up is important and the monitor should be positioned directly in front of the endoscopist, with instruments within easy reach.

In some situations, the use of saline infusion can be especially helpful, especially when dealing with a hollow viscus (e.g., gastro-intestinal tract, bladder, cloaca, etc), and is also preferred for aquatic animal coelioscopy because of the potential for effects of residual insufflation gas on post-operative buoyancy.

The 1.7-mm (5 Fr) grasping forceps are useful for manipulating tissues, debridement and retrieving foreign objects or parasites. The fine aspiration/injection needle can be used for the aspiration of fluid from cystic structures where biopsy may be contraindicated due to post-sampling leakage. The needle can also be used for irrigation, and drug administration. The flexible biopsy forceps are used to harvest tissue samples for histopathology and microbiology. The small sample size usually permits the collection of several biopsies for multiple laboratory tests, and serial biopsies to monitor disease progression over time to assess response to treatment. To take a tissue sample, the biopsy forceps are inserted down the operating channel and into the field of view (Fig 1B). It is much easier to advance and manipulate the sheath-telescope-instrument as a single device than to try and keep the sheath-telescope still and independently move the biopsy forceps back and forth. With the biopsy forceps held open, the sheath-telescope-instrument is advanced to the tissue of interest and when tissue enters the biopsy cup, the handle is released. These instruments are delicate and the biopsy handle is only required to open the biopsy jaws. The handle's spring mechanism is usually sufficient to take a soft tissue biopsy without additional manual pressure, as long as the instrument is sharp. Clamping down on the handle will damage the forceps and increase biopsy crush artifact. Some organs may be protected by a more fibrous membrane. The fixed blade of the scissors is inserted at a shallow angle through the membrane, and the sheath-telescope-scissors are advanced as a single unit, cutting the membrane as they proceed. The scissors can then be replaced by the biopsy forceps to take a sample through the capsular incision.

Biopsies are delicate and measure approximately 1.3 mm^3 in size when collected using a 1.7-mm instrument. Handling and other histologic artifacts are reduced by gently dislodging the tissue from the biopsy cups into a small volume of sterile saline, before then decanting into a biopsy cassette, and submitting in 10% neutral buffered formalin. Picking the biopsy out of the instrument using a needle will cause severe damage, and even moistened cotton-tipped applicators have been shown to cause tissue alteration. Biopsies for microbiology are best submitted in sterile saline for immediate processing. Alternatively, if mailing to a laboratory, they should be submitted in appropriate transport media. For the submission of samples for toxicology or parasitology, it is wise to consult with the laboratory prior to sample collection.

A definitive diagnosis is important to maximize treatment success, and many exotic animal cases are unsuccessfully managed simply because of a failure to identify the specific problem. Definitive diagnoses rely upon the demonstration of a host pathologic response (by histopathology, less reliably cytology, or using paired rising antibody titers), and, if infectious/toxicologic, demonstration of the causative agent (by microbiology, parasitology, or toxicology). There are few reliable serologic tests available for most birds, and those that are available require a minimum of 6 weeks for ectotherms to demonstrate the necessary two-fold increase in titers. It is therefore clear that tissue samples offer the most expedient means to a diagnosis of many diseases, and endoscopy offers the least invasive ante-mortem method to collect such material. Consequently, clinicians have discovered that diagnostic endoscopy offers an unparalleled ability to facilitate ante-mortem diagnosis, and maximize treatment success in many cases.

Diagnostic Procedures

Typical considerations for anesthesia and surgery are required and include fasting in accordance with patient size and diet, pre-emptive analgesia and sedation (e.g., butorphanol and midazolam for most companion psittacines), thermal support, vascular access, endotracheal intubation, inhalant maintenance, and dedicated monitoring. The general handling techniques used in avian endoscopy are similar to those employed for domesticated animals; however, the fact that most companion birds weigh < 1kg requires careful control with the base of the telescope, eyepiece, and camera supported using the superior hand and the terminal end held by thumb and forefinger of the inferior hand.[8]

Tracheoscopy

The lower respiratory tract can be evaluated using the left and right approaches to the airsacs and lungs (see coelioscopy); however, to complete the respiratory examination an oral approach to the choana, trachea and syrinx, and an external approach to the nares are required. Parrots that are severely dyspneic, or suddenly lose their voice and present in acute respiratory distress must first be stabilized. Oxygen and, when necessary, an airsac tube to provide an alternative airway should be considered. Gas anesthesia can be delivered by an airsac tube leaving the mouth and trachea clear for endoscopy, biopsy and debridement. In larger birds the 3.5-mm protection sheath should be used and will enable tracheoscopy of birds over 400-500 g (e.g., Amazons, African grays, macaws and cockatoos). In smaller birds, a 1.0-mm semi-rigid endoscope or 1.9-mm telescope is required. The 1.9-mm and 2.7-mm telescopes can be used without any sheath; however, the advantages of reduced diameter should be weighed against increased risks of telescope damage. The bird is positioned in dorsal recumbency with the head and neck extended, and the telescope can be inserted through the glottis and advanced down the trachea. A surgical plane of anesthesia is required to prevent coughing, but irritation and mucosal damage can be reduced by raising the leading edge of the 30° telescope above the mucosal surface while advancing down the trachea. The complete tracheal rings, syrinx and sometimes even the proximal primary bronchi can be examined. Even where tracheal diameter prevents the use of a sheathed telescope, retrieval and biopsy forceps can be inserted alongside the telescope for retrieval of foreign bodies, debridement and sample collection.

Ingluvioscopy and gastroscopy

Examination of the oral cavity, esophagus, crop, proventriculus, and ventriculus are possible in most birds under 400-500 g using the 2.7-mm telescope and 4.8-mm operating sheath, via an oral approach. In larger birds, the ventriculus can only be reached using flexible or by the introduction of a rigid telescope through a temporary ingluviotomy. Gas insufflation can be used to dilate and examine the esophagus, crop and proventriculus for foreign bodies; however, warm (39-40°C, 102-104°F) sterile saline irrigation provides better visualization, superior mucosal detail, and helps dilate the tract as the sheath-telescope is advanced. All birds should be intubated and placed at a 30-45° incline to reduce the risks of aspiration. Foreign bodies can be removed using retrieval forceps or wire basket devices.

Cloacoscopy

Cloacoscopy using warm saline irrigation is very rewarding and preferred for the evaluation of cloacal papillomas, coprodeum, urodeum (including the openings to the ureters and, in females, the distal uterus), bursa of Fabricius, and proctodeum. Excessive cloacal fluid administration can result in oral regurgitation, and so intubation remains important. Proliferative lesions can be easily biopsied but care is required not to penetrate the cloacal wall.[9] Cloacal neoplasia can be ablated using radiosurgical or diode laser probes introduced via the instrument channel of the sheath.

Coelioscopy

There are four basic approaches to the coelomic cavities; left, right, ventral, and interclavicular. Physical examination, diagnostic imaging (including radiography and ultrasonography), and clinical pathology should be conducted to identify the most appropriate approach. For example, the spleen is best visualized via a left approach, the psittacine pancreas from the right, and both liver lobes can only be seen from a ventral approach to the left and right ventral hepatoperitoneal cavities. Most approaches to the avian coelom involve entry into and examination from within the airsacs. Therefore, it is important to keep sheath ports closed to avoid anesthetic compromise. In addition, left and right approaches should be avoided in birds with ascites because fluid leakage into the airsac system is almost unavoidable; however, in such cases, a ventral midline approach is practical.

The most commonly employed procedure involves a left approach into the airsac system because male and female reproductive organs can always be seen (only a few species have bilateral ovaries). The bird is positioned in right lateral recumbency with wings secured dorsad over the bird's back using self-adhesive tape. The left pelvic limb is pulled craniad and secured to the neck, again using self-adhesive bandage, to expose the left flank. The entry site is located immediately behind the last rib, and just ventral to the flexor cruris medialis muscle as it courses from caudal stifle to ischium. Very few feathers, if any, need to be plucked prior to aseptic preparation of the area. Following a 2- to 4-mm skin incision, straight hemostats, directed in a slight craniodorsal direction, are used to bluntly dissect between the thin subcutaneous tissues and enter the left caudal thoracic airsac. The hemostats are replaced by the sheathed telescope, and correct position within the caudal thoracic airsac is confirmed by the identification of lung (straight ahead), cranial thoracic airsac (left), abdominal airsac (right), caudal edge of liver and proventriculus (ventral), and ribs and intercostal muscles (dorsal).

Exploration of adjacent airsacs is accomplished by pressing the tip of the telescope against the airsac membrane and advancing the telescope in a sweeping motion until the airsac membranes are breached. Normal membranes are transparent and tissues in the adjacent airsac can be visualized and avoided. Great care is required when breaking through thickened, opaque airsacs because vision is impaired and visceral trauma can occur if the telescope is blindly advanced. Lung, liver, heart and associated great vessels can be examined from the cranial thoracic airsac, while urogenital, intestinal, splenic, adrenal, and associated vasculature can be visualized from the abdominal airsac.

There is no need to repair the small holes punctured in the airsac membranes as they generally heal within 5-10 days. Post-operative subcutaneous emphysema may be seen in some birds when only skin closure is performed. Therefore, either a single absorbable (e.g., poliglecaprone 25) suture that incorporates both muscle and skin, or a two-layer closure is recommended.

Biopsy technique

One of the greatest benefits of endoscopy is that when an abnormal structure or pathologic lesion is observed, biopsies can be taken under direct visual control. Biopsies can be harvested from the kidneys, gonads, liver, spleen, pancreas, lung, fat, airsac, coelomic musculature, and, in general, any abnormal soft tissue structure. It is important to examine as much of the target structure as possible to determine whether pathology is focal, multifocal or diffuse. In cases of diffuse renal or hepatic disease (e.g., tubulonephrosis, nephrocalcinosis, hepatic lipidosis, hepatitis), two or three biopsies taken from the most convenient sites are generally diagnostic.

Ultrasound-guided and blind-percutaneous biopsy techniques may be equally effective in diagnosing diffuse disease; however, poorer visualization of closely associated structures makes iatrogenic trauma more likely. Most diagnostic failures occur because of poor tissue selection for biopsy, and this is especially true when dealing with focal (e.g., abscess, neoplasm, cyst) and multifocal diseases (e.g., pyogranulomata, mycobacteriosis). In these cases, direct endoscopic visualization offers the best chance of sampling the most appropriate area(s). In cases of focal or multifocal disease, single or multiple discrete lesions are visible and biopsies should ideally be harvested from the edge of the lesion taking normal and abnormal tissue in the same biopsy sample for both microbiology and histology. Alternatively, and technically easier, small biopsies can be collected from the abnormal and normal areas and submitted together for comparison. Focal disease deep within an organ showing no surface lesions, although rare, may be missed on endoscopic examination. It is important to correlate biopsy histopathology and microbiology with clinicopathologic data when dealing with organ disease. It is often surprising how biopsy results provide a definitive diagnosis even in the face of unremarkable clinicopathologic data.

When confronted by a potentially cystic lesion or abscess, it is safer to first attempt drainage using fine needle aspiration, rather than risk leakage into the airsacs or coelom. When attempting to take a liver or lung biopsy, it is preferable to first incise the airsac and serosal membranes using endoscopic scissors. This provides better access to the tissue parenchyma and yields biopsies of superior histologic quality with minimal artifacts. Biopsies of the spleen, kidney, testis, adrenal and most pathologic lesions can usually be taken without the use of scissors. Post-sampling hemorrhage tends to be minor and inconsequential thanks to the avian extrinsic coagulation pathway, and in particular, tissue-associated thromboplastin.[9,10]

Complications

The major complications encountered are typically associated with anesthesia, and the advanced disease state of many birds at the time of presentation. The importance of stabilization, and a thorough pre-operative evaluation cannot be overemphasized. Endotracheal intubation, ventilation, intravenous or intraosseous catheterization with perioperative fluid support, and warm air/water blankets are important. Minor hemorrhage following tissue biopsy is common but generally insignificant. Most endoscopy issues are related to operator error until experience and ability have been gained. In general, the ability to examine birds internally and collect tissue samples greatly aids diagnosis and improves treatment success.

Post-operative care

Birds should be closely supervised on recovery as anesthetic compromise can ensue following extubation and cessation of cardiorespiratory support. Fluid therapy and nutritional support should continue, with psittacines and passerines eating within an hour of recovery. Meloxicam is used routinely post-operatively, although opiates and local anesthetics could prove useful as part of a balanced approach to analgesia. Typically birds return to normal function quickly following single entry diagnostic procedures compared to more traditional, invasive surgery. Sutures are removed at 7-10 days if still present.

Outcome

Return to normal behaviors and weight gain are often the most useful indicators for improvement. Serial clinicopathology and imaging can also be useful if abnormalities were detected preoperatively. Serial endoscopic evaluations can be used to monitor patients, especially as some diseases (e.g., hepatic lipidosis, glomerulonephrosis) may have a protracted course.

How to Add Endoscopy to My Avian Practice or Service

While the clinical virtues of diagnostic and surgical endoscopy in exotic animal practice are obvious, appropriate management including a fiscally responsible fee structure appears to be more challenging. The costs of a basic rigid endoscopy system for exotic animal practice can range between $10,000 and $20,000, and is not an insignificant capital investment, and therefore a suitable fee structure is required to help recoup costs associated with purchase (or lease), use, repair, replacement, technician time, and practice facilities. From discussions with various colleagues, it became obvious that there was widespread variation in endoscopy practice management. In addition, concerns regarding fee structures were frequently raised by attendees at endoscopy courses. Consequently, I designed and circulated a short survey in the summer of 2014 in an attempt to obtain basic information on endoscopy practice management in Europe and the United States. This 2014 internet-based survey was completed by 35 veterinarians experienced with exotic animal endoscopy, of which 21 were from the United States or Canada, 10 from Europe, three from Australia or New Zealand, and one from Africa. The general demographics of their respective practices varied considerably; however, overall the mean exotic pet caseloads were avian 31%, reptile/amphibian 21%, small mammals 23%, fish 2%, with the remainder being domesticated animals. In the following sections, I have tried to indicate the general themes and commonalities, and have provided numerical mean values where appropriate.

Step 1: Case audit

It can be particularly rewarding and surprising to go through your exotic caseloads and identify the types of cases that you are seeing and how endoscopy could be utilized for such cases. This is important for several reasons:

a) It highlights what you are actually seeing rather than what you think you are seeing.

b) It will indicate equipment priorities.

c) It will indicate which specific practice cultures and verbal deliveries must change to integrate endoscopy, including specific marketing ploys.

Step 2: Select equipment

It can be disheartening when looking at equipment displays or lists knowing that you cannot have everything. But remember that endoscopy systems are modular, and you can buy a basic system today and expand that system as your caseload increases and evolves. You will need a camera, monitor, light source, light guide cable, and CO_2 insufflator if you plan on doing any laparoscopy/coelioscopy in non-aves. Also don't forget that endoscopy is not species based, and should be integrated for ALL species in a practice, birds, herps, small mammals, dogs and cats. In general, the 2.7-mm system has greatest versatility in exotic companion animal practice (there are even multiple uses in dogs and cats); however, if the majority of your patients are small birds then the 1.9-mm system might be a better choice to start. The basic 2.7-mm rigid endoscopy system can be purchased for around $15000 (considerably less on the second hand market), or leased for approximately $300 per month. Most survey respondents were experienced exotic animal endoscopists, and had obviously expanded their endoscopy facilitates from this basic starting point. Nevertheless, their continued preference for a 2.7-mm system was obvious with the majority (86%) owning this system. In addition, standard definition cameras and monitors were twice as common (66%) compared to high definition systems (34%), and xenon light sources were favored (60%) over cheaper halogen sources (34%). These differences probably represent a delay between new technologic developments and the need to replace older items. I expect high definition and xenon to continue to gain in popularity in the future. The majority of veterinarians prefer to keep their equipment on a mobile endoscopy cart (or endoscopy tower).

Step 3: Training

Training is essential because it prevents damage to patient and equipment, improves ability and efficiency, and most important of all, it breeds confidence that is infectious and obvious to clients. So training is essential and fortunately, there are a variety of training opportunities within United States and Europe, and short 4-hour workshops are frequently available at veterinary conferences (e.g., ARAV, AAV, AEMV, AAZV, ICARE). These are especially useful as a first step to try and see if this is something that you are comfortable with; some of us are more like physicians, and others are more like surgeons. If your initial experience was positive then a 2-day intensive course would be the next step. Such weekend courses are held on reptile/avian and small mammal endoscopy at the University of Georgia (http://vet.uga.edu/ce/calendar). After 8 hours of hands-on endoscopy time, the goal is to provide sufficient confidence to scope a client owned animal the next day!

Step 4: Charges and fees

Equipment charges and ancillary fees: Back to the survey, there was considerable variation in the charging mechanism for the set up and use of endoscopy equipment. Only half the veterinarians surveyed specifically charged for use of an endoscopy or operating room ($150), sterilization, use, and cleaning of endoscopy equipment ($122), or standard surgical pack including traditional instruments, drape, cap, mask, and gloves ($71). Only 25% charged a specific fee for technician or nurse time ($31). No doubt, many veterinarians that do not charge the above fees separately, but probably incorporate them into their endoscopy procedural fees; however, there was almost universal consistency in applying separate charges for anesthesia ($184), histopathology ($210), and microbiology ($154).

Endoscopy procedural fees: There was also variation in how veterinarians actually charged for their endoscopy procedures. Approximately one third of veterinarians used a tiered fee structure (e.g., level 1-6, Table 2), while another third used an individual fee for every specific procedure (e.g., reptile coelioscopy, avian tracheoscopy, small mammal otoscopy).

Table 2. The tiered endoscopy fee structure used at the University of Georgia in May 2016 by all services including zoological medicine.

Endoscopy fee level	Fee ($)[a]	Examples of exotic animal procedures
1	94.05	Basic stomatoscopy, tracheoscopy, otoscopy, rhinoscopy
2	209.67	Complicated stomatoscopy, tracheoscopy, otoscopy, rhinoscopy. Basic gastroscopy, cloacoscopy
3	305.79	Complicated gastroscopy, cloacoscopy Basic coelioscopy with biopsy
4	402.38	More complicated coeliotomy, bilateral entries, basic endosurgery (e.g., granuloma removal)
5	498.72	More complicated endosurgery, multiple ports, surgical assistant
6	58.97	Avian sex identification fee per animal belonging to the same owner (in addition to a $500 set-up fee)

[a]These fees represent only the procedure; anesthesia, equipment, and laboratory tests are additional.

The advantage of a tiered fee structure is that it provides flexibility to increase or decrease fees depending upon difficulty, is simple and easily utilized by staff and doctors, and avoids numerous fee codes and individual descriptors. A major advantage of using detailed descriptors is that electronic medical record searches can be more targeted; however, this is at the expense of inputting considerably more data into a computerized accounting system.

Surprisingly, the final third of the survey population admitted to using no pre-determined fee structure, and calculated the fee to be charged on a case-by-case basis. While this provides the greatest flexibility, general business management dogma suggests that it may result in under charging, especially by those not directly invested in the fiscal wellbeing of the practice.

Table 3 lists a number of exotic animal endoscopy procedures and the mean fees charged from the survey responses. It must be appreciated that, just like any fee structure, there was considerable variation between low and high levels (approximately 50-150% of stated means) that was probably related to geographic location and associated costs of living.

Table 3. Mean fees charged in 2014 for specific avian endoscopy procedures performed by 35 experienced clinicians.

Procedure	Fee ($)
Oculoscopy	27
Tracheoscopy	198
Tracheoscopy with biopsy or debridement	280
Esophagus/proventriculus/ventriculus	245
Esophagus/proventriculus/ventriculus with biopsy or debridement or foreign body removal	400
Cloacoscopy	173
Cloacoscopy with biopsy or debridement	294
Coelioscopy for reproductive evaluation of healthy birds (fee per bird in a group)	194
Single-sided coelioscopy for disease investigation	304
Single-sided coelioscopy with biopsy or debridement	338
Multiple entry endosurgery in a bird (e.g., orchidectomy, salpingohysterectomy, mass removal)	508

Step 5: Marketing

Whenever a practice develops a new service area, appropriate marketing is essential. The ability to capture endoscopy images and video for documentation, makes it relatively simple to show case the procedure to the client or referring veterinarian. It is important that the name and telephone number of the practice accompanies any image or video, as clients will frequently show their friends and family. Past materials can also be used in the examination room to demonstrate a proposed diagnostic plan to a new client. These images should also be incorporated into letters sent to other veterinarians regarding cases that they referred to you, or for referral brochures. Some practices have also discovered the benefits of holding an open day, where the public can take behind the scenes tours, and see demonstrations including use of endoscopy equipment.

When marketing endoscopy to clients, verbal delivery is critical. Take the following as an example;

> "*Mrs Smith your African grey appears to have respiratory disease, most likely fungal but could be bacterial? We could anesthetize for a radiograph, and even perform endoscopy, or we could start treatment today with both antibiotics and antifungals.*"

Compare the above delivery with the following and consider which would be more (a) convincing for client and (b) medically appropriate.

> "*Mrs Smith your African grey appears to have respiratory disease, but there are several possible causes and different treatments. We need to stabilize before briefly anesthetizing for radiographs, and then consider endoscopy and biopsy to determine the best approach to treatment. Precise diagnosis will maximize treatment success.*"

I hope you agree that the second delivery is more convincing, and avoids the inappropriate use of antimicrobials (which if rife within the veterinary profession, is almost out of control in exotic animal practice!); however, it does depend on the veterinarian having confidence in his or her endoscopic abilities. There are various ways to practice endoscopy. One way that I found particularly effective in private practice was to offer a minimally invasive necropsy. If a beloved pet exotic died, clients were sometimes resistant to permit a necropsy for aesthetic reasons. Using the telescope to see inside a recently dead patient provides a no risk learning opportunity, a mini-necropsy examination that a) could yield biopsies for histopathology and cultures, and be another fee generating service. Also any wildlife that comes in and needs to be euthanatized could be yet another practice opportunity. The use of plastic Tupperware containers, green peppers, and other items can be used to help maintain eye-hand-telescope-instrument coordination when there is down time. The trick with all these practice options is to keep the equipment non-sterile. We as veterinarians often have an innate resistance to open sterile pack for a non-fee generating fun and games. So keep it unsterile and you will be much more likely to play with the equipment, and if its needed in a hurry it can be cold sterilized and ready to go in 20 minutes.

References

1. Roth L. Comparison of liver cytology and biopsy diagnoses in dogs and cats: 56 cases. *Vet Clin Pathol.* 2001;30(1):35-38.

2. Falcone RE, Wanamaker SR, Barnes F, Baxter CG, Santanello SA. Laparoscopic vs. open wedge biopsy of the liver. *J Laparoendosc Surg.* 1993;3(4):325-329.

3. Rau B, Hunerbein M. Diagnostic laparoscopy: indications and benefits. *Langenbeck's archives of surgery/ Deutsche Gesellschaft fur Chirurgie.* 2005;390(3):187-196.

4. Mejia-Fava J, Divers SJ, Jimenez DA, et al. Diagnosis and treatment of proventricular nematodiasis in an umbrella cockatoo (*Cacatua alba*). *J Am Vet Med Assoc.* 2013;242(8):1122-1126.

5. Divers SJ. Endoscopy equipment and instrumentation for use in exotic animal veterinary medicine. *Vet Clin North Am Exot Anim Pract.* 2010;13(2):171-185.

6. Taylor M. Endoscopic examination and biopsy techniques. In: Ritchie BW, Harrison GJ, Harrison LR, eds. *Avian Medicine: Principles and Application.* Lake Worth, FL: Wingers Publishing;1994:327-354.

7. Chamness CJ. Equipment for the Avian and Exotic Endoscopist. *Semin Avian Exot Pet Med.* 1999;8: 102-106.

8. Divers SJ. Avian diagnostic endoscopy. *Vet Clin North Am Exot Anim Pract.* 2010;13(2):187-202.

9. Hernandez-Divers SJ, Wilson GH, Lester VK, Hernandez-Divers SM, Latimer KS, Ritchie BW. Evaluation of coelioscopic splenic biopsy and cloacoscopic bursa of Fabricius biopsy techniques in pigeons (*Columba livia*). *J Avian Med Surg.* 2006;20(4):234-241.

10. Hanley CS, Wilson GH, Latimer KS, Frank P, Hernandez-Divers SJ. Interclavicular hemangiosarcoma in a double yellow-headed Amazon (*Amazona ochrocephala oratrix*). *J Avian Med Surg.* 2005;19(2):130-137.

Section 10

AAV Research

Joanne Paul-Murphy, DVM, Dipl ACZM,
Dipl ACAW
Moderator

Pharmacokinetics of a Single Dose of Oral and Subcutaneous Enrofloxacin in Caribbean Flamingos *(Phoenicopterus ruber ruber)*

Melissa R. Nau, DVM,
James W. Carpenter, MS, DVM, Dipl ACZM,
Butch KuKanich, DVM, PhD, Dipl ACVCP,
Matt Warner, MS

Session #170

Affiliation: From the Departments of Clinical Sciences (Nau, Carpenter) and Anatomy and Physiology (KuKanich, Warner), College of Veterinary Medicine, Kansas State University, 1800 Denison Avenue, Manhattan, KS 66506, USA.

Enrofloxacin is a fluoroquinolone antibiotic that is widely used in veterinary medicine due to its bactericidal activity and safety in a broad range of species. Caribbean flamingos, a member of the order Phoenicopteriformes, are popular in zoological collections and suffer from a variety of conditions which can result from or lead to bacterial infection. However, there are currently no pharmacokinetic studies to guide antibiotic use in this species. In this study, two groups of 7 adult, unknown gender, captive Caribbean flamingos received a single dose of 15 mg/kg enrofloxacin, administered either orally or subcutaneously. Plasma concentrations of enrofloxacin and its metabolite, ciprofloxacin, were measured using liquid chromatography mass spectrometry. Pharmacokinetic parameters were determined using noncompartmental analyses. The pharmacokinetic parameters for both routes of administration were similar, with a mean peak plasma concentration (C_{max}) of 5.25 and 5.77 µg/ml, a mean time to peak plasma concentration (T_{max}) of 1.49 and 1.1 h, a mean area under the curve (AUC) of 49.9 and 47.3 h·µg/ml, and a mean terminal half-life ($T_{1/2}$) of 5.83 and 6.46 h for oral and subcutaneous dosing, respectively. Conversion to ciprofloxacin was minimal, with the AUC of ciprofloxacin representing < 3% of the enrofloxacin AUC for both routes of administration. Based on the results of the present study, a dose of 15 mg/kg enrofloxacin administered either orally or subcutaneously in the Caribbean flamingo every 24 hr is recommended for susceptible bacterial pathogens with an MIC ≤ 0.25 µg/ml.

Intravenous Catheter Use in Avian Species

Daniel Calvo Carrasco, LV Cert. AVP (ZooMed), MRCVS,
Naomi Shimizu, DVM, MRCVS,
Graham Zoller, DVM, IPSAV (Zoological Medicine),
Minh Huynh, DVM, Dipl ECZM (Avian), MRCVS

Session #201

Affiliation: Great Western Exotics, Vets Now Referrals Swindon, Unit 10 County Business Park, Shrivenham Rd, Swindon, SN1 2NR UK (Calvo Carrasco), Clinique Vétérinaire Universitaire, University of Liège, Quartier Vallée 2, Avenue de Cureghem, 3, Sart-Tilman, 4000 Liège (Shimizu) and Centre Hospitalier Vétérinaire Frégis, 43 Avenue Aristide Briand, 94110 Arcueil, France (Zoller, Huynh).

Guidelines for human[1] and small mammals suggest replacing intravenous (IV) catheters every 72 to 96 hours.[2,3] It is believed that IV catheters are not well tolerated in avian species and significant or even fatal hemorrhage may occur if the patient removes or chews it.[4,5] Tolerance and complications have not been evaluated in avian species. The objective of this study was to evaluate the use of IV catheters in different avian species.

Data collected from hospitalized patients of three referral hospitals included species, body weight, indication for IV catheter placement, duration of catheter patency, catheter size and reason for removal (i.e., discharge of the animal, catheter-related complication or patient-related complication). Fluid therapy rates, type of fluid used, and therapeutic agent used also were recorded.

Birds (n = 139) from six different orders with IV catheters placed were included. The main orders were Psittaciformes (52.5%), Accipitriformes (19.4%), Galliformes (12.2%) and Falconiformes (7.2 %). The mean body weight was 1102.24g ($CI_{95\%}$=[882.79-1321.69]).The main site for IV catheter placement was the ulnar or basilica vein (93.02%), followed by the medial metatarsal vein (6.97%). The mean time of IV catheter patency was 42.42hrs ($CI_{95\%}$=[34.24-50.60]). The catheter size was 22 gauge in 14.96%, 24 gauge in 56.69% and 26 gauge in 28.34%. The reasons for removal were discharge (43.18%), patient-related complications (pulled out/ chew it) (12.12%) or catheter-related complications (kink/blocked extravasation) (17.42%). A portion of patients (26.51%) died while the IV catheter was still in place due to unrelated reasons. A linear statistical model evaluated differences among groups.

References

1. Rickard CM, Webster J, Wallis MC, Marsh N, McGrail MR, French V, Foster L, Gallagher P, Gowardman JR, Zhang L, McClymont A, Whitby M. Routine versus clinically indicated replacement of peripheral intravenous catheters: a randomised controlled equivalence trial. *Lancet*. 2012;22:1066-1074.

2. Spencer KR. Intravenous catheters. *Vet Clin North Am Small Anim Pract*. 1982;12(3):533-543.

3. Weil AB. Intravenous indwelling catheters: use and care. Clinician's brief. 2006;53-58. Available at: http://www.cliniciansbrief.com/columns/51/intravenous-indwelling-catheters. Accessed July 6, 2016.

4. Beaufrère H. Parenteral fluid therapy. Medical cursing and cosmetic procedures. In: Samour J, ed. *Avian Medicine*. 3rd ed. St. Louis, MO: Elsevier;2016:209-215.

5. Lichtenberger M, Lennox A. Critical care. In: Speer BL, ed. Current Therapy in Avian Medicine and Surgery. St. Louis, MO: Elsevier;2016:582-588.

Effects of High Inspiratory Gas Flow Rates on Avian Post-Intubation Tracheal Stenosis

Katherine A. Imig,
Peter J. Pascoe, BVSc, Dipl ACVAA, Dipl ECVAA,
Michael K. Keel, DVM, PhD, Dipl ACVP,
Michelle G. Hawkins, VMD, Dipl ABVP (Avian)

Session #070

Affiliation: From Veterinary Student (Imig), Department of Medicine & Epidemiology (Hawkins), Department of Surgical & Radiological Science (Pascoe), Department of Pathology Microbiology & Immunology (Keel), School of Veterinary Medicine, University of California Davis, 1 Garrod Drive, Davis, CA 95616, USA.

Endotracheal tubes are used in avian anesthesia to protect the airway and facilitate intermittent positive pressure ventilation (IPPV). Tracheal stenosis is a reported complication following avian intubation, however the cause(s) is currently unknown. A high inspiratory gas flow rate during IPPV was hypothesized to increase the occurrence of post-intubation tracheal stenosis in birds. Twelve rock doves (*Columba livia*) were anesthetized with isoflurane, orotracheally intubated with uncuffed endotracheal tubes attached to a non-rebreathing ventilation circuit, and ventilated using IPPV with a high inspiratory gas flow rate (0.5-1 second inspiratory time) at 5-15 breaths per minute. Group 1 (n=6) was ventilated for 3 hours maintaining an end tidal carbon dioxide of 35-40 mmHg. Due to the lack of histologic evidence of tracheal stenosis in group 1, a second group (n=6) was ventilated with IPPV for 6 hours using the same parameters but maintaining an end tidal carbon dioxide of 20 mmHg. All birds were observed for clinical signs of respiratory distress for 14 days post-intubation, then humanely euthanized to evaluate for gross and microscopic evidence of tracheal injury. During the monitoring period, clinically abnormal respiratory signs were observed in 3 birds from group 1 and 3 birds from group 2. Necropsy revealed that 1 bird from group 2 had mild histologic evidence of tracheal injury, however there was no gross or histologic evidence of significant tracheal stenosis in either group. Additional factors need to be examined to determine the etiology of avian post-intubation tracheal stenosis.

Tear Production in Red-Tailed Hawks With and Without Ocular Disease

Price Dickson, and
Julia K. Whittington, DVM

Session #160

Affiliation: From Veterinary Clinical Medicine, University of Illinois, 1008 W. Hazelwood Dr., Urbana, IL 61802, USA.

Lubrication of the cornea in birds is achieved through an interaction between the nictitating membrane, its associated gland, and the lacrimal gland. The nictitating membrane lies deep to the eyelids and moves freely in a nasal-to-temporal direction, spreading secretions from the gland over the cornea. The lacrimal gland, situated in a temporal position, also contributes to the tear film.[1] Decreased tear production, in addition to pathology of the nictitating membrane and adnexa, is likely to result in a loss of corneal integrity.

Tear production is routinely measured by the Schirmer Tear Test and the Phenol Red Thread Test (PRTT) with values established for a variety of raptor species.[2,3] Fifteen free-living red-tailed hawks (*Buteo jamaicensis*) were given full ophthalmic examinations, including PRTT measurements. Eyes were evaluated independently. The mean measurement for eyes without detectable pathology was 12.15 mm/min., (Standard Deviation: 12.19 mm). Eyes with damage to the internal or external ocular structures had a mean measurement of 3.80 mm/min, (Standard Deviation: 1.99 mm). This data establishes a reference range for tear production in normal eyes of red-tailed hawks and indicates that ocular pathology has a pronounced impact on tear production in this species.

Various drugs, as well as inhalant anesthetic gases, negatively impact tear production in dogs, cats, and humans.[4] Reduction in tear production may result in vision-threatening issues, such as the development of corneal ulcers. In raptors, the effect of inhalant anesthetic administration on tear production has not been investigated and will be the focus of future data collection.

References

1. King AS, McLelland J. Special sense organs. In: *Birds: Their Structure and Function*, East Sussex, England: Baillière Tindall; 1984:284-314.

2. Barsotti G, Briganti A, Spratte JR, Ceccherelli R, Breghi G. Shirmer tear test type I readings and intraocular pressure values assessed by applanation tonometry (Tonopen XL) in normal eyes of four European species of birds of prey. *J Vet Ophthalol.* 2013;5:365-369.

3. Beckwith-Cohen B, Horowitz I, Bdolah-Abram T, Lublin A, Ofri R. Differences in ocular parameters between diurnal and nocturnal raptors. *Vet Ophthalmol.* 2015;Suppl 1:98-105.

4. Peche N, Köstlin R, Reese S, Pieper K. Postanaesthetic tear production and ocular irritation in cats. *Tierarztl Prax Ausg K Kleintiere Heimtiere.* 2015;43:75-82.

Lead Concentrations in Tissues from Free-Ranging North American Vultures

Jesse A. Fallon, MS, DVM,
Shannon Behmke, MS,
Todd Katzner, PhD

Session #293

Affiliation: From Virginia Tech, Cheatham Hall, RM138, Virginia Tech, 310 West Campus Drive, Blacksburg, VA 24061, USA (Fallon), School of Natural Resources, 322 Percival Hall, PO Box 6125, Morgantown, WV 26506-6125 USA (Behmke), United States Geological Survey, 970 Lusk Street, Boise, ID 83706 USA (Katzner).

Lead poisoning is a threat to wild birds, especially scavenging birds of prey.[1-4] To understand the degree to which wildlife populations are exposed to lead, we quantified lead tissue concentration in black vultures (*Coragyps atratus*; BLVU) and turkey vultures (*Cathartes aura;* TUVU), two species that are useful as environmental sentinels in eastern North America. Every individual sampled ($n = 108$) had bone lead levels suggestive of chronic exposure to anthropogenic lead (BLVU: x = 36.99 ± 55.21 mg Pb/kg tissue (±SD); TUVU: x = 23.02 ± 18.77 mg/kg). We found limited correlation between lead concentrations in soft tissue (liver, kidneys, muscle) and bone. Isotopic ratios suggested multiple potential sources of lead including ammunition, gasoline, coal-fired power plants, and zinc smelting, but not lead-based paint. Black and turkey vultures range across eastern North America, from Quebec to Florida, and individuals may traverse thousands of kilometers annually. The extent to which vultures are exposed suggests that anthropogenic lead likely permeates eastern North American ecosystems. Lead exposure in such widespread and common species and the failure of soft-tissue sampling to diagnose this pattern has implications for understanding modern wildlife and human health concerns.

References

1. Scheuhammer AM, Templeton DM. Use of stable isotope ratios to distinguish sources of lead exposure in wild birds. *Ecotoxicol.* 1998;7:37-42.

2. Church ME, Gwiazda R, Risebrough RW, Sorenson K, Chamberlain CP, Farry S, Heinrich W, Rideout BA, Smith DR. Ammunition is the principal source of lead accumulated by California condors re-introduced to the wild. *Environ Sci Technol.* 2006;40:6143-6150.

3. Pain DJ, Fisher IJ, Thomas VG. A global update of lead poisoning in terrestrial birds from ammunition sources. In: Watson RT, Fuller M, Pokras M, Hunt WG, eds. *Ingestion of Lead from Spent Ammunition: Implications for Wildlife and Humans.* Boise, ID: The Peregrine Fund;2009:99-118.

4. Pokras MA, Kneeland MR. 2009. Understanding lead uptake and effects across species lines: a conservation medicine based approach. In: Watson RT, Fuller M, Pokras M, Hunt WG, eds. *Ingestion of Lead from Spent Ammunition: Implications for Wildlife and Humans.* Boise, ID: The Peregrine Fund;2009:1-16.

Manual Restraint and its Physiologic Effects in Birds

Grayson Doss, DVM, and
Christoph Mans, Dr. med. vet., Dipl ACZM

Session #107

Affiliation: From the University of Wisconsin Veterinary Teaching Hospital, Madison, WI 53706, USA.

Manual restraint is commonplace in veterinary medicine and is routinely employed for physical examinations, diagnostic sample collection, and administration of nutrition and therapeutics in avian species. Despite its frequent employment, manual restraint is often a stressful process, with debilitated birds being at risk of serious complications. The avian stress response is a complex neuroendocrine process involving the sympathetic nervous system and the hypothalamic-pituitary-adrenocortical axis.[1,2] Research in birds shows that restraint triggers immediate physiologic effects through activation of the "fight or flight" response, and chronic stressors may result in long-term deleterious effects. In a clinical setting, monitoring physiologic variables is the most practical approach to evaluating patient stress. Although the short-term physiologic changes seen with restraint have been studied in a handful of bird species, little information exists in birds of prey, particularly owls. To further investigate the effect of restraint in this group of raptorial birds, nine adult barred owls (*Strix varia*) were restrained for 15 minutes. Heart rate, respiratory rate, and cloacal temperature were monitored at regular intervals. Over the restraint period, both heart rate and respiratory rate decreased significantly, while cloacal temperature increased significantly. This is the first documentation of stress hyperthermia in an owl species. Physiologic changes during restraint were comparable to those in the red-tailed hawk (*Buteo jamaicensis*) while distinct from findings in parrots of the genus *Amazona*. Further research is needed to examine the short-term stress response secondary to manual restraint in birds in order to improve patient welfare.

References

1. Siegel HS. Physiological stress in birds. *Bioscience.* 1980;30:529-534.

2. Blas J. Stress in birds. In: Scanes CG, ed. *Sturkie's Avian Physiology*. 6th ed. San Diego, CA: Academic Press, 2015;769-810.

Anti-Ganglioside Antibodies Are Not Associated with the Development of Proventricular Dilatation Disease

Jeann Leal Arujo, DVM, and
Ian Tizard BVMS, PhD, ACVM, DSc

Session #074

Affiliation: From The Schubot Center, Veterinary Pathobiology, Texas A&M University, College Station, TX 77845, USA.

The identification of parrot bornaviruses (PaBV) in psittacine birds with proventricular dilatation disease (PDD) has not been sufficient to explain the pathogenesis of this fatal disease since not all infected birds develop clinical signs. One hypothesis suggests that PaBV could trigger the production of autoantibodies targeting neuronal gangliosides. These are major neuronal antigens, and PDD might therefore resemble Guillain-Barré syndrome (GBS) in its pathogenesis. Experimental inoculation of pure bovine gangliosides and chicken brain-derived ganglioside extracts were used in two different immunization studies. A preliminary study on seven healthy chickens (*Gallus gallus domesticus*) was performed using a group of four chickens inoculated with a brain ganglioside extract in Freund's complete adjuvant (FCA) and a control group comprised by three chickens inoculated only with phosphate-buffered saline (PBS). A second study with five healthy quaker parrots (*Myiopsitta monachus*) was comprised of three groups. Two quaker parrots received purified gangliosides in FCA, two received a crude brain extract in FCA, and one control quaker parrot received FCA alone. Birds were observed for 50 days following inoculation. They were then necropsied and subjected to histologic examination. In the preliminary study, one chicken developed ataxia and weakness. This is consistent with a motor deficit observed in chickens with Guillain-Barré syndrome. None of the quaker parrots had any clinical signs that could resemble PDD or GBS. None of the chickens or quaker parrots presented any gross lesions. The chicken with clinical signs had a perivascular and perineural lymphocytic infiltrate in the proventriculus. Two of the quaker parrots (one from each treatment group) developed mild lymphoplasmacytic encephalitis and myelitis. Guillain-Barré syndrome can be induced experimentally in birds but its lesions and clinical consequences do not resemble PDD. Our results suggest that autoantibodies against gangliosides in birds are not associated with a condition resembling PDD.

Consistency of Behavioral Responses of Grey Parrots Across Different Contexts

N. Thie, MSc,
C.M. Vinke, PhD;
N.J. Schoemaker, DVM, PhD, Dipl ECZM (Small Mammal), Dipl ECZM (Avian), Dipl ABVP (Avian),
Y.R.A. van Zeeland, DVM, MVR, PhD, Dipl ECZM (Avian), Dipl ECZM (Small Mammal), CPBC

Session #262

Affiliation: From the Division of Zoological Medicine, Department of Clinical Sciences of Companion Animals (Thie, Schoemaker, van Zeeland), and the Department of Animals in Science and Society, Faculty of Veterinary Medicine, Utrecht University, Yalelaan 108, 3584 CM Utrecht, the Netherlands (Vinke).

Behavioral consistency has been reported in numerous animal species and for a variety of behavioral traits.[1-7] Despite reports on behavior tests in psittacine birds,[8-11] information regarding their behavioral consistency during such tests is lacking. The current study therefore aimed to investigate behavioral consistency in parrots during behavior tests. For this purpose, 94 privately-owned grey parrots were subjected to an open field test (OFT), novel object test (NOT) and manual restraint test (MRT), which were performed one year apart in a veterinary clinic (unfamiliar environment) and at the parrots' homes (familiar environment). In addition, the caretakers' perceptions of their parrots' responses in similar, day-to-day (non-experimental) settings were collected by means of a survey.

Results revealed low agreement between the behavioral responses across the different contexts, indicating low behavioral consistency. For the OFT, responses in the home environment corresponded best with those in the veterinary clinic, whereas caretakers perceived their parrots to respond more calm and confident compared to the aforementioned situations ($P < 0.05$). In contrast, for the NOT and MRT parrots responded less calm and confident in the clinic environment compared to the other contexts ($P < 0.05$). Findings of this study provide further insight in reactivity of grey parrots during behavior tests and underline the influence of context on the observed responses. Moreover, results suggest a role for the familiarity of the environment and presence of a caregiver as an explanation for the found differences in the parrots' reactivity in the different contexts, although further studies will be needed to confirm these findings.

References

1. Verbeek MEM, Drent PJ, Wiepkema PR. Consistent individual differences in early exploratory behavior of male great tits. *Anim Behav.* 1994;48:1113-1121.

2. Spoolder HAM, Burbidge JA, Lawrence AB, et al. Individual behavioural differences in pigs: intra-and inter-test consistency. *Appl Anim Behav Sci.* 1996;49:185-198.

3. Dingemanse NJ, Both C, Drent PJ, et al. Repeatability and heritability of exploratory behaviour in great tits from the wild. *Anim Behav.* 2002;64:929-938.

4. Svartberg K. A comparison of behaviour in test and in everyday life: Evidence of three consistent boldness-related personality traits in dogs. *Appl Anim Behav Sci.* 2005;9:103-128.

5. Svartberg K, Tapper I, Temrin H, et al. Consistency of personality traits in dogs. *Anim Behav.* 2005;69:283-291.

6. Bell AM, Hankison SJ, Laskowski KL. The repeatability of behaviour: a meta-analysis. *Anim Behav.* 2009;77:771-783.

7. David M, Auclair Y, Cézilly F. Assessing short- and long-term repeatability and stability of personality in captive zebra finches using longitudinal data. *Ethol.* 2012;118:932-942.

8. Meehan CL, Mench JA. Environmental enrichment affects the fear and exploratory responses to novelty of young Amazon parrots. *Appl Anim Behav Sci.* 2002;79:75-88.

9. Luescher AU, Sheehan K. Rearing environment and behavioral development of psittacine birds. In: *Current Issues Res Vet Behav Med.* 2005;35-41.

10. Fox RA, Millam JR. The use of ratings and direct behavioural observation to measure temperament traits in cockatiels (*Nymphicus hollandicus*). *Ethol.* 2010;116:59-75.

11. Van Zeeland YRA, van der Aa MMJA, Vinke CM, et al. Behavioural testing to determine differences between coping styles in grey parrots (*Psittacus erithacus erithacus*) with and without feather damaging behaviour. *Appl Anim Behav Sci.* 2013;148:218-231.

Welfare Effects of Feeding Naturally Sized Biscuit Diet to Orange-Winged Amazon Parrots

Tamsen Polley, MS,
Kate Gustavsen, PhD, DVM,
Johnathan Perez, BA,
Kirk Klasing, PhD,
Joanne Paul-Murphy, DVM, Dipl ACZM Dipl ACAW

Session #171

Affiliation: From the Department of Medicine and Epidemiology at the School of Veterinary Medicine (Polley, Gustavsen, Paul-Murphy), Department of Neurobiology, Physiology, and Behavior (Perez), and Department of Animal Science (Klasing) at the University of California - Davis, One Shields Avenue, Davis, CA 95616, USA.

Welfare is a conceptual summation of physical and psychological health. It is difficult to define, and more difficult to quantify. Podomandibulation, manipulating food items with a foot and beak, is a critical foraging behavior in large psittacine birds. Wild parrots spend 37-60% of waking time foraging, versus 4-10% in captive parrots. Foraging is highly motivated, inherently rewarding, and positively impacts psittacine welfare. Formulated psittacine diets are significantly smaller than wild-type diets. We hypothesized that a nutritionally complete, "oversized" biscuit diet that promotes podomandibulation would lead to improved welfare in orange-winged Amazon parrots (*Amazona amazonica*) over a 5-month period compared to a control group receiving a standard size pelleted diet. Welfare was measured by behavioral observation, food interaction, hematology, plasma biochemistry, avian haptoglobin, plasma corticosterone, fecal corticosterone metabolites, and plumage evaluation. The oversized biscuit diet group did not significantly change their foraging time budgets, however the birds provided the control pellets decreased the time spent on podomandibulation. Foraging enrichment, measured by mean ± SD food interaction value, was 5-fold higher and podomandibulation increased 1.3-fold, both in the biscuit group compared to baseline. Podomandibulation and stereotypy were negatively correlated in all birds regardless of group. Plasma and fecal corticosterone decreased in all birds throughout the study. Juvenile parrot feather quality increased throughout the study, indicated by increased total feather quality score. The provision of oversized biscuits had a positive impact on welfare based on quantitative improvements in the food interaction value and more engagement in self-enrichment behavior through tossing and foraging for biscuits.

Tibiotarsal Fractures in Pet Birds: A Retrospective Study of 86 Cases

Louden Wright, DVM,
Christoph Mans, Dr. med. vet., Dipl ACZM,
Grayson Doss, DVM, Geoff Olsen, DVM, Dipl ABVP (Avian),
Brian Speer, DVM, Dipl ABVP (Avian), Dipl ECZM,
Gerd Britsch, Dr. Med. Vet.,
Jill Heatley, DVM, Dipl ACZM, Dipl ABVP (Avian),
Jane Christman, DVM,
Michael Jones, DVM, Dipl ABVP (Avian)

Session #108

Affiliation: From the University of Wisconsin Veterinary Teaching Hospital, Madison, WI 53706, USA (Wright, Mans, Doss), The Medical Center for Birds, Oakley, CA 94561, USA (Olsen, Speer), Vogel und Reptilienpraxis Dr Britsch, 76139 Karlsruhe, Germany (Britsche), Texas A&M College of Veterinary Medicine and Biomedical Sciences, College Station, TX 77845, USA (Heatley, Christman), and University of Tennessee College of Veterinary Medicine, Knoxville, TN 37996, USA (Jones).

Tibiotarsal fractures are a common orthopedic injury in small companion birds.[1] Because of the body size and lifestyle of small pet bird species, external coaptation is often used as an alternative to surgical repair. Tape splints are a commonly utilized form of external coaptation in avian medicine but little information regarding their efficacy exists. The goal of this study was to retrospectively evaluate cases of tibiotarsal fractures in small pet birds weighing < 200 grams. Records from 5 institutions were examined. Eighty-six cases met the inclusion criteria. Information obtained included species affected, fracture classification, method of fixation, case progression, complications, and case resolution. Budgerigars and cockatiels were the most common species encountered. The majority of fractures were mid-diaphyseal and closed. Tape splints alone were used in the majority of birds, while an intramedullary pin was used adjunctly in a few cases. Case resolution was classified as either acceptable or unacceptable based on seriousness of side effects. An acceptable outcome was a case in which function was maintained with the potential for some lameness. Unacceptable outcomes were defined by serious side effects including severe lameness, necrosis of the limb necessitating amputation, or death. The majority of cases (> 90 %) had an acceptable outcome. Results from this study demonstrate that the tape splint is an appropriate method of external coaptation for tibiotarsal fractures in small, pet birds weighing less than 200 g.

References

1. Helmer P, Redig PT. Surgical resolution of orthopedic disorders. In: Harrison GJ, Lightfoot TL, eds. *Clinical Avian Medicine*. Vol 2. Palm Beach, FL: Spix Publishing, Inc.;2006:761-774.

Section 11

AAV Clinical

Thomas N. Tully, Jr., DVM, MS,
Dipl ABVP (Avian), Dipl ECZM (Avian)
Moderator

Comparison of Methods for Determining Prevalence of *Macrorhabdus ornithogaster* in a Flock of Captive Budgerigars *(Melopsittacus undulatus)*

Patrick J. Sullivan, DVM,
Cheryl Greenacre, DVM, Dipl ABVP (Avian), Dipl ABVP (Exotic Companion Mammal),
Edward Ramsay, DVM, Dipl ACZM,
Michael Jones, DVM, Dipl ABVP (Avian)

Session #009

Affiliation: From the Department of Small Animal Clinical Sciences, Avian and Zoological Medicine Service, University of Tennessee, Knoxville, TN 37996, USA.

Macrorhabdus ornithogaster, known as avian gastric yeast, is a yeast organism that infects multiple species of birds, both captive and wild, although it appears to disproportionately affect Australian psittacine species. Diagnosis in vivo has historically proven difficult, as shedding can be intermittent, and not all positive animals are clinically ill. Ill animals can show a myriad of symptoms, including regurgitation, loose stool with undigested food material, marked weight loss, and acute death. In this study we compared fecal gram's staining (FGS) and polymerase chain reaction (PCR) of cloacal swabs to evaluate prevalence of active shedding of *M. ornithogaster* in a captive flock of budgerigars (*Melopsittacus undulatus*). Of the 100 birds sampled, 34 were positive for *M. ornithogaster* by FGS and 57 were positive by PCR. All FGS-positive results correlated to a positive PCR, which helped to act as an internal control for this study. Based on the PCR results, the prevalence of *M. ornithogaster* shedding in this flock was 59.6%. One bird that had previously tested negative on both FGS and PCR, and found to be in good body condition at sampling, was found dead 26 days post testing. The patient was emaciated, and necropsy revealed abundant *M. ornithogaster* at the isthmus, as well as marked koilin disruption with multifocal to coalescing lymphoplasmacytic and granulocytic infiltrates and gram-positive bacterial colonization. Cloacal swab PCR is a more sensitive predictor of the presence of *M. ornithogaster* compared to FGS in budgerigars, but negative PCR does not appear to preclude infection.

Unusual Presentations of Chlamydiosis in Psittacine and Other Birds

H. L. Shivaprasad, BVSc, PhD, Dipl ACPV

Session #315

Affiliation: From California Animal Health and Food Safety Laboratory System - Tulare branch, 18830 Road 112, Tulare CA 93274, School of Veterinary Medicine, University of California, Davis, CA 95616, USA.

Avian chlamydiosis is a naturally occurring contagious systemic disease of various species of birds including psittacines caused by *Chlamydia psittaci.*[1] This bacterium was previously named *Chlamydophila psittaci.*[2] *Chlamydia psittaci* belongs in the family Chlamydiaceae within the order Chlamydiales. Chlamydia is a zoonotic agent and has been reported in more than 460 species of birds and in 30 orders, including psittacine birds, pigeons, doves, rheas, raptors, geese, passerine birds (perching birds), poultry including turkeys, chickens, ducks and wild birds.[1,3,5] Avian strains of *C. psittaci* are currently divided in to 13 genotypes designated, A, B, C, D, E, F and E/B, 1V, 6N, Mad 16, R54, YP84, and CPX0308 based on analysis of outer membrane protein A (*omp*A).[4] These strains predominately infect different avian hosts, although cross transmission occurs. In addition to *C. psittaci*, other species of *Chlamydia* that have been known to infect birds include *C. abortus, C. muridarum, C. suis, C. pecorum*, and *C. trachomatis*. Recently, new species of *Chlamydia* have been isolated from birds: *C. gallinacea* from chickens, *C. avium* from pigeons and *C. andidatus ibidis* from ibis.[6,7]

Avian chlamydiosis is most common in psittacine birds that are the predominate hosts for genotype A. Clinical signs due to chlamydiosis in psittacine birds include greenish-yellow diarrhea, respiratory signs, lethargy, ocular discharge, emaciation, dehydration, and occasionally neurologic signs. Lesions frequently observed are airsacculitis, hepatitis, splenitis, pericarditis, and pneumonia.[1,8] Diagnosis of chlamydiosis can be made by serology, or cytology of impression smears of liver, spleen, air sac, or other organs and demonstration of small round organisms in the cytoplasm of macrophages by special stains.[1,9,10] Other tests for diagnosing *Chlamydia* include immunofluorescence (FA), isolation of the organism, polymerase-chain reaction (PCR), histopathology, and immunohistochemistry (IHC).[9] *C. psittaci* can be further typed by multilocus sequence variable number of tandem repeat analysis, multilocus sequence typing and sequencing of *omp* gene.[4]

Cases of chlamydiosis in psittacines and other birds submitted to the California Animal Health and Food Safety Laboratory System, Fresno branch were reviewed. Several unusual cases were identified. Two cases of otitis media, two cases of encephalitis with chlamydia located in the endothelial cells, several cases involving the bursa of Fabricius, and two cases of nephritis in which chlamydia were not detected in the air sac, liver and spleen. Four unusual cases of *Chlamydia* involving the nasal glands in commercial turkeys were included. Data were reviewed for clinical signs, pathology, FA and IHC. In two cases PCR was performed on 23S rRNA gene specific for known Chlamydiaceae and in two cases multilocus sequencing was performed.

Case 1. Otitis Media Associated with *Chlamydia*

Many juvenile and a few adult cockatiels (*Nymphicus hollandicus*) experienced lethargy, loss of weight, respiratory and ocular discharge along with whitish diarrhea. A few birds experienced occasional neurologic-like symptoms. There was increased mortality especially among the juveniles in the aviary. Three juveniles submitted for necropsy revealed moderate to severe hepatosplenomegaly. Two out of three cockatiels had otitis media associated with chlamydia in addition to other lesions.

Case 2. Chlamydiosis in the Endothelial Cells Associated with Meningoencephalitis

Two psittacines, one year-old male blue-headed pionus (*Pionus menstruus*) and an 8 wk-old, female sun conure (*Aratinga solstitialis*), from different aviaries had chlamydial organisms identified by IHC in the endothelial cells of the brain and other organs. Pionus had died suddenly and the sun conure had diarrhea and loss of weight. Pathology included hepatitis, splenitis, pulmonary congestion and edema, nephritis, and meningoencephalomyelitis associated with vasculitis. In addition to IHC, the liver in the Pionus was analyzed by PCR and sequencing. The liver was positive for *C. psittaci* by PCR and sequencing of the organism identified it as *C. psittaci* and not *C. pneumoniae. C. pneumoniae* has been associated with atherosclerosis in people.

Case 3. Chlamydia in the Bursa of Fabricius

Several psittacines; a 4-week-old green wing macaw (*Ara chloropterus*), two juvenile cockatiels (*Nymphicus hollandicus*) and two baby cockatiels had inflammation of the bursa of Fabricius associated with chlamydial organisms. The bursa of Fabricius of the macaw was severely enlarged and had pale foci of necrosis. Clinical history included diarrhea, respiratory signs and ocular discharge. Most of these psittacines also had conjunctivitis, airsacculitis, hepatitis, splenitis, nephritis, enteritis, etc.

Case 4. Chlamydia in the Kidneys But Not in the Liver, Spleen, Air Sac

Two psittacines, an immature female yellow-shouldered Amazon parrot (*Amazona barbadensis*) and a 12- week-old immature female cockatiel (*Nymphicus hollandicus*) were diagnosed with Chlamydiosis. These psittacines were from different aviaries having increased mortality in weaned birds and chicks respectively. Clinical signs included diarrhea and loss of weight. Airsacculitis, nephritis, hepatitis, and splenitis were observed in both birds at necropsy. Air sacs, livers, and spleens were negative for chlamydia by FA and IHC but kidneys were positive by IHC.

Case 5. Chlamydiosis Involving the Nasal Glands in Turkeys

Between 1991 and 2015, 4 cases of unusual presentation of chlamydiosis involving the nasal glands were diagnosed in 8- to 14-week-old commercial turkeys based on FA and immunohistochemistry. The only clinical sign observed were unilateral or bilateral swelling above the eye. The swelling was due to the inflammation of the nasal glands present in the dorsolateral aspect of the extraorbital region of the nasal cavity. Except for slight drop in feed and water consumption the birds did not exhibit respiratory signs or increased morbidity or mortality.

References

1. Vanrompay D. Avian chlamydiosis (psittacosis, ornithosis). In: Swayne D, ed. *Disease of Poultry*, 13th ed. Ames, IA: Wiley-Blackwell Publishers;2013:1055-1073.

2. Sachse K, Bavoil PM, Kaltenboeck B et al. Emendation of the family Chlamydiaceae: proposal of a single genus, *Chlamydia*, to include all currently recognized species. *Syst Appl Microbiol.* 2015;38:99-103.

3. Kaleta EF, Taday EMA. Avian host range of *Chlamydophila* spp. based on isolation, antigen detection and serology. *Avian Pathol.* 2003;32:435-462.

4. Shivaprasad HL, Carnaccini S, Bland M, et al. An unusual outbreak of chlamydiosis in turkeys involving the nasal glands. *Avian Dis.* 2015;59:315-322.

5. Knittler MR, Sachse K. C*hlamydia psittaci*: update on an underestimated zoonotic agent. *Pathog Dis.* 2015;73:1-15

6. Laroucau K, Vorimore F, Aaziz R, et al. Isolation of a new chlamydial agent from infected domestic poultry coincided with cases of atypical pneumonia among slaughterhouse workers in France. *Infect Genet Evol.* 2009;9:1240-1247.

7. Sachse K, Laroucau K, Riege K *et al.* Evidence for the existence of two new members of the family Chlamydiaceae and proposal of *Chlamydia avium* sp. nov. and *Chlamydia gallinacea* sp. nov. *Syst Appl Microbiol.* 2014;37:79-88.

8. Mohan R. Epidemiologic and laboratory observations of *Chlamydia psittaci* infection in pet birds. *J Am Vet Med Assoc.* 1984;185:1372-1374.

9. Graham DL. Histopathologic lesions associated with chlamydiosis in psittacine birds. *J Am Vet Med Assoc.* 1989;195:1571-1573.

10. Suwa T, Touchi A, Hirai K, Itakura C. Pathological studies on chlamydiosis in parakeets (*Psittacula krameri manillensis*). *Avian Pathol.* 1990;19:355-370.

Clinical Use of Melatonin in Birds

Lauren V. Powers, DVM, Dipl ABVP (Avian), Dipl ABVP (Exotic Companion Mammal)

Session #289

Affiliation: From the Avian and Exotic Pet Service, Carolina Veterinary Specialists, Huntersville, NC 28078, USA.

Abstract: Recent studies in birds have shown that melatonin stimulates the release of hypothalamic gonadotropin-inhibitory hormone (GnIH), a newly discovered neuropeptide hormone that down-regulates GnRH, inhibiting the synthesis and release of LH and FSH. In addition to the brain, melatonin receptors are found elsewhere in the avian body, including the gonads. In the author's practice, three of four birds treated for aggression, one out of three birds treated for undesired hormonal behavior, and two out of four birds treated for anxiety appeared to respond to treatment based on client feedback. No clinical improvement was observed in any of 14 birds treated for feather destructive behavior. Further investigation into clinical use of melatonin in avian species is warranted.

Introduction

Reproduction is a highly complex function in birds, orchestrated by the hypothalamic-pituitary-gonadal (HPG) axis.[1] Most avian species breed during a restricted time of the year regardless of the latitude at which they are found in the wild. The most important factor that determines reproductive activity in birds is the availability of food and water. For birds not breeding close to the equator, seasonal changes in day length provide important predictive clues about breeding conditions.[1] Species that respond to changes in the photoperiod are termed "photoperiodic" species. In photoperiodic birds such as a the Japanese quail (*Coturnix japonica*) and the domestic chicken (*Gallus gallus domesticus*), prolonged exposure to a short photoperiod induces a physiologic state termed "photosensitivity."[2] When photosensitive birds experience a critical day length, the HPG axis becomes stimulated and initiation of reproductive activity and behavior occurs.[2] After several weeks of photostimulation by long days, these birds then become photorefractory, meaning that the reproductive system becomes inactive, the gonads regress, and reproductive behaviors cease, even while days are still increasing in length.

Melatonin

Melatonin (N-acetyl-5-methoxytryptamine) is an indole hormone synthesized from serotonin by the pineal gland, retina, and other tissues in birds.[3] Birds exhibit daily and seasonal changes in the duration of pineal melatonin secretion, similar to mammals. Melatonin is secreted by the pineal gland at highest concentrations at night or in darkness.[2] In avian species, melatonin is involved with the regulation of circadian rhythm, feeding patterns, thermoregulation, neuroendocrine function, and reproductive function.[3]

The elimination half-life of melatonin in domestic chickens is short, less than 30 minutes in one report. When administered in feed, plasma concentrations were maintained in the "nocturnal range" for 12 (ethanol washed feed) to 24 hours. Transdermal delivery of melatonin in Eucerin cream at 6 μg and 20 μg in European stonechats (*Saxicola torquata rubicola*) and 40 mg in garden warblers (*Sylvia borin*) resulted in elevated circulating concentrations of melatonin within 30 to 60 minutes and for at least 180 minutes after application.

Melatonin receptors

Melatonin regulates various biological functions in birds through three different receptor subtypes (Mel-1a, Mel-1b, and Mel-1c).[3] Melatonin receptors are found throughout the brain, including the pars tuberalis of the pituitary gland, and elsewhere in the body of avian species including the gonads.[3,4] In ducks, 2-[^{125}I]iodomelatonin binding was shown to be homogeneous throughout the testes, and in the ovary it appeared to be concentrated around the ovarian follicles, particularly in the granulosa cells.[4] All three melatonin receptor subtypes are expressed in the ovary of the domestic chicken, with various receptor type expression in the granulosa and theca layers of preovulatory and postovulatory follicles.[3] These subtypes are identical to the brain receptors.[3]

Gonadotropin-Inhibitory Hormone (GnIH)

Gonadotropin-inhibitory hormone (GnIH) is a recently discovered hypothalamic neuropeptide hormone comprised of 12 amino acids.[5] GnIH was originally isolated from quail brain tissue and described in 2000.[5] This report was the first demonstration of a hypothalamic neuropeptide hormone inhibiting gonadotropin release in any vertebrate species.[1,5] The presence of GnIH has since been documented in a number of avian species, including domestic chickens, house sparrows (*Passer domesticus*), and European starlings (*Sturnus vulgaris*), with various amino acid differences between species.[5]

Within the brain, neuron cell bodies containing GnIH have only been found in the posterior paraventricular nucleus (PVN). However, nerve terminals containing GnIH are found in the median eminence, suggesting that GnIH is released into the hypophysial portal vasculature.[1] GnIH fibers also directly contact GnRH-1 (gonadotropin releasing hormone-1) neurons, suggesting that GnIH may regulate the synthesis and release of GnRH-1 at the cell body level.[1] Both GnRH-1- and GnRH-2-containing neurons in the hypothalamus possess receptors for GnIH. GnIH likely down-regulates the synthesis and release of GnRH, inhibiting the synthesis and release of the gonadotropins LH and FSH.[2]

GnIH receptors

GnIH receptors (GnIHRs) are found elsewhere throughout the brain, and may directly influence certain behaviors such as reproductive behavior and territorial aggression in birds.[1] For example, GnIH fibers are observed in the preoptic area and the periaqueductal central grey (PAG), areas of the brain that regulate socio-sexual behavior such as aggression and sexual behavior. GnIH has been shown to inhibit aggressive and sexual behaviors in quail by directly activating aromatase and increasing neuroestrogen synthesis in the brain.[6] In addition to the brain, GnIH and its receptors are also expressed in the avian testes and ovary.[2] GnIH has not yet been detected in the circulation of any species studied to date, and so is more likely to be produced locally and exhibit a local effect (paracrine effect).

Effects of melatonin on GnIH production

GnIH neurons possess receptors for melatonin, including Mel-1c.[2] Melatonin increases GnIH mRNA expression and release in a dose-dependent fashion from quail hypothalamic explants.[7] Melatonin administration to quail after removal of the pineal gland and eyes resulted in a dosage-dependent increase in GnIH mRNA expression and production of GnIH.[2]

Effects of Melatonin and GnIH on Gonadal Activity in Birds

Melatonin and GnIH may also act peripherally to control reproduction.[2] In chickens, ducks, and quail, the testes and ovary (granulosa cells) bind 2-[^{125}I]iodomelatonin. GnIH and GnIHRs are expressed in avian testes and ovaries in addition to the brain. GnIH reduces testosterone secreted from cultured testes in a dose-dependent manner. Melatonin and GnIH both decrease testosterone release from gonadotropin-stimulated testes *in vitro* before, but not during, the breeding season in European starlings.[2] LH-stimulated progesterone production by hen granulosa cells is markedly reduced with melatonin coincubation. Conversely, melatonin injected at 0.01 mg IM in quail exposed to long day lengths did not inhibit long day-induced gonadal growth and may have resulted in minimal, but significant, reproductive stimulation.[8] The reasons for these effects were unknown, but provide further evidence that the relationship between melatonin and reproductive timing is quite complex.[3,8]

Effects of Melatonin on Reproductive Activity in Birds

Melatonin delayed sexual maturity in growing pullets by 6 to 11 days compared with controls when administered at 25 mg/kg of feed at 7 hours at the end of the photoperiod.[8] Wild female great tits (*Parus major*) implanted with melatonin-filled 10-mm silastic implants prior to the breeding season started their first clutch of the season significantly later than females carrying an empty implant.[9]

Other Effects of Melatonin in Birds

Immune function

Melatonin is capable of affecting both cell-mediated and humoral immunity in birds.[2] Melatonin enhances cell-mediated immunity in European starlings.[2] Continuous or transient administration in the drinking water at 50 mg/ml significantly increased humoral and cellular immunity in Japanese quail.

Anxiety and distress

Melatonin (0.10 to 1.0 mg/kg by intraperitoneal injection) reduced isolation-distress vocalizations in young (2 to 12 days of age) domestic chickens in a dose-dependent manner.[10]

Antioxidant effects

Melatonin is a potent antioxidant. At 10 mg/kg PO and 40 mg/kg in feed (separate study), melatonin greatly reduced the adverse effects caused by aflatoxin in chicks such as hepatic vacuolar degeneration, necrosis, and bile duct hyperplasia.

Human Use and Formulations

Melatonin is used in humans for a wide variety of disorders, including insomnia, jet lag, headache, gastroesophageal reflux disorder, and irritable bowel syndrome. Human dosages vary widely. For example, oral doses from 2 to 5 mg qHS (nightly at bedtime) are often advised for disorders such as insomnia and jet lag.

Melatonin is categorized by the US Food and Drug Administration (FDA) as a dietary supplement and is widely available over-the-counter in the USA in a variety of formulations, including tablets, capsules, and liquids for oral and sublingual use. Melatonin is also available in a prescription sustained release 2-mg tablet (not FDA approved) for use for insomnia (Circadin, Neurim Pharmaceuticals, Tel-Aviv, Israel) and as a subcutaneous sustained-release implant for use in domestic ferrets (*Mustela putorius furo*) (Ferretonin, 5.4-mg) for the treatment of adrenocortical disease, and in domestic dogs (Dermatonin, 18.0-mg) for "Alopecia X" and "Seasonal Flank Alopecia" (Melatek, Middleton, WI, USA). Melatonin is soluble in water at 0.10 mg/ml. Sublingual liquid preparations are generally formulated at lower concentrations impractical for clinical use in birds. Several liquid formulations with higher concentration (10 mg/ml) are commercially available (eg. Puritan's Pride, Oakdale, NY, USA; Solgar, Leonia, NJ, USA; SISU, Burnaby, BC, Canada).

Clinical Use in Birds and Response to Treatment

Clinical use of melatonin in companion and aviary birds has not been reported. In the author's practice, 25 birds were treated with melatonin between 2002 and 2016, 18 of which had adequate documented treatment response based on client feedback (Table 1). Indications for use included human-directed and conspecific-directed aggression, undesired sexual behavior and activity, anxiety, and feather damaging behavior (FDB). Treatment plans were usually multi-modal. Three out of four birds treated for aggression responded to treatment. One out of three birds treated for undesired sexual behavior and activity appeared to respond to treatment. Two out of four birds treated for anxiety responded to treatment. None of 14 birds with FDB responded to treatment with oral melatonin at the dosages, frequencies, and routes of administration used. Mild to moderate sedation was reported in four birds at dosages ranging from 1.1 to 13.3 mg/kg PO. No other adverse effects were reported by clients.

Table 1. Clinical use of oral melatonin and response to treatment in companion birds.

Species	Gender	Age (yr)	Indication(s)	Administration	Response to treatment
Grey parrot (*Psittacus erithacus*)	Male	7	FDB	1 mg/8 oz drinking water (0.0042 mg/ml)	No improvement in FBD.[a]
Grey parrot	Female	18	FDB	3.3-13.3 mg/kg PO q24hr in morning	No improvement in FBD. Bird calmer, especially with restraint. Higher dosages resulted in sedation.
Grey parrot	Female	23	FDB	1 mg/8 oz drinking water (0.0042 mg/ml)	No effect on FBD, resulted in sedation and incoordination, client discontinued.
Indian ringneck parakeet	Male	8	FDB	1 mg/8 oz drinking water (0.0042 mg/ml)	No improvement in FBD. Bird subjectively calmer.
Blue and gold macaw (*Ara ararauna*)	Female	Adult	FDB	1 mg/kg PO q24hr	Uncertain effect on FBD

Species	Gender	Age (yr)	Indication(s)	Administration	Response to treatment
Hahn's macaw (*Diopsittaca nobilis nobilis*)	Male	9	FDB	Unknown	Initial improvement but FBD soon recurred.
Umbrella cockatoo (*Cacatua alba*)	Male	Adult	FDB	Unknown	Difficult to administer, questionable FBD response.
Blue and gold macaw	Male	Adult	FDB, anxiety	1 mg/8 oz drinking water (0.0042 mg/ml)	No improvement in FBD or anxiety.
Green cheeked conure (*Pyrrhura molinae*)	Unknown	Adult	FDB, anxiety	1 mg/8 oz drinking water (0.0042 mg/ml)	No improvement in FBD.
Grey parrot	Female	5	FDB, anxiety	0.64 mg/kg PO (evenings) Dissolved in fruit juice	Uncertain effect on FBD. Appeared to have calming effect.
Blue and gold macaw	Female	24	FDB, seasonal aggression	Unknown	No effect on FBD. Dramatic improvement in seasonal aggression. Beneficial calming effect.
Eclectus parrot (*Eclectus roratus*)	Male	Adult	FDB, human-directed aggression	Variable	Higher dosages effective but resulted in heavy sedation.
Eclectus parrot	Female	8	FDB, salpingitis, egg binding	0.56 mg/kg PO q24hr before bed every night	No improvement in FBD. 1.1 mg/kg in food made bird drowsy.
Umbrella cockatoo	Female	19	Cloacal prolapse, FDB	Unknown	Initial temporary improvement but both problems quickly recurred.
Pekin duck (*Anas platyrhynchos*)	Female	Adult	Megacloaca, excessive egg laying	0.12 mg/kg PO q24hr	No further egg laying, megacloaca persisted.
Pekin duck	Male	1	Sexual aggression towards other ducks	0.48 mg/kg PO q24hr	Remarkable effect according to client, calmer, still interested in breeding, especially younger ducks.
Moluccan cockatoo (*Cacatua moluccensis*)	Male	19	Human-directed aggression	Unknown	Minimal to no improvement. Responded to deslorelin acetate implant.
Green winged macaw (*Ara chloroptera*)	Female	19	Anxiety	1-3 mg/kg PO PRN	Dramatic improvement in anxiety. Sedation at higher dosages.

[a]FDB = Feather damaging behavior.

Response to treatment likely depends in part on dosage and frequency of administration. A multi-modal approach to therapy is likely necessary for successful treatment of these challenging conditions in birds. Further studies are warranted in companion bird species, including pharmacokinetic evaluation and clinical response to treatment.

References

1. Bédécarrats GY, McFarlane H, Maddineni SR, et al. Gonadotropin-inhibitory hormone receptor signaling and its impact on reproduction in chickens. *Gen Comp Endocrinol.* 2009;163(1):7-11.

2. McGuire NL, Kangas K, Bentley GE. Effects of melatonin on peripheral reproductive function: Regulation of testicular GnIH and testosterone. *Endocrinol.* 2011;152(9):3461-3470.

3. Sundaresan NR, Leo MM, Subramani J, et al. Expression analysis of melatonin receptor subtypes in the ovary of domestic chicken. *Vet Res Commun.* 2009;33(1):49-56.

4. Ayre EA, Wang ZP, Brown GM, et al. Localization and characterization of [^{125}I] iodomelatonin binding sites in duck gonads. *J Pineal Res.* 1994;17(1):39-47.

5. Tsutsui K, Saigoh E, Ukena K, et al. A novel avian hypothalamic peptide inhibiting gonadotropin release. *Biochem Biophys Res Commun.* 2000;275(2):661-667.

6. Ubuka T, Tsutsui K. Gonadotropin-inhibitory hormone inhibits aggressive behavior of male quail by increasing neuroestrogen synthesis in the brain beyond its optimum concentration. *Gen Comp Endocrinol.* 2014;205:49-54.

7. Chowdhury VS, Yamamoto K, Ubuka T, et al Melatonin stimulates the release of gonadotropin-inhibitory hormone by the avian hypothalamus. *Endocrinol.* 2010;151(1):271-280.

8. Lewis PD, Middleton BA, Gous RM. Exogenous melatonin modifies rate of sexual maturation in domestic pullets. *Poult Sci.* 2006;85(1):117-122.

9. Greives TJ, Kingma SA, Beltrami G, et al. Melatonin delays clutch initiation in a wild songbird. *Biol Lett.* 2012;8(3):330-332.

10. Nelson E, Panksepp J, Ikemoto S. The effects of melatonin on isolation distress in chickens. *Pharmacol Biochem Behav.* 1994;49(2):327-333.

Spironucleus in Cockatiels *(Nymphicus hollandicus)*

Fern Van Sant, DVM

Session #282

Affiliation: From For the Birds, 1136 S De Anza Blvd. San Jose, CA 95129, USA.

A novel motile trophozoite was consistently noted in an outbreak of enteritis in cockatiels (*Nymphicus hollandicus*) housed at a sanctuary that focused on older and debilitated birds. The outbreak escalated quickly with many birds showing clinical disease and mortality rates approaching 15%. The enteritis was characterized by anorexia, green diarrhea, melena and profound weight loss. Severely affected birds were noted to quickly lose their pectoral muscle mass. Necropsy with histopathology failed to identify the pathogen. Because the motility pattern of the trophozoite was so unique and matched descriptions of *Spironucleus meleagridis* as reported by Levy and Powers,[1] samples were submitted to the parasitology research lab at North Carolina State University and identified by polymerase chain reaction (PCR). A commercial PCR was subsequently developed at the Infectious Disease Laboratory at University of Georgia.

Anti-protozoal drugs including parenteral metronidazole, oral metronidazole and oral carnidazole were tried but no antiprotozoal drug was effective at decreasing the number of trophozoites shed or diminishing symptoms. Based on current recommendations for treatment of cats with tritrichomonas infections resistant to metronidazole, ronidazole was started.[6] A dose of 15 mg/kg (50 mg/ml, Roadrunner Pharmacy, AZ 85085) for 7 days was well tolerated and effective at addressing shedding of protozoa and symptoms.

Spironucleus meleagridis (formerly *Hexamita meleagridis*) is a flagellated diplomonad protozoa that causes infectious catarrhal enteritis. The literature reflects potential for significant economic losses in commercial game bird farming.[3,4] *Spironucleus* has been linked to significant disease in both wild living and companion psittacine birds.[1,5,6]

References

1. Levy MG, Powers LV, Gore KC, Marr HS. *Spironucleus meleagridis*, an enteric diplomonad protozoan of cockatiels (*Nymphicus hollandicus*): preliminary molecular characterization and association with clinical disease. *Vet Parasitol.* 2015;208(3-4):169-73.

2. Gookin JL, Copple CN, Papich MG et al. Efficacy of ronidazole for treatment of feline Trichomonas foetus infection. *J Vet Intern Med.* 2006;20(3):536-543.

3. Lloyd S, Gibson JS. Hematology and biochemistry in healthy young pheasants and red legged partridges and effects of spironucleus on these parameters. *Avian Pathol.* 2006;35(4):335-340.

4. Cooper GL, Charlton BR, Bickford AA, Nordhausen R. *Hexamita meleagridis* (*Spironucleus meleagridis*) infection in chukar partridges associated with high mortality and intracellular trophozoites. *Avian Dis.* 2004;48:706-710.

5. Doneley B. Weight loss syndrome in juvenile free-living galahs (*Eolophus roseicapillus*). *Proc Annu Conf Assoc Avian Vet Austr Comm.* 2012.

6. Philby AW, Andrews PL, Gestier AW, Reece RL. Arzey KE. Spironucleus in Australian king parrots (*Alisterus scapularis*). *Aust Vet J.* 2002;80:154-160.

Angiocentric Immunoproliferative Lesions as a Cause of Pruritus and Automutilation in Parrots

Y.R.A. van Zeeland, DVM, MVR, PhD, Dipl ECZM (Avian), Dipl ECZM (Small Mammal), CPBC,
M.J.L. Kik, DVM, PhD, Dipl ECZM (Herpetology),
N.J. Schoemaker, DVM, PhD, Dipl ECZM (Small Mammal) Dipl ECZM (Avian), Dipl ABVP (Avian)

Session #264

Affiliation: From the Division of Zoological Medicine, Department of Clinical Sciences of Companion Animals (van Zeeland, Schoemaker), and the Division of Pathology, Department of Pathobiology, Faculty of Veterinary Medicine, Utrecht University, Yalelaan 108, 3584 CM Utrecht, the Netherlands (Kik).

Two female blue-fronted Amazon parrots (*Amazona aestiva*) and a female grey parrot (*Psittacus erithacus*) were presented with a history of progressive pruritus, feather plucking and automutilation for a duration of 0.5 to 3 years. Dermatologic examination revealed dry, erythematous skin with multiple crusts, erosions and feather loss, particularly on the ventral side of the wings, chest, and legs. Cytology of tape-strip samples revealed a (secondary) bacterial and/or yeast infection. Following treatment (fluconazole, TMP/S and meloxicam systemically for 3-6 weeks), skin biopsies were collected for histopathologic examination. These revealed presence of hyperkeratosis, acanthosis and extensive perivascular lymphohistiocytic infiltrates in the dermis, consisting of small-to-medium-sized pleiomorphic lymphocytes, histiocytes and some heterophils. Immunohistochemical staining demonstrated a mixed T and B cell population. Based on the aforementioned criteria, angiocentric immunoproliferative lesions (AIL) were diagnosed in two birds. In the third bird, angiocentric lymphoma was diagnosed as blastic cells were the predominant cell-type. This bird responded poorly to treatment with prednisone (0.1 mg/kg PO q24h, given in combination with antibiotics and antimycotics), whereas the other birds showed good clinical improvement for at least 1 year.

In humans, AIL is a rare Epstein-Barr-virus-associated lymphoproliferative disorder,[1] which can histologically be classified into lymphocytic vasculitis (AIL-I); lymphomatoid granulomatosis (AIL-II); and angiocentric lymphoma (AIL-III).[2-4] Treatment can consist of corticosteroids, antiviral therapy, chemotherapy and/or rituximab.[5-8] Reports in veterinary medicine are scarce,[9-11] with this case series being the first to diagnose AIL in avian species. Moreover, this report demonstrates that long-term management with corticosteroids can be achieved in birds without detrimental effects.

References

1. Marks IL, Kurtin PJ, Katzenstein AL, et al. Lymphomatoid granulomatosis: evidence of immunophenotypic diversity and relationship to Epstein-Barr virus infection. *Am J Surg Pathol.* 1995;19:1300-1312.

2. Liebow AA, Carington CRB, Friedman PJ. Lymphomatoid granulomatosis. *Hum Pathol.* 1972;4:457-558.

3. Jaffe ES. Pathologic and clinical spectrum of post-thymic T-cell malignancies. *Cancer Invest.* 1984;2: 413-426.

4. Lipford EH, Margolick JJB, Longo DL, et al. Angiocentric immunoproliferative lesions: A clinicopathologic spectrum of postthymic T-cell proliferations. *Blood* 1988;72:1674-1681.

5. Wilson WH, Kingma DW, Raffeld M, et al. Association of lymphomatoid granulomatosis with Epstein-Barr viral infection of B lymphocytes and response to interferon-alpha 2b. *Blood* 1996;87:4531-4537.

6. Rao R, Vugman G, Leslie WT, et al. Lymphomatoid granulomatosis treated with rituximab and chemotherapy. *Clin Adv Hematol Oncol.* 2003;1:658-660.

7. Tagliavini E, Rossi G, Valli R, et al. Lymphomatoid granulomatosis: A practical review for pathologists dealing with this rare pulmonary lymphoproliferative process. *Pathologica* 2013;105:111-116.

8. Roschewski M, Wilson WH. Lymphomatoid granulomatosis. *Cancer J.* 2012;18:469-474.

9. Smith KC1, Day MJ, Shaw SC, et al. Canine lymphomatoid granulomatosis: an immunophenotypic analysis of three cases. *J Comp Pathol.* 1996;115:129-138.

10. Magi GE1, Di Pasquale S, Renzoni G. Cutaneous lymphomatoid granulomatosis (angiotropic lymphoma) in a dog: immunophenotyping analysis. *Vet Res Commun.* 2009;33:145-147.

11. Shimazaki T, Nagata M, Goto-Koshino Y, et al. A case of canine lymphomatoid granulomatosis with cutaneous lesions. *J Vet Med Sci.* 2010;72:1067-1069.

Immunization of Poultry

Bruce A. Singbeil, DVM

Session #206

Affiliation: Bellevue Crossroads Vet, 15600 NE 8th St, Ste 0-6, Bellevue WA 98008, USA.

Abstract: This presentation provides insight into the considerations of vaccination of poultry. Given the broad diversity of species and diseases, only *Gallus* species are represented. The list of diseases currently vaccinated for is provided, but is considered not absolute at publication. Examples of various viral genetics for demonstration are provided between enveloped, non-enveloped and immunosuppressive diseases. Genetics, both virus and species, for immunization are presented.

The immunization of poultry is a complex and ever-evolving subject.[1] This is because virus genetics and breed genetics are always in a state of flux. Serotypes and pathotypes differ among the same viruses. Geographic differences are common. Current molecular genetics are a great aid for understanding viral mutations/differentiations but do not always provide answers for the right immunization approach. This often requires laboratory testing for cross-protection studies with different viral strains.

An appreciation for immunology, avian genetics, and molecular virology for the protection of poultry from disease is needed to be successful to perform immunization; vaccination is just a step in the process. The strain of pullet for egg production varies widely in the acceptance for vaccination. The same is true for meat-type poultry. There are many pharmaceutical companies that produce a multitude of products for immunization. All are licensed for poultry but incorrect strain or timing of administration will lead to illness and mortality. I, personally, have seen more illness from improper vaccination than the diseases themselves. To demonstrate immunization in poultry, we will cover 4 viruses as examples of immunizations and the influences of breeder vaccination programs, timing, routes and administration.

Immunization in poultry is performed using attenuated live, chemically mutated, gene deleted, recombinant or adjuvant mediated killed pathogens (Table 1). Methods of administration include in-ovo, gel droplet, water, eye drop, spray, wing web and needle injection.

Table 1. Poultry vaccines available in the United States and diseases.

Category	Vaccine
Viral	Marek's, infectious bursal, Newcastle disease, infectious bronchitis, infectious laryngotracheitis, fowl pox/pigeon pox, avian encephalomyelitis and chicken infectious anemia
Bacterial	*E. coli*, *Salmonella*, *M. gallisepticum*, cholera *(Pasteurella)*, coryrza (*Haemophilus* species)
Protozoal	Coccidiosis

Marek's vaccination is the most ubiquitous in distribution and need for vaccination. This enveloped herpes virus causes lymphoproliferative disease in many forms. These are dermal lesions seen in broilers, paralysis (most common), neurologic, ocular, degenerative inflammatory and immunosuppression, often are presented in multiple forms in the same bird. Immunization needs to be performed in the hatchery via in-ovo or injection, as once exposed to field virus the vaccine is ineffective. While the infection is ubiquitous, the disease is not. This speaks to the genetics of both the virus and the breed. Some avian breeds have genetic resistance; others do not.

Transmission of the virus is not vertical (hen to chick). It is transmitted via inhalation of fluff at hatch, fomites and people working in the environment. There is a genetic and age resistance to this virus. Passive immunity, (maternal antibody from hen to chick), does not occur as "sterile immunity," in that it lasts a very short while but progeny are still susceptible. Many backyard breeders are of the opinion that if the breeders are vaccinated, chicks are protected. Current vaccines are of serotype 1 such as SB1, RB1B and HVT. Serotype 2 virus from the Netherlands has been used to control problem areas in the United States, using CVI988 strains, but can induce losses to A strain avian leukosis virus B cell tumors. Immunization in poultry is complex.

Infectious bronchitis virus (IBV) is the most common serologic-positive disease that I observe, whether it is causing symptoms or not. IBV is a coronavirus, enveloped, positive sense RNA virus, which lends itself to translational and transcriptional mutations even amongst the vaccine viruses. Incubation periods as short as 18 hours are possible. Most commonly, this causes respiratory problem, eliminating the protective tracheal lining, which leads to secondary infections and pneumonia. In addition to respiratory disease, it is notorious for causing reproductive declines and egg shell problems. Renal disease from exposure causes increased mortality and production losses. Corona viruses in most species are intestinal problems. In poultry, successful isolation is from cecal tonsil. Early onset may be found in tracheal swabs but the cecal tonsil is consistent. Knowing the strain for your geographic area in order to immunize for this disease is paramount. There are currently 3 strains in the United States that are not controlled by current vaccines. Immunization is best using breeder vaccinations for passive immunity and administration in hatchery and field exposure.

Infectious bursal disease virus (IBDV) is a non-enveloped birnavirus double-stranded RNA virus, which allows for transformation of trimeric surface binding proteins. Vertical transmission of this virus has not been demonstrated. There are many serologic and pathogenic types of this virus. Clinical signs may be seen in non-apparent changes in flock development, mild non-pathologic immunosuppression, and major mortality due to vvIBDV (very virulent). Immunosuppression is the hallmark of this disease due to the depletion of B cells in the bursa. Its counterpart, chicken infectious anemia (CIA), is a positive sense DNA non-enveloped virus. This virus affects the thymus T cells and is transmitted horizontally and vertically. Both cause major immunosuppression and due to their non-enveloped structure are very difficult to control by sanitation. The question to ask in any poultry disease is: Is there an underlying cause for the current infection? Both diseases are best addressed as part of the breeder vaccination program, which allows for passive transfer of maternal antibodies. CIA is only addressed this way. IBDV is used in hatchery and field vaccination programs in both broilers and laying flocks as well as breeder programs.

References

1. Swayne DE, Glisson JR, McDougald LR, Nolan LK, Suarez DL, Nair VL. *Diseases of Poultry*, 13th ed. Oxford, UK: Blackwell Publishing;2013.

A Novel Technique for Cloacal Removal of Mature Retained Eggs in Birds

Angela M. Lennox, DVM Dipl ABVP (Avian), Dipl ABVP (Exotic Companion Mammal), Dipl ECZM (Small Mammal)

Session #195

Affiliation: From the Avian and Exotic Animal Clinic, 9330 Waldemar Road, Indianapolis, IN 46268, USA.

Reproductive disease is common in pet species, including psittacines, passerines and even pet poultry. Clinical signs can be obvious (e.g., a palpable egg, or egg retained in the cloaca); however, signs also can be vague in some cases appear unrelated to the reproductive tract. Examples include increased respiratory rate and effort, leg lameness and feather plucking. Reproductive disease in the female bird can involve the ovary, oviduct, and/or cloaca, and include ovarian cysts, neoplasia, infection, inflammation, uterine rupture, retained eggs and prolapse. Treatment is dependent upon the exact diagnosis and can include behavioral modification, GnRH agonists, antimicrobial therapy, analgesics and surgery. Surgery of the reproductive tract can be challenging, and may include partial ovariectomy, removal of ovarian cysts, removal of the oviduct and in some cases egg products within the oviduct or free in the celomic cavity, and treatment of cloacal/uterine prolapse.

During normal oviposition, the infundibulum catches the egg after ovulation and directs it to the uterus. Aberrations in anatomy or function can result in an "ectopic" yolk deposited directly into the celomic cavity (internal laying). In some cases the ovum is reabsorbed, or can result in yolk peritonitis. The oviduct consists of the magnum (addition of albumin), isthmus (deposition of inner and outer shell membranes) and shell gland (plumping, or addition of water and electrolytes, and actual calcification); after passage of the egg through the oviduct the egg passes through the vagina to the cloaca and out the vent.

The author and others have encountered cases of ectopic fully formed and shelled eggs within the celomic cavity; potential causes include retrograde movement of the formed egg up and out the infundibulum, or rupture via the wall of the oviduct due to disease of the oviduct, or iatrogenic manipulation of the egg.

Retained Egg and Dystocia

Treatment of dystocia and retained eggs must include careful patient evaluation and stabilization prior to beginning primary therapy. Supportive care includes fluid therapy, nutritional support, increased ambient temperature, and if warranted, administration of calcium, antibiotics and analgesics. In some cases, supportive care and warmth alone may allow the bird to pass an egg present within the oviduct or cloaca naturally. Drug therapy can benefit when the uterus is intact and functional and there are no adhesions or obstruction. PGE2 (Prepadil Pharmacia and Upjoh, Kalamazoo, MI, USA) relaxes the uterovaginal sphincter while inducing uterine contractions, and is administered .025 ml/100 g.

Several techniques have been described for delivery of a retained egg non-responsive to medical therapy; these include manual delivery, ovocentesis and celiotomy. Manual delivery must be performed with care, and is only considered for shelled eggs in the distal oviduct/vagina or cloaca. With the bird under general anesthesia, digital pressure is applied to the proximal portion of the egg, and the cloaca lubricated and gently dilated. As the egg begins to present into the cloaca, the vagina is identified via the appearance of white shell through the thin wall, and the vaginal opening identified and dilated with a blunt probe to allow passage of the egg. Complications of this procedure include uterine trauma and/or rupture, collapse of the egg, hemorrhage, and respiratory compromise secondary to celomic pressure.

Some authors have described ovocentesis, either through the cloaca or through the body wall. Ovocentesis through the body wall cannot be recommended due to inability to confirm that the egg is actually in the reproductive tract and not ectopic, and due to the risk of damage to other organs, in particular the GI tract. In one case of attempted ovocentesis using this method, the needle produced intestinal perforation and leakage of GI contents into the ceolom (Nemetz, personal communication).

The Lonestar retractor is an excellent tool for visualization of eggs located within the distal oviduct and the cloaca. Anesthetize the bird routinely, using appropriate pre-medications including an analgesic. Apply a local block circumferentially to the lips of the vent where the retractor stays will be placed (lidocaine/bupivacaine 2 mg/kg each combined). Careful traction allows excellent visualization of the entire cloaca; if the egg is retained in the caudal oviduct, gently caudal pressure on the egg allows it to be seen "crowning" through the vagina. Puncture the shell carefully with a hemostat or 18 g needle, and carefully remove the shell piece by piece, taking care not to damage the oviduct with sharp portions of the shell. This can be time consuming in cases where the egg is adhered to the wall of the oviduct. Collect diagnostic samples if desired (cytology, specimens for culture/sensitivity), and flush the cloaca. Confirmation of removal of all portions of the shell can be obtained radiographically.

While the author has not seen complications related to use of the retractor and manual delivery using this method, it should be kept in mind that birds with dystocia often represent again for the same problem at the next attempt to lay. For this reason, birds with reproductive disease should be worked up carefully for underlying factors contributing to dystocia, and treated appropriately.

References

1. Bowles H. Evaluating and treating the reproductive system. In: Harrison G, Lightfoot T, eds. *Clinical Avian Medicine.* Palm Beach, FL: Spix Publishing Inc.;2006:519-539.

2. Joyner KL. Theriogenology. In: Ritchie B, Harrison G, Harrison L, eds. *Avian Medicine: Principles and Applications.* Lake Worth FL: Wingers Publishing;1994:748-804.

3. King AS, McLelland J. Female reproductive system. In: King AS, McLelland J (eds) *Birds: Their Structure and Function.* 2nd ed. Philadelphia, PA: Bailliere Tindal;1984:145-165.

4. Pollock CG, Orosz S. Avian reproductive anatomy, physiology and endocrinology. *Vet Clin North Am Exot Anim Pract.* 2002;5(3):441-474.

Coxiella-like Infection in a Juvenile Eclectus Parrot (*Eclectus roratus*): Diagnosis and Management

Claudia Kabakchiev, DVM,
Hugues Beaufrère, Dr. med. vet., PhD, Dipl ACZM, Dipl ABVP (Avian), Dipl ECZM (Avian),
Britta Knight, DVM, MSc, PhD,
Robert Foster, BVSc, PhD, MANZCVS, Dipl ACVP,
Hugh Cai, DVM, MSc, DVSc,
Dale A. Smith, DVM, DVSc

Session #130

Affiliation: From the Health Sciences Centre (Kabakchiev, Beaufrère), the Department of Pathobiology (Knight, Foster, Smith), Ontario Veterinary College, and the Animal Health Laboratory (Cai), University of Guelph, Guelph, N1G 2W1, ON, Canada.

A 9-month-old male eclectus parrot presented for decreased appetite, weight loss, and lethargy. A mild non-renegerative anemia and heterophilic leukocytosis were noted on the complete blood count. There was also a hyperproteinemia, hypercholesterolemia, and an elevated amylase level on biochemical evaluation. Radiographs identified splenomegaly, hepatomegaly, and an enlarged proventriculus. Coelioscopy was performed to collect biopsy samples from the spleen and liver; biopsies were also collected from the air sacs, the left kidney, and the crop. The splenic and hepatic tissues showed evidence of chronic and ongoing systemic inflammation. A PCR test on the splenic tissue was positive for a *Coxiella*-like organism previously identified as a cause of hepatitis in birds. The eclectus parrot was started on azithromycin therapy (40 mg/kg orally every 24 hours) and an improvement was noted in its appetite and energy level. Serial hematologic examinations demonstrated a steady decrease in the white blood cell count and normalization of the hematocrit. The patient developed diarrhea soon after completing the azithromycin course. Antimicrobial therapy was re-instituted, this time using doxycycline (25 mg/kg orally every 12 hours) which is the primary treatment used in mammals infected with *Coxiella burnetii*. Two 4-week courses of doxycycline were used on this patient. The patient was doing well clinically and according to hematologic and biochemical evaluation at the end of the antimicrobial treatment and one month after discontinuation. This is the first known report of *Coxiella*-like infection diagnosed ante-mortem and successfully treated in a psittacine bird.

Experiences with Free Flight

Kenneth R. Welle, DVM, Dipl ABVP (Avian)

Session #138

Affiliation: From University of Illinois, College of Veterinary Medicine, Urbana, IL 61802, USA.

Abstract: Free unrestrained flight of parrots is a small but growing trend. Inspired by videos of free-flying birds, the practice has some benefits for pet birds, but also some serious dangers. The author has undertaken this hobby with two young macaws. This paper will describe the general process used and some of the risks and benefits.

Introduction

There is a YouTube video of a couple hiking with their seven macaws.[1] They weren't carrying these birds around with harnesses or anything of the sort. The birds just flew around and periodically landed on one or the other of the people. It was a spectacular thing to see. One can click on a link and find a whole web page dedicated to this very practice. There also are free flight clubs, chat rooms, and other groups that allow enthusiasts to communicate about and practice this activity. Avian veterinarians may encounter individuals who do, or are planning to, engage in free flight with their birds. A basic knowledge of the activity may be helpful in advising these clients.

Advantages of Free Flight

It's very clear from seeing the aging population of pet birds, that exercise is something that most birds are lacking. While there are certainly other exercises that parrots can engage in, flight is the most natural and vigorous. Exercise allows helps maintain cardiac, vascular, muscular, and mental health.

Enrichment is currently a big topic in animal welfare. One of the challenges of keeping any intelligent species in captivity is providing for their psychological needs. While there are many ways to do this, expanding a bird's environment to include nearly unlimited spaces, a much higher degree of enrichment becomes possible. Exposure to fresh air and natural sunshine is an added benefit.

One advantage of free flight is more self-serving. These birds are spectacular to watch in flight. Most people who love birds enjoy watching their antics. Free flight expands the range of available activities to watch. As scary as it seems to see your parrot flying away from you, it doesn't match up to the joy of watching them fly back and make a perfect landing on your hand.

Flight requires a fairly significant amount of mental activity as well as physical activity. Navigation, wind adjustments, and other fine motor skills are partly mental. Allowing these activities can promote mental development that may not be allowed to develop otherwise.

One of the problems with the traditional manner in which birds are kept is that someone with an active lifestyle will have to leave behind their bird when engaging in activities. Free flighted birds can be involved in daily life. They can go for hikes, hang out in the yard during a picnic, or other outdoor activities.

Risks of Free Flight

There are certainly risks in free flight. There is risk inherent in all activities. Sending a child to school, allowing them to drive, or date, or leave the home at all, will expose them to risks. However, to eliminate all risk is also to eliminate opportunities, enjoyment, and any real semblance of a life. Most people choose to teach their children to live in our world and then trust them to do so. The same process can be applied to pet birds.

The first thought most people will get when thinking about free flight is that the bird will escape. Most people seem to think that pet birds dream of escaping and will do so at the very first opportunity. Certainly, the escape of a bird is always devastating. It is also the last thing that most pet birds want to happen. Most parrot escapes are really the result of the bird's inability to navigate, turn, and land correctly. An unskilled bird can gain altitude but may have difficulty slowing, turning, landing accurately, and performing other fine motor skills. This often results in the bird flying straight away from the owner and landing only from exhaustion in a tree somewhere. Then they lack the ability to return. One of the essential components of successful free flight is allowing a bird to learn these skills in an area where they will not get lost. Some of that is indoors, later it is outdoors in large open expanses with little to land on besides the familiar owner or a perch. The skills build from there. In fact, birds that are well trained to this behavior have far less risk of escape than an unskilled, untrained bird that accidentally gets out the door.

The risk of predation also is very real. While this can be true for many pets, the small size of most companion birds makes them susceptible to a fairly wide variety of predators. In particular, we worry about aerial predators, raptors. Raptors can prove to be a risk when a bird is in the air and on the ground or perch. This risk cannot be entirely prevented. It can be mitigated by several things. Flying very large parrots narrows the range of predators that would attack them. Few Cooper's hawks will try to attack a macaw that is three times its own weight. Second, the flying skills of the bird will reduce the chances that they will be caught by a raptor. The birds can also learn safe perching strategies, flock flying, and other skills that can help them evade predation. The presence of the owner can serve as a deterrent to predators as well. Often just walking toward a hawk perched on a telephone pole will cause it to vacate the area.

A fast flying bird is certainly going to be prone to certain injuries. This is a risk even wild birds are prone to. Anyone that has worked in a wildlife hospital has seen broken wings, broken legs, concussive head injuries, and other casualties. These cannot always be prevented. The best way to mitigate this risk is to control where birds are flown. The complexity of the flying environment can be increased as the flight skill increases. Some environments, such as those with large plate glass windows, wind turbines, transformers, and other dangerous structures are best avoided.

Avian veterinarians should promote responsible pet ownership. The last thing that should be encouraged is anything that could contribute to the establishment of feral populations of non-native species. Many of the pet bird species could not survive in a cold climate. If they can, it should not be assumed that they would be harmless to the local native species. To prevent this from happening, only well trained birds should be flown outdoors without restraint. Additionally, these birds should be brought into a safe enclosure at night and whenever they can't be supervised. Ideally, someone in addition to the primary owner should be able to handle the birds. In the event of the owner's death, the bird should not be left to fend for itself.

Bird Selection

Ideal candidates for free flight are birds that, you guessed it, can fly. Flying clumsily across the living room may not be adequate qualification however. Free flight will require the ability to turn, slow down, move up and down, and otherwise navigate in three dimensions. Full wings are a requirement. Trimming the wings will only hinder the development of the required skills. Injuries could also inhibit adequate flight skills.

A free flying bird will need to have a very close bond to the handler. Although independent behavior is usually a good thing, a certain dependence on the owner or handler is critical to successful free flight. The bird should seem somewhat uncomfortable with being left behind. Although food reinforcers are used, a bird that will fly to the owner just for social interaction will be much more successful.

Young birds are recommended for free-flight training. Birds may learn flight skills best immediately upon fledging. This would be the time when wild birds learn these skills. Birds will also have an innate need to stay with their perceived parent during this time. This allows them to develop outdoor skills while still having a mental "leash" that keeps them close to the owner.

Birds that are easy to find also are helpful. Large, loud, brightly colored birds are easier to spot in a tree than small, quiet, green birds. Larger birds present less temptation to hawks. In the author's area, the only aerial predators larger than the macaws would be large female red-tailed hawks, eagles, or large great horned owls. None of these species preferentially hunt birds. A cockatiel or conure would be more subject to attacks from Cooper's hawks, merlins, and other falcons. The other factor that will reduce the risk of predation is the skill level of the bird. Flying skills, vigilance in looking for danger, and other skills required practice. These skills can be honed while flying closer to the handler which will serve as somewhat of a deterrent to most raptors. Flock flying can also be helpful. A group of birds is oddly more difficult for a predator to successfully attack.

Training the Trainer

The first step that must be taken is to properly train the person that will be handling the birds. Classes are recommended before undertaking free flight. There are few trainers that will encourage this practice, due to the liability. The author took the course taught by the same trainer that made the video of the flying macaws. Although there are probably other approaches to learning how to train birds for free flight, this discussion will center on the ones used in the course taken by the author.[2] Much of what must be learned are the basics of bird training. Traditional training methods that have been used in parrots are somewhat coercive. Positive reinforcement training is a form of operant conditioning is a time and scientifically tested method of modifying behavior. Instinctive drift will allow some modification of behaviors learned through operant conditioning. The instincts of the animal may cause the behavior to change to match natural adaptive behaviors of the species. This may be either deleterious or beneficial to the desired behaviors. For free flight, instinctive drift allows the bird to develop flight skills without the specific behavior being taught. For instance, learning to fly in a loop instead of straight to the handler, to avoid predators, and other similar skills would be very difficult to specifically train. However, by allowing the bird to fly in a safe area they can develop through instinctive drift. At the same time, the basic behavior of flying to the hand remains intact. Location selection is another very important factor in learning to free fly birds. When skills are poor, it is critical to maximize the chances for successful repetitions of the desired behavior. Indoor flight is a great starting point. This allows the bird to develop the basic skill of flying to the hand. If a suitably large site is available, or with smaller species, almost all of the required skills can be learned indoors. For large species like macaws, however, the amount of space needed is rarely available.

The first outdoor level should afford very few opportunities for mistakes. There should be no trees, open water, extreme terrain or other areas for the bird to fly that would make them difficult to retrieve. The weather should also be favorable with little wind, no rain, and mild temperatures. The first outdoor flights should be identical to the indoor flights. Flying from a perch to the handler and back is the initial exercise. Eventually the bird will miss the mark and end up flying out of the straight line. Then they will have to learn the skill of coming from any direction to their desired landing spot. Most birds will learn that it's easier to land when flying into the wind than with a tailwind. When the bird becomes skillful at that level, the environment can be changed to include increasing levels of challenges. For the author, trees are the primary factor that affects a level. If working with more than one bird, they should initially be flown separately. This is because one bird may lead another astray. Once both birds achieve an appropriate skill level, they can be flown together. This method also leaves an anchor bird if one bird lands somewhere difficult to retrieve. The anchor bird is a strong motivating resource to help recover a bird from an inaccessible site.

Basic Bird Skills to Teach

Flight to the hand is the most basic and essential skill for a free flight bird. It is equivalent to the step up behavior that every bird should learn. It is also one of the easiest to teach to the bird. Using a target or a treat lure, the bird is encouraged to fly to the hand. Initial flights may really be a step. Then they increase in length. Ideally, the bird should also be taught to fly back to the perch. This is more difficult to teach. After flying to the hand, the bird is bridged and rewarded. They are returned to the perch to consume the reward. At first this is from just a step from the perch. Once they learn the routine, they can be launched back to the perch. It becomes a game. Flight skills are generally learned through practice and instinctive drift. Likewise, navigation must be learned by moving from easy areas to more difficult areas. The handler can call to the bird to help them locate the site to land. Some birds may initially have difficulty slowing down for a landing. This can be a difficult skill initially, especially when there is significant wind. Negotiating wind currents is more advanced. Days with minimal wind should be chosen for early flights. However, a strong flyer will often play in the wind currents, rising and dipping, flying almost in place when against the wind. Evasive maneuvers are also initially done in play. These playful turns and dives can then be used when needed for evading predators on the wing.

Take a Deep Breath!

The first outdoor flight is likely to be a nerve-wracking experience. Regardless of how skilled of an indoor flyer the bird seems, there is a heart sinking feeling the first time a bird flies past the owner and then away. This feeling is matched by the incredible satisfaction of seeing them flying back toward them and ultimately landing on the hand.

Tribulations

Rarely does anything go entirely as planned. There will usually be some bumps in the road to free flight. Usually this will be having a bird land somewhere difficult to retrieve, or worse completely out of sight. Anyone planning to engage in free flight should have a recovery plan in place. Recovery of a bird that flies away is very likely to be successful when the handler is prepared. Some tools to have ready include a phone, binoculars, transportation, a cage or carrier, and preferably an anchor bird, a recording of a species-matched bird, or similar lure. A pouch of treats should be ready as well. The first goal when a bird is flying away from the handler is to keep visual contact with the bird. If the location of the parrot is known, and it has had some training, recovery is very likely. The binoculars can help to maintain that visual contact. If visual contact is lost, the bird is likely to be panicking as much as the owner. Periodically calling out will often prompt a contact call from the bird. This is generally how a lost bird is located. It's surprising how easy it is to miss a large red macaw in a tree until the contact vocalization is made. If contact cannot be established within a short period of time, various contacts should be made to get help in re-establishing contact. Anchor birds are invaluable in the recovery of lost birds, especially when the lost bird is socialized to other birds.

Usually, birds are not going to continue to fly further away unless something frightens them. Often they will already be frightened by the ordeal. They may be reluctant to launch from the tree or other site initially. The presence of the owner may actually cause them to relax. Constantly calling to them will not be helpful. Avian eyesight is excellent and holding some of their favorite food items or even water will sometimes be enough to spark some interest in moving. They should be cued to come down when they appear receptive and ready to move. Sometimes walking away from them will stimulate them to fly after their handlers.

Although it is not entirely new, free flight of parrots is increasing in popularity and avian veterinarians may encounter owners that are or are planning to engage in this activity. A knowledge of the methods, risks, and benefits of this practice will help the practitioner provide good advice.

References

1. Biro C. Hiking with seven macaws. Available at: https://www.youtube.com/watch?v=hD4w_5-LRpw. Accessed July 6, 2016.

2. Liberty Wings. Avian training and flight instruction. Available at: www.libertywings.com. Accessed July 6, 2016.

The Chicken Foot: When is a Bumble Not a Bumble?

Marli Lintner, DVM

Session #336

Affiliation: From Avian Medical Center, 15952 SW Quarry Road, Lake Oswego, OR 97035, USA.

Abstract: The term "bumble foot" referred to a specific swelling and necrotic core formation in the metatarsal pad of the avian foot. Many other foot ailments have been erroneously lumped into this category by the pet chicken owner. Diagnosis and treatment of bumble foot and the most common foot issues will be discussed. These include bony issues, infectious synovitis, toe jam, scale overgrowth, frostbite, gangrene, and gout.

Diseases of the Chicken Foot

Introduction

It is important not to lump a large variety of diseases and injuries affecting the foot of the pet chicken into the diagnosis of "bumble foot". This can be very problematic when well-meaning owners, armed with advice found on the Internet, attempt to treat their birds. Armed with pocket knives and herbal mixtures containing cayenne pepper, olive oil and apple cider vinegar, they will carve away at bumbles that in fact are arthritis, fractures, gout or just dead skin buildup. An accurate diagnosis from a veterinarian will save time, money, pain to our patients, and hopefully result in a more informed, if not better, outcome.

Anatomy

The gallinaceous foot is composed of 4 toes with toes 2, 3, and 4 directed forward and the number 1 toe smaller and directed backwards. While not as extensive as in the duck family, chickens do have a moderate web structure extending about half the distance of the first phalange. The base of the tarsometatarsus forms the base of the foot. A distinct cartilaginous structure, the metatarsal pad, protects and cushions this base. The majority of the phalangeal structure is bone, joint, ligament and long tendon insertions.[1] Virtually no muscle is present in the gallinaceous foot.

The scaled skin of the foot presents some challenges for the practitioner. The limited stretch and flexibility of the skin can make closing large wounds difficult if not impossible. Scabbing and constriction of wounds can impinge on blood supply. Response to trauma, especially burns and frostbite, may be slow to manifest visually. And loss of scale pattern with resulting exposure of thin dermis can take months to rescale.

Spurs can occasionally be present in hens as well as roosters. The spur is a bony growth arising medially from mid metatarsal bone. A vigorous, often sharp, horny covering progressively grows from this bone. The length, sharpness and curvature can interfere with the bird's normal gait and cause trauma to the opposing tarsometatarsus. Trimming of spurs should take into account the bony structure and concurrent blood and nerve supply. Aggressive trimming or trauma can deglove the spur leaving exposed bone.

Bumble Foot Issues

The backyard chicken will traverse a variety of substrates in its daily travels. Backyards can consist of lawn, bark dust, gravel, cement, wood, areas of brambles, garden, pine needles, etc. A large variety of litters are used in coops including sand, shavings and straw. The chicken foot is well adapted for this and all of the above have produced chickens without foot issues. Current theory incites thorns or sharp substrates puncturing the foot and allowing bacteria to invade the foot pad. In my experience, foot issues are closely related to chronic pressure and bruising of the metatarsal pad. Roost height in the coop (or patio decks) is the most common cause; the higher the roost the more issues. Roosts with sharp edges (2 × 4's) and/or hard, slick flooring also increase the chance of foot and leg issues. In the evening, as the birds fly up to roost, significant jockeying for position is normal. On sharp edged perches such as 2 × 4 lumber, bruising of the foot pad can result. In the morning, a mad scramble to be the first hen out of the coop is also typical behavior. Landing on hard, slick surfaces such as plywood, linoleum, or cement, even if padded with straw or shavings, can also result in bruising of the metatarsal pad. Finally, walking up or down steep "ladders" where the foot slides till a rung is encountered can also cause bruising.

Bruising of the metatarsal pad results in a focal area of necrosis, much in the same way as a decubital ulcer or bed sore forms. As the scab and core form, secondary infection can occur, but sterile abscesses are possible. Continued weight bearing on this core continues the damage and blood supply compromise.

Treatment modalities have been well documented.[2] Most consist of a variety of debridements of the bumble, antibiotic treatments and doughnut-type padding to keep pressure off the lesion as it heals. Aggressive debridement, when the lesion is in the cellulitis stage, should be avoided and can cause additional damage to the metacarpal pad. At this stage, anti-inflammatory treatment may be helpful. When wrapping, immobilize the metatarsal joint to decrease plantar tendon motion. This will help limit tendonitis and the spread of infection up the leg. Environmental changes should be initiated to limit additional damage to the feet, and a weight loss program started if obesity is a factor.

Non Bumble Foot Issues

Bony issues

So when is bumble foot not actually bumble foot? One of the most common causes of sores on the bottoms of feet and toes comes from bony enlargement as seen in osteoarthritis, fractures or dislocations. Swollen, hard joints or bony calluses will add pressure and wear to the skin, even on soft surfaces. The resulting sores often become infected and resemble bumble foot lesions. The large amount of hard, bony reaction can make it surprisingly difficult to assess with palpation alone. Radiographs are very helpful. These lesions are best padded and completely immobilized. Simple fractures heal fairly quickly, but secondary infections leading to osteomyelitis are a long-term problem and require a lengthy treatment and guarded prognosis.

Infectious synovitis

Infectious synovitis caused by *Mycoplasmas synoviae* can cause extremely painful swelling of the footpads. In most cases, swelling of the hock and flexor tendon sheaths is also present. The swellings contain thick, clear, viscous material as compared to the harder caseous cores of a bumble. Other bacteria, including *Staphylococcus, Escheridria coli, Pasteurella,* and *salmonellae* can also cause synovitis.[3] Antibiotics can rid the joints of bacterial infections, but often irreparable damage has resulted.

Toe jam

Chickens have a very small foot web, much less developed than a duck, but this tissue can still cause some problems. Dirt, feces, and dead skin and debris can build up from underneath, forming nodules in the webs. These appear as swellings between the base of the toes, and can become red and sore. Treatment is simple, by exerting mild pressure on the swelling, push down and turn the web inside out, thus pushing out the built up debris. The resulting "toe jam" is easily removed. Owners should be instructed to monitor for reoccurring build ups.

Scale overgrowth

The cranial aspect of the tarsometatarsus is covered with a thick, overlapping scale pattern. Due to the sharp angle and mobility of the tarsometatarsus with the phalanges, these scales can put pressure and even embed into the top of the foot causing pain and swelling. A variety of conditions will increase the likelihood of this occurring. Knemidocoptes mite infections can distort the scale pattern, elevating the scutes so they dig into the foot.

Well-meaning owners believe lubricating the scales will alleviate the scale issue. Unfortunately, more moisture swells the scales and the skin and makes the problem worse. The addition of herbals, especially tea tree oil, can increase the irritation. Mineral oil makes a greasy mess and olive oil will eventually go rancid on the skin.

Pet chickens have a much longer life expectancy than their commercial counterparts. Skin issues in the older hen (we are seeing hens over 10 years old) are not well documented. But, it is apparent the ingrown scale issues increase with age. Scaled skin is healthiest when dry and clean. Scales are designed to exfoliate naturally, but if they are building up, assistance may be needed. Pulling them off should be done with caution as the underlying skin can be torn. Using rongeur forceps to split the scales up the center will naturally help them to flake off.

Frostbite

Frostbite lesions on the foot can mimic the appearance of bumble foot lesions. In the early stages, frostbite causes pain, edema and swelling. The toes are affected most commonly (and most severely), but the pad can also be damaged. It may take days, or even weeks, before the tissues show the characteristic black scabbing and necrosis. Weather conditions usually have changed by the time lesions appear, complicating the diagnosis.

Dry gangrene

While dry gangrene of the toe tips is characteristic of frostbite, any injury to the blood supply of the foot or toes can result in hypoxia, necrosis and death of the tissue. Strangulation injuries with twine or wire are common. Lacerations that result in scab constricture around a digit need to be monitored and possibly split to avoid blood supply compromise. When dry gangrene occurs, there is usually a very clear demarcation between live and dead tissue. The hard, dry, black tissue can be trimmed back with rongeur forceps to leave a smooth, flush scab. No anesthesia is necessary as there is no feeling in the dead tissue. Scab formation is an important part of wound healing and scabs should be left in place when possible, but debulking and scab decompression can be critical to saving digits.

Gout

Gout is seen in two forms in commercial poultry, visceral and articular. Articular gout results from urate deposition within the joints, most commonly those of the feet.[3] Swellings can be huge, disfiguring and very painful. Pressure and abnormal weight bearing can cause sores on toes and pads. Secondary infections add to the joint damage. Studies with commercial poultry have implicated high protein diets and genetic predilections. Pet poultry are often fed diets high in protein, animal fat and cholesterol. In the author's experience, diets high in dark greens, such as kale, spinach, and chard, can also cause gout. Dehydration from lack of water, or unpalatable water due to heavy medication dosing, is another concern. Opening the lesions may relieve some pressure, but increases the chance for joint infection. Nutritional changes can keep the disease from progressing, but will not significantly change the damage done to the joints. Anti-inflammatories and pain medication should be used with caution as they can be nephrotoxic, especially with compromised renal function.

Conclusion

The backyard chicken will expose their feet to surfaces and situations undreamed of by those in commercial poultry operations. A thorough history and physical examination will go far in helping the practitioner diagnose and treat diseases of the foot.

References

1, King AS, McLelland, J. *Birds: Their Structure and Function.* East Sussex, England: Balliere Tindall; 1984;1-5:9-22.

2. Greenacre CB, Morishita TY. *Backyard Poultry Medicine and Surgery: A Guide for the Veterinary Practitioner.* Ames, IA: John Wiley & Sons; 2015.

3. Calnek BW, Barnes HJ, Beard CW, McDougald, LR, et al. *Diseases of Poultry.* Ames, IA: Iowa State University Press;1997.

Section 12

AAV Case Reports

Geoff Olsen, DVM, Dipl ABVP (Avian)
Moderator

Stifle Disarticulation as a Pelvic Limb Amputation Technique in a Cockatiel *(Nymphicus hollandicus)* and a Northern Cardinal *(Cardinalis cardinalis)*

Sarah Ozawa, DVM, and
Christoph Mans, Dr. med. vet., Dipl ACZM

Session #052
AAV 2016 Student Manuscript Competition Award Recipient

Affiliation: From the Department of Surgical Sciences, School of Veterinary Medicine, University of Wisconsin-Madison, 2015 Linden Drive, Madison, WI 53706, USA.

In avian species, the most common sites for pelvic limb amputation are transection of the proximal tarsometatarsus and mid-femur.[1,2] In human orthopedics, stifle disarticulation, or through the knee amputation, is an established technique of pelvic limb amputation.[3-6] The advantages of this amputation method in humans include decreased surgical and anesthesia time, simpler surgical procedure, reduced blood loss, less trauma to soft tissue structures and the lack of required bone and muscle sectioning. We describe the use of stifle disarticulation in 2 birds as an alternative to transfemoral limb amputation. Case 1 was a 1-year-old northern cardinal that presented for soft tissue swelling and fracture of the 3rd digit of the left limb with secondary bacterial infection and necrosis. Case 2 was a 25-year-old male cockatiel that was diagnosed with distal extremity necrosis following a non-union of a closed tibiotarsal fracture. In both cases amputation was recommended due to the poor prognosis for healing and vascular compromise. The stifle disarticulation method was elected in the two cases because both birds were poor surgical candidates for traditional methods of amputation due to patient size or age. Stifle disarticulation surgery for amputation of the pelvic limb was performed successfully in both cases and recovery was unremarkable. The outcomes of these two cases suggest that stifle disarticulation may be used as an alternative to transfemoral amputation in birds where prolonged surgery, anesthesia or patient size precludes traditional amputation techniques.

References

1. Martin H, Ritchie BW. Orthopedic surgical techniques. In: Ritchie BW, Harrison GJ, Harrison LR, eds. *Avian Medicine: Principles and Application.* Lake Worth, FL: Wingers Publishing Inc;1994:1138-1162.

2. Helmer P, Redig PT. Surgical resolution of orthopedic disorders. In: Harrison GJ, Lightfoot TL, eds. *Clinical Avian Medicine.* Palm Beach, FL: Spix Publishing Inc;2006:761-773.

3. Pinzur MS, Bowker JH. Knee disarticulation. *Clin Orthop Relat Res.* 1999;361:23-38.

4. Stark, G. Overview of knee disarticulation. *J Prosthet Orthot.* 2004;16:130-137.

5. Cull DL, Taylor SM, Hamontree SE, et al. A reappraisal of a modified through-knee amputation in patients with peripheral vascular disease. *Am J Surg.* 2001;182:44-48.

6. Pinzur MS. Knee disarticulation: Surgical procedures. In: Bowker JH, Michael JW, eds. *Atlas of Limb Prosthetics: Surgical, Prosthetic and Rehabilitation Principles*. Rosemont, IL: Mosby Year Book Inc;1992:479-486.

7. Penn-Barwell JG. Outcomes in lower limb amputation following trauma: A systematic review and meta-analysis. *Injury.* 2011;42:1474-1479.

Osteomyelitis in Two Aquatic Avian Species, A Mute Swan *(Cygnus olor)* and a Ring-Billed Gull *(Larus delawarensis)*, Caused by Novel Bacteria: *Chryseobacterium indologenes*

Shachar Malka, DVM, Dipl ABVP

Session #340

Affiliation: From The Humane Society of New York, 306 East 59th Street, New York, NY 10022, USA.

Chryseobacterium indologenes is a soil and water gram-negative bacteria infrequently diagnosed as a human opportunistic pathogen and rarely mentioned in veterinary literature. Its pathogenicity has been reported only in fish, and has been sporadically mentioned as a culture result in a tree frogs and map turtle species with unknown significance. In human medicine, *Chryseobacterium indologenes* was described as the cause of various infections, including meningitis, pneumonia and implant failure. In this report, clinical presentation, diagnosis, treatment and outcome is described in one Anseriforme species and one Charadriiforme species. Both birds were wild born and brought into veterinary care by the same wildlife center. The mute swan was presented initially with an infection in its maxillary bill. The bird shared a rehabilitation swimming pool with the ring-billed gull, a long term resident of the wildlife center, that later developed osteomyelitis and osteoarthritis of its leg. Both birds responded well to antibiotic therapy. The mute swan was released back to a reservation lake from where she originally came and the gull was kept in captivity due to its original injury. Water culture from the wildlife rehab center's pool confirmed the same bacteria that was cultured from the patients. To the author's knowledge, this is the first report of *Chryseobacterium indologenes* infection in an avian species.

Normal Circadian Tear Production in Two Steppic Avian Species, the Houbara Bustard and the Stone Curlew

P. Azmanis, DVM, Dr. med. vet, Dipl ECZM (Avian),
S. Gough, RVN NCert. (A&CC) Cert Exotic, Dipl CABT,
D. O'Donovan, Dipl H. Ed., BSc, Pg Dipl, MSc, CBiol, MSB

Session #240

Affiliation: From Dubai Falcon Hospital, PO Box 23919, Dubai, UAE (Azmanis, Gough), Wadi Al Safa Wildlife Center, Dubai, UAE (O' Donovan).

Circadian tear production in avian species is not well studied and only a few reports exist in other species.[1,2] In healthy mammals the normal circadian tear production depends on various factors including the integrity of ocular anatomic structures, the corneal surface area and the eye size, the blinking rate, the time zone (photoperiod), the environmental humidity and temperature, the species and possibly the sex and the eye (left or right). We hypothesized that steppic species need to conserve moisture and therefore may have reduced tear production compared to other species. Additionally nocturnal and diurnal steppic species may differ in tear production due to the fact that the moisture in arid environment is higher at night and that nocturnal species have usually larger eyes. We measured the tear production with Schirmer's Tear Test (STT I) in 4 adult houbara bustards and 4 adult stone curlews (2 males and 2 females from each species) in 5 different time zones (dawn, morning, noon, dusk, night) for 3 subsequent days during winter time. Birds had no ocular pathology and were kept/fed in outdoor pens. Cornea size was measured with a micrometric caliber, and the blinking rate (nicticans stroke/min) observed. Although the project is under process the preliminary results suggest that the mean daily tear production in bustards was 8.16 mm/60s (OD) and 7.3 mm/60s (OS) and in stone curlews 16.31 mm/60s (OD) and 15.3 mm/60s (OS). There was no significant difference in the left and right eye tear production or between males and female houbara bustards. Further results will be presented and discussed.

References

1. Berger SL, King VL. The fluctuation of tear production in the dog. *J Am Animal Hosp Assoc*. 1998;34: 79-83.

2. Piccione G, Giannetto C,Fazio F, Giudice G. Daily rhythm of tear production in normal horse. *Vet Ophthalmol*. 2008;11:57-60

Highly Pathogenic Avian Influenza in Juvenile Black-Capped Chickadees *(Poecile atricapillus)*

Renee Schott, DVM,
Anibal Armien DVM, MSc, Dipl ACVP,
Leslie Reed, DVM,
Mia Kim Torchetti DVM, MS, PhD,
Arno Wünschmann Dr. med. vet., Dipl ACVP

Session #158

Affiliation: From the Wildlife Rehabilitation Center of Minnesota, 2530 Dale St N, Roseville, MN 55113, USA (Schott, Reed), University of Minnesota, 1365 Gortner Ave, St Paul, MN 55108, USA (Armién, Wünschmann), US Department of Agriculture, 1920 Dayton Ave, Ames, IA 50010, USA (Torchetti).

During the 2015 highly pathogenic avian influenza (HPAI) outbreak in Minnesota, several thousand samples from wild birds were tested, however only 3 total wild birds were diagnosed as having HPAI.[1] Between June 11 and 26, 2015, nine juvenile black-capped chickadees (*Poecile atricapillus*) with abnormal neurologic signs were submitted for necropsy. All were found by concerned citizens in the Minneapolis/Saint Paul metropolitan area and were taken to the Wildlife Rehabilitation Center of Minnesota. Seven individuals presented with severe neurologic signs including torticollis, nystagmus, and log-rolling; these individuals were euthanized upon admission due to grave prognosis for recovery; two others developed intention tremors after two weeks in care. All were submitted to the Minnesota Veterinary Diagnostic Laboratory for necropsy and diagnostic testing. One individual was positive for Influenza A on immunohistochemistry. Another individual was suspect on Influenza A Matrix (USDA) PCR. This sample was then attempted to be subtyped H5 or H7 (USDA) via PCR but was negative for both. The National Veterinary Services Laboratory (Ames, IA) found this sample positive for avian influenza H5 and avian influenza H5 icA via rRT-PCR. Virus isolation on both samples failed. Histologic descriptions for the two HPAI positive individuals consisted of non-suppurative, multifocal encephalitis. The remaining chickadees had the following diagnoses: Ascarid larval migration encephalitis (2), encephalitis of unknown origin (2), and no significant microscopic lesions (3). The two HPAI positive chickadees are the first confirmed case of a natural HPAI infection of a passerine bird during the 2015 Eurasian HPAI outbreak.

References

1. US Department of Agriculture, US Geological Survey. Wild Bird Highly Pathogenic Avian Influenza Cases in the United States: 2015. Available at: https://www.aphis.usda.gov/wildlife_damage/downloads/WILD BIRD POSITIVE HIGHLY PATHOGENIC AVIAN INFLUENZA CASES IN THE UNITED STATES. pdf. Accessed July 6, 2016.

Unilateral Complete Obstruction of the Primary Bronchus in Two Juvenile Macaws

Anna Martel-Arquette, DVM, and
Christoph Mans, Dr. med. vet., Dipl ACZM

Session #169

Affiliation: From School of Veterinary Medicine, University of Wisconsin-Madison, WI 53715, USA.

An 11-month-old green-winged macaw *(Ara chloropterus)* (case 1) and an 8-month-old military macaw *(Ara militaris)* (case 2) were independently presented for acute onset of dyspnea. Both birds had no previous medical history. Survey radiographs in case 1 showed no abnormalities. Computed tomography was performed in both cases and complete obstruction and dilation of one primary bronchus was present, with extension of hyperattenuating material into the syrinx in both cases. No air sacculitis was present in either case. Coelioscopy was performed in case 1 and revealed no abnormalities. Tracheoscopy was performed in both cases and debridement of the syrinx was performed in case 1. Case 2 was euthanized on the day of diagnosis. Treatment with oral and intratracheal antifungal therapy was attempted in case 1 but the bird died 12 days after the initial diagnosis. Complete necropsies were performed in both cases. Fungal infections predominately obstructing a single primary bronchus was diagnosed in both cases. These two cases demonstrate the value of computed tomography in avian medicine, as bronchial obstruction cannot be diagnosed by radiography, tracheoscopy, or coelioscopy. Obstruction of a primary bronchus should be a differential for birds presenting with acute dyspnea.

Intraosseous Lipoma in a Rose-Breasted Cockatoo *(Eolophus roseicapilla)*

Graham Zoller, DMV, IPSAV,
Jerome Cavoizy, DMV,
Lauriane Devaux, DMV,
Alexandra Nicolier, DVM, Dipl ECVP,
Minh Huynh, DVM, Dipl ECZM

Session #186

Affiliation: From Centre Hospitalier Vétérinaire Frégis, 43 avenue Aristide Briand 94110 Arcueil, France (Zoller, Devaux, Huynh), ExoVet, 17 Rue de la Bergerie 77181 Coutry, France (Cavoizy) and Vetdiagnostics, 14 avenue Rockefeller 69008 Lyon, France (Nicolier).

A 9-year-old, female, rose-breasted cockatoo (*Eolophus roseicapilla*) was presented for an open humeral fracture noticed an hour earlier. Upon presentation the bird was in shock and markedly obese. A right humeral open fracture was palpated associated with soft tissue swelling. The left wing was drooping and a swelling of the left intertarsal joints was noticed. Subcutaneous adipous masses were observed ventrally to the celomic cavity. Radiographs showed an oblique, simple and cranio-medially displaced right humeral fracture. Heterogenic and invasive intramedullary lesions of both humeri, ulnas and distal tibiotarsi were noticed. Bloodwork results were compatible with inflammation and muscular trauma associated with the fracture. Triglycerides and bile acids concentrations were within reference interval. A surgical reduction of the fracture was attempted after a 24-hour period of stabilization. The bird went into cardiorespiratory arrest postoperatively and did not respond to resuscitation. Complete necropsy was performed. Histopathology findings were consistent with a severe multifocal osseous and subcutaneous lipomatous infiltration. Focally extensive osseous necrosis, fibrin deposition and acute hemorrhages were observed in the right humerus, consistent with an acute focal osseous fracture. Histologic examination of the liver revealed a diffuse hepatic lipidosis. This is the first reported case of multiple intraosseous lipoma in a bird and the first case of associated fracture.

Lateral Body Wall Herniation Secondary to Chronic Oviductal Disease in Two Psittacine Birds

Anna Martel-Arquette, DVM,
Courtney Wong, DVM,
Brian Speer, DVM, Dipl ABVP (Avian),
Christoph Mans, Dr. med. vet., Dipl ACZM

Session #073

Affiliation: From the School of Veterinary Medicine, University of Wisconsin-Madison, WI, 53715, The Medical Center for Birds, Oakley, CA 94561, USA.

Lateral body wall herniation with involvement of the oviduct were diagnosed in a nanday conure *(Aratinga nenday)* (case 1) and a cockatiel *(Nymphicus hollandicus)* (case 2). Both cases had a history of chronic egg laying. Case 1 presented with a mass on the left lateral body wall, which was diagnosed as an egg within the herniated oviduct protruding from the lateral body wall. Case 2 presented with a chronic fluid-filled hernia on the right flank. A cystic, persistent right oviduct and an impacted left oviduct were found during surgical exploration. Both cases were successfully treated by surgical correction of the hernia and salpingectomy. Chronic reproductive activity in psittacines can predispose to body wall herniation due to elevated levels of reproductive hormones. Lateral body wall herniation with oviductal involvement in pet birds has not previously been reported in the literature and should be considered as a differential for birds presenting with masses on the lateral body wall.

Diabetes Insipidus in a 4-Year-Old Congo African Grey Parrot *(Psittacus erithacus erithacus)* Positive for Avian *Bornavirus*

Delphine Laniesse, DMV, DVSc,
Hugues Beaufrère, DMV, PhD, Dipl ACZM, Dipl ABVP (Avian), Dipl ECZM (Avian),
Laura Bourque, DVM,
Dale A. Smith, DVM, DVSc

Session #275

Affiliation: From the Department of Clinical Studies (Laniesse, Beaufrère), and the Department of Pathobiology (Laniesse, Bourque, Smith), Ontario Veterinary College, University of Guelph, 50 Stone Road, Guelph, ON, N1G 2W1 Canada.

A 4-year-old African grey parrot was presented for polyuria and polydipsia (PUPD): it was drinking 2.5 L of water/kg/day. Blood work revealed hypernatremia and increased plasma osmolality. A modified water deprivation test was performed: plasma osmolality increased significantly while the specific gravity of urine remained low, and the bird rapidly lost weight. Arginine vasopressin (AVP, mammalian antidiuretic hormone), given conjunctivally at 20 µg/kg, did not decrease plasma osmolality but caused a 50% increase in urine osmolality (43-62 mmol/L). The degree of PUPD, the hyperosmolar plasma despite free access to water, and the marked dehydration in response to minimal water deprivation suggested a diagnosis of diabetes insipidus (DI). Central DI was suspected based on the response to AVP administration, thus the bird was prescribed AVP (20 µg/kg intranasally q12h). No reduction in PUPD was identified in response administration of AVP; therefore, the treatment was changed to arginine vasotocin, the avian antidiuretic hormone. An empirical dosage of 0.5 µg/kg arginine vasotocin intranasally q12h was initially prescribed, and was gradually increased to 4 µg/kg q12h over 2 weeks. Water consumption decreased progressively (up to 38%) over several months, blood osmolality decreased and urine osmolality increased slightly, but the response to therapy was variable. The bird died 2 years after initiation of therapy. Avian *Bornavirus* (ABV) infection was diagnosed on post-mortem examination based on histopathology and positive rtRT-PCR. A cause-effect relationship between ABV infection and diabetes insipidus was suspected.

Autoimmune-Like Disease Affecting the Skin and Feathers in a Goshawk *(Accipiter gentilis)*

H. L. Shivaprasad, BVSc, PhD, Dipl ACPV, and
Victoria Joseph, DVM

Session #316

Affiliation: From California Animal Health and Food Safety Laboratory System-Tulare branch, 18830 Road 112, Tulare CA 93274, School of Veterinary Medicine, University of California, Davis, CA 95616, USA (Shivaprasad) and 3985 Foot Hills Blvd, Roseville, CA 95747, USA (Joseph).

An approximately three-year-old male goshawk (*Accipiter gentilis*) had a one-year history of gradual loss of feathers all over the body except for the tail and primary feathers in the wings. Feathers on the head were sparse and appeared immature. The bird was euthanatized and submitted for necropsy. The bird was in fair postmortem condition, mildly emaciated and weighed 1.40 kg. The bird had generalized loss of feathers as described above. There were no gross lesions of diagnostic significance in any internal organs. Tissues from various organs especially numerous feathers along with the skin were collected in 10 % neutral buffered formalin for histopathology. Culture of the skin revealed coagulase negative *Staphylococcus* spp. and liver was negative for aerobic bacteria and intestine were negative for *Salmonella* spp. The bird was negative for avian influenza and West Nile virus by PCR.

Histopathology of various organs revealed significant lesions only in the skin and feathers. There was infiltration of mostly lymphocytes and a few plasma cells in the epithelium of the feather follicles extending in to the wall as well as to the epidermal collar. There was mild to moderate necrosis of the epithelial cells with occasional formation of bulla that contained a few acantholytic cells. Other changes in the feather follicles consisted of necrosis of the feather follicle epithelium, fibrinosuppurative inflammation and giant cell granuloma's. There were many feather follicles that were filled with small to large amounts of keratin. There was hyperkeratosis of the epidermis. An attempt to demonstrate IgG in the skin or the feather follicles in the hawk by the use of chicken anti IgG antibody was inconclusive.

Various infectious and non-infectious diseases affecting the skin and feather have been described in birds.[1-5] Except for two autoimmune skin diseases in chickens; scleroderma and vitiligo, no similar condition has been described in birds.[6] The histologic lesions in the present case are suggestive of an autoimmune skin disease. But IgG in the feathers could not be demonstrated equivocally by the use of anti-chicken IgG which may not cross react with accipiter species. One of the authors (HLS) has seen a few similar cases of autoimmune-like skin disease in psittacines over the years but was unable to confirm an autoimmune skin disease due to lack of proper reagents as anti-canine IgG did not work.

References

1. Perry RA, Gill J, Cross GM. Disorders of the avian integument *Vet Clin North Am Sm Anim Pract.* 1991;21:1307-1327.

2. Koski MA. Dermatologic diseases in psittacine birds: an investigational approach. *Semin Avian Exot Pet Med.* 2002;11:105-124.

3. Hudelson S, Hudelson P. Dermatology of raptors: a review. *Semin Avian Exot Pet Med.* 1995; 4:184-194.

4. Gill JH. Avian skin diseases. *Vet Clin North Am Sm Animal Pract.* 2001;4:463-492.

5. Shivaprasad HL. Integumentary. In: Fletcher O. ed. *Avian Histopathology*. 3rd ed. Jacksonville, FL: Am Assoc Avian Pathologists. 2008;393-427.

6. Wick G, Andersson L, Hala K, et al. Avian models with spontaneous autoimmune diseases. *Adv Immunol.* 2006;92:71-117.

Infraorbital Keratin Cyst in an Umbrella Cockatoo *(Cacatua alba)*

Jennifer E Graham, DVM, Dipl ABVP (Avian), Dipl ABVP (Exotic Companion Mammal), Dipl ACZM,
Shelley Hahn, DVM,
Abigail Mariano, VMD,
Rob McCarthy, DVM, Dipl ACVS

Session #036

Affiliation: From the Departments of Clinical Sciences (Graham, Mariano, McCarthy) and Pathology (Hahn), Cummings School of Veterinary Medicine at Tufts University, 200 Westboro Rd, North Grafton, MA 01536, USA.

A 10-year-old female umbrella cockatoo presented for evaluation of a grape-sized mass noted at the right commissure of the beak. The mass had been noted by the owner approximately 4 weeks prior to presentation. Computed tomography revealed a soft tissue attenuating, peripherally contrast enhancing mass measuring 2.5 cm in diameter within the right infraorbital sinus and extending into the right retrobulbar space. The mass extended between the right suborbital arch and mandible with a 1.8cm portion of the mass protruding externally. The bird was taken to surgery in an attempt to remove the mass. Debridement of the mass and its caseous components was not possible from the initial surgical incision at the caudolateral aspect of the eye so the incision was extended around the bony orbit to mobilize the eye. The eye was retracted with 2 stay sutures in the lower lid and another lobe of the cyst was visualized. Blunt dissection around two lobes of the cyst medial to the eye facilitated removal. Histopathologic examination consisted of a large cyst, lined entirely by keratinizing squamous epithelium, occasionally thrown into folds resulting in a slightly papillary appearance. The cyst was filled entirely with amorphic and lamellar keratin. The wall of the cyst was composed of a moderately cellular fibrous capsule. Also represented in the biopsy was the wall of the infraorbital sinus, lined by non-keratinizing squamous epithelium and ciliated pseudostratified columnar epithelium (corresponding to caudal and cranial sinuses) with no associated histological changes. This is the first known case describing diagnosis and management of an infraorbital keratin cyst in a bird.

References

1. Stiles J, Greenacre C. Infraorbital cyst in a white cockatoo (*Cacatua alba*). *J Avian Med Surg*. 2001;15(1): 40-43.

Esophageal Squamous Cell Carcinoma in Two Amazon Parrots

Anne Staudenmaier, VMD,
Elisabeth Simone-Freilicher, DVM, Dipl AVBP (Avian),
Brendan Noonan DVM, Dipl AVBP (Avian),
Patty Ewing DVM, MS, Dipl ACVP

Session #187

Affiliation: From the Avian and Exotic Department (Staudenmaier, Simone-Freilicher, Noonan) and the Department of Pathology (Ewing), Angell Animal Medical Center, 350 South Huntington Avenue, Boston, MA 02130 USA.

Esophageal squamous cell carcinoma was diagnosed in a 28-year-old blue-fronted Amazon parrot (*Amazona aestiva*) that was presented for a repetitive regurgitation-like motion of the neck and a 34-year-old yellow-naped Amazon parrot (*Amazona auropalliata*) that was presented for regurgitation and crop distention. A discrete, firm, non-moveable, subcutaneous mass was palpable near the crop on physical exam of the blue-fronted Amazon, and severe esophageal thickening was found on radiographs and CT scan in the case of the yellow-naped Amazon parrot. A diagnosis of squamous cell carcinoma was confirmed on histopathology of the esophagus in both patients.

Though esophageal squamous cell carcinoma is a not uncommon neoplasia in human medicine, being the eighth most commonly diagnosed cancer in people, it is a relatively rare diagnosis in veterinary medicine. Most avian cases are in poultry, and for the one psittacine case previously reported in the literature, the bird presented for severe systemic signs and without a gross mass palpable on physical examination. The cases presented here illustrate the need to consider esophageal carcinomas in psittacine birds as differential diagnoses in a parrot presenting with upper gastrointestinal signs or gastrointestinal motility issues and suggest that esophageal squamous cell carcinoma may not be as rare in parrots as previously thought.

References

1. Anderson WI, Steinberg H. Primary glossal squamous-cell carcinoma in a Spanish cochin hen. *Avian Dis*. 1989;33:827-828.

2. Higuchi K et al. Current management of esophageal squamous-cell carcinoma in Japan and other countries. *Gastrointest Cancer Res*. 2009;3(4):153-161.

3. Khaki F et al. Cervical type AB thymoma (mixed) tumour diagnosis in a mynah as a model to study human: clinicohistological, immunohistochemical and cytohistopathological study. *Diagn Pathol*. 2013;8:98-104.

4. Murtaugh RJ et al. Squamous cell carcinoma of the esophagus in an Amazon parrot. *J Am Vet Med Assoc*. 1986;188(8):872-873.

5. Strafuss AC, Ladds PW. Fibrosarcoma in a young chicken. *Avian Dis*. 1970;14(2):406-409.

Section 13

AAV Practical Information

M. Scott Echols, DVM, Dipl ABVP (Avian)
Moderator

Extreme Enrichment for the Modern Macaw

Kenneth R. Welle, DVM, Dipl ABVP (Avian)

Session #137

Affiliation: From University of Illinois College of Veterinary Medicine, 3505 Veterinary Medicine Basic Sciences Building, 2001 S Lincoln Ave., Urbana, IL 61802, USA.

Abstract: Captive birds are protected from some of the dangers they would face in the wild. In addition, they are usually guaranteed the resources they need for survival. The restriction of behaviors imposed by captivity, however, can result in some undesirable effects. While this is not entirely preventable, providing enrichment can help avoid some of these problems. Enrichment involves changes to animals' environments, diets, social arrangements, and human interactions to improve their quality of life. This paper will outline some methods that can be used to enrich the lives of captive parrots.

Introduction

Taking an animal into the home carries a great deal of responsibility. While the basic needs of food, water, and shelter are obvious, it is also necessary to provide for the more abstract needs of companion animals. Psittacine birds are intelligent and socially complex animals that can suffer physical and behavioral consequences when these needs are not met. Captivity, by necessity, restricts some natural behaviors and, in response, creates some new behaviors (displacement behaviors) to fill the void. Enrichment is defined as the act of improving the quality of something, or making it better. In the case of captive animals, this means making their life better. Enrichment in the broadest sense involves creating a life that provides for all of the physical, psychological, and emotional needs of an animal. The larger the behavioral spectrum an animal has, the more challenging this enrichment will be; likewise the larger the behavioral spectrum, the more critical this enrichment will be. This paper will describe some methods of enrichment that can be used for companion parrots. Some will involve trying to replicate natural behavioral opportunities for these birds, while others will involve replacement behaviors that can benefit the bird rather than creating further stresses. [1-5]

Time Budgets

One of the challenges of enrichment is the drastic alteration of the time budget for a captive animal compared to the time budget of a wild animal.[6,7] Although time budgets for wild animals are not usually consistent, some things are reasonable to assume. For instance, tropical birds would typically have a light cycle approximating 12 hours of light and 12 hours of dark. This narrows the range of time that birds will be active. The remaining time may be utilized feeding, socializing, performing self-maintenance, moving to and from feeding areas, and playing. Of course, many of these usages overlap significantly. Time cannot always be strictly budgeted into one category. Feeding, self-maintenance, and play are all social activities, so in reality, the total time budget would exceed 24 hours. In captivity, some of these activities are necessarily restricted. This void will sometimes force a shift toward other activities or create pathologic displacement behaviors such as stereotypies, apathy, or self-damaging behaviors. Much of the goal of enrichment is to manage the time budget to recreate the activities of a wild bird or to replace proportions of them with healthy, stimulating, and appropriate behaviors.[8-10]

Environmental Complexity

The problem with captivity is, simply put, captivity. Restricting the movements of a bird will limit the extent of exposure to the environment. How restrictive this is depends highly on the environmental complexity available to the bird. A good analogy would be incarceration of prisoners. Placing a prisoner in solitary confinement is far more restrictive than housing one in the general prison population, which is in turn, far more restrictive than someone confined on "house arrest" with a monitoring anklet. None of us like to think of our birds as prisoners, but many pets are given environments far smaller and less complex than an average prisoner is afforded.

Size is one of the factors involved in environmental complexity. Generally, larger spaces will allow a greater diversity of activity than smaller spaces. This can be affected somewhat by other factors, but extremely small spaces will necessarily restrict the available choices for activity. While it is true that some birds seem to become bewildered by an extremely large cage, this is likely to be due to early experiences or lack thereof. No parrot has a small cage in the wild, and their aversion to open space is a pathological response, born from developmental stunting. It is relatively easy to find information about the minimum cage size for birds. This may not be the best way to look at the equation. A better way would be to determine the maximum space that can be provided. Once that is determined, then the bird can be chosen to match the available space. Guidelines stating that a cage should be large enough to allow for the wings to extend without touching is grossly inadequate to allow normal behavior. The majority of birds spend 12-22 hours in the cage; it is impossible to allow normal behaviors to occur in a space barely large enough to turn around in. Whether or not a bird is flighted, a flight cage, open bird-room, or aviary is highly recommended in order to provide ample opportunity for activity. In addition to the raw space, the shape and orientation of the cage is also important. Two cages of equivalent volume may have very different amounts of "effective" space. The older cages for small birds with a vertical, narrow, cylindrical shape had no more usable space for the bird than a cage of half the height. Generally rectangular cages oriented horizontally provide the most usable space for birds.

Exploratory opportunities are another feature of avian enclosures that can influence the complexity. Parrots have the unique capability among birds to climb using their feet and beaks. For birds with clipped wings, and those with wing conditions preventing flight, this is the primary mode of locomotion. Provision of furnishings to the cage that encourage this activity will enhance the ability to explore. Wire cages provide this opportunity inherently. However, climbing excessively on the wire of the cage can have deleterious effects on the remiges and rectrices. The feathers often push between bars bending the barbs back and ultimately breaking them. Natural branches are a good addition to most cages. In addition to providing climbing opportunities, they allow for beak activity. These branches can be oriented horizontally or vertically, depending on the bird's abilities and the configuration of the branches. They can also be hung rigidly between the cage sides or can swing more freely. Very few perches in nature are rigid. If natural branches are limited, wooden dowels, cactus spines, rope perches, commercial swings, and polyvinyl chloride (PVC) pipe can be used. PVC can be roughened to improve grip or can be wrapped in self adherent bandages or natural twine or rope. Ropes can be used alone as well. The diameter should be appropriate for the size of the bird. The ways the ropes are hung are dependent on the climbing abilities of the bird. Ropes should generally be natural fibers in case the bird chews off fragments and swallows them, and to prevent strands from constricting toes. Ladders offer a simple climbing opportunity as well and can be rigid or constructed of rope. Large manila cargo nets are excellent climbing furniture for large birds.

One type of item that should be avoided in most situations is any nesting structure. Most species of parrot are quite territorial about their nests, and will display aggressive responses to people or even other birds that they would normally tolerate. This is because most species will separate from the flock for breeding. This is when the nest structure is usually utilized. For most parrot species, the nest is a hollow tree cavity. Any hollow item large enough to climb into could be perceived as a nesting structure. One possible exception to this rule may be quaker parrots. These birds are communal nesters, and nest building is not isolated to breeding activity. They also build nests, which is unique among psittacids. Providing these birds with small twigs will allow natural nest-building activity to occur.

Visual and auditory stimuli will also affect the environmental complexity. The activity and scenery surrounding a bird's cage should be interesting but not stressful to the bird. A cage usually should be located in an area with a great deal of household activity. This allows non-contact interaction with people in the household. Verbal interaction should be relatively constant when people are present. In households where no one is home through most of the day, alternative stimuli should be provided. Television, radio, other pet birds, a view of a bird feeder outside, or other interesting stimuli may be useful. Care must be taken that the stimuli do not evoke fear responses, at least on a constant basis. One may think that watching an outdoor bird feeder would be a great stimulus but seeing birds snatched up by Cooper's hawks may be more like a horror movie to some birds. This is not to say that brief exposure to threatening stimuli are harmful to a psychologically stable bird, but this will depend on prior experience and the individual personality of the bird. The olfactory sense of birds has long been considered weak, largely due to the small olfactory lobe of the brain in most species. There is, however, some evidence that olfactory cues are sometimes used by birds. This can be encouraged simply by feeding fragrant vegetables (e.g., sweet peppers) or using a small amount spices (e.g., cinnamon, nutmeg) on chew toys.

Toys can provide a great deal of activity for parrots. There are numerous types of toys that are appropriate for parrots. These will provide different types of activities. They should be selected specifically for the species involved. Foot toys simply allow the bird to manipulate the item with the foot. This is usually combined with beak activity or foraging. Birds that would feed in trees in the wild are more likely to use foot toys than ground feeding species like cockatiels or grass parakeets. Chew toys are critical for all parrots. These should be constructed of natural, non-toxic materials and should be easy enough to destroy that the bird does not give up on the project. Branches with the bark still on it, cardboard, and thin, soft woods are good for birds with low to moderate destructive potential. Harder woods are appropriate for very beaky birds. Puzzles allow the bird to work out a problem to achieve some goal, whether that is a food item, another toy, or just a sound. Puzzles should be simple at first, gradually increasing the complexity as the bird learns and gains confidence. Like humans, birds will become bored with continued exposure to the same toys. How many of us have home gyms that are collecting dust or serving as a clothing rack? Rotation of the toys will help keep the interest up. Care must be taken to avoid frightening timid birds with new toys also. Some birds will react adversely to new items being placed in the cage. In these cases, toys should be introduced outside of the cage in a non-threatening way (i.e., allow the bird to approach the toy on its own time schedule).

No matter how interesting a home is, it is always desirable to get away from it once in a while. Even if it seems stressful at first, having exposure to varying environments is critical for pet parrots. The truth is that very few pet parrots will live their entire lives in one home. With the long life expectancy of parrots, having the owner move, die, or lose interest in the bird will necessitate adapting to a new environment. Birds that are more accustomed to various locations will be better adapted to this change. For well-adjusted parrots, the variable environment will be pleasant stimulation. Play stands are a good alternative site for parrots to spend time. There are numerous arrangements that can be useful. Branches, PVC pipe structures, and many other stands are commercially available or can be readily built to provide a play area for psittacids of varying sizes and activity levels. Multiple cages can be helpful in some cases. A sleep cage will allow the primary cage to be in an area of high activity, but still have a dark quiet place to sleep at night. The sleep cage can be smaller and minimally furnished. Otherwise, having another cage in another room may just provide a change of scenery. An outdoor aviary allows for exposure to natural sunlight and fresh air. Commuter birds that go to the workplace with the owner have the advantage of staying with their human flock most of the time. The non-contact interaction with various people through the day is ideal for well-socialized birds. Travel can be either highly stressful or highly rewarding depending on the individual bird and the methods used. Generally, travel will involve more close confinement and sometimes motion sickness, dehydration, and other issues. These factors need to be taken into account for travel to be a positive experience.

Photoperiod

As previously mentioned, most psittacids come from tropical to subtropical latitudes. The photoperiod in these areas vary minimally and will always be close to 12 hours of light and 12 hours of dark. By comparison, most pet birds have prolonged day length. This affects the time budget significantly by reducing the amount of dark, quiet time when the bird would be sleeping. As in humans, it can be assumed that inadequate sleep would be a significant stress to these birds. Additionally, it creates a greater void in the time budget that must be filled in some way. Although photoperiod is highly critical for signally reproductive cycles in temperate species, in tropical species these effects will be variable. Other parameters such as food availability, rain, and nest site availability may have a larger effect in these cases. When there is photoperiod dependence, it is primarily dependent on changes in the light cycles rather than the actual day length. Whenever possible, a consistent cycle of 12 hours of light and 12 hours of dark should be maintained. Unfortunately, this rarely is practical for those with typical work schedules. However, the goal should be to stay as close to this schedule as possible.

Nutrition

Avian veterinarians have always concerned themselves with nutrition. Provision of all of the required nutrients is critical to the physical health of any animal and the effects of imbalanced diets are familiar to the profession. Nutritional balance can be achieved by giving appropriately formulated commercial diets. With extruded or pelleted diets, each bite generally has all of the required nutrients. If anything, there are more of some nutrients (e.g., vitamin D, vitamin A, calcium) than is necessary for maintenance in many birds. Other formulated diets take whole seeds and/or pellets and bind them together into a ball or cake with added supplements. Still other diets are seed mixtures with supplements concentrated in nuggets. With these "fortified seed mixtures" all of the components of the diet must be consumed to provide these same nutrients. This rarely happens since most owners feed far more of the diet than the bird will consume. This can lead to the same nutritional deficiencies generally seen with all seed diets. In order to best provide for nutritional needs of the bird, the author recommends feeding a primarily pelleted or extruded diet as the base with a defined small amount of unsupplemented grains or seeds to slightly dilute the nutrients. Vegetables and fruits are added to provide vitamin precursors and other phytonutrients. The total amount of food provided should be calorically appropriate for the bird. It is unnecessary for birds to have unlimited food in front of them 24 hours a day. This practice has led to obesity, constant reproductive stimulation, and birds that are poorly responsive to training techniques. The amount a bird is eating should be determined, and if the body weight is appropriate, the food should be trimmed to exactly what is being consumed, with little or none left. There are certainly times in a bird's life when this will not be appropriate. Birds that are growing, raising offspring, or laying eggs may need additional calories or specific nutrients (i.e. calcium for egg laying).

Food is important for more than nutrition. In humans food is a part of every culture. Likewise, food for parrots is more than just nutrition. Feeding takes up a large part of the day. Most birds feed socially. To fill a bowl with a uniform nugget may not provide the same opportunities for these birds. To improve the psychological benefits of the diet, a number of different things can be done. Variation of the diet is one method of making it more interesting. Even with a diet composed of 100% formulated nuggets a pretty wide variety is available. Using several varieties not only provides some variation, but may be a good safety precaution. There have been a small number of recalls of formulated diets over the years where a mixing error has caused dangerous excesses of some nutrients in a diet. Using a variety of types would cut the total amount of any one diet that a bird eats. Extruded diets come in many forms and shapes. These may provide the ability of the bird to hold the food in the foot and manipulate it. There are a variety of sizes of most brands of pellets as well. Even the flavors and scents of the diets vary. Some have food coloring added. Vegetables can be given whole or cut into different sized and shaped pieces. Most wild birds will have variable diets depending on the season, so temporal variation of the diet may also be beneficial. This can be readily done by securing seasonal produce for the birds.

Presentation of the food is the other main way that it can provide enrichment. This will be discussed in more detail in the section on foraging.

Socialization

Parrots are obligately social animals.[11,12] While it is common to remark about birds being "spoiled" because they demand constant attention, it is unnatural for most parrots to be alone. The only time most parrots are ever alone in the wild is when they are incubating eggs. At other times, they would be with their parents, offspring, clutch mates, juvenile flock, adult flock, or their mate. Being alone is completely unnatural. It is not just "loneliness" that results from this isolation. Flocking is primarily a protective mechanism to reduce the pressures of predation. Being within a group allows for multiple sentinels looking for signs of danger. A group of birds can "mob" a predator and chase it away. Even thinking less altruistically, a bird in a group of 10 birds has a 10% chance of being the target of a predator while an individual bird has 100% chance. Just as most people feel more vulnerable when alone, parrots will also feel stressed and vulnerable when kept alone.

Despite their importance, social skills must be learned. Early socialization starts with parental care. Even before hatching, the parent birds start interacting with a chick. Like mammals, the parents will stimulate defecation and carry the waste out of the nest initially. Parental care is fairly constant and involves feeding, preening, vocalization, and other interactions. This is not readily replicated in incubator hatched and hand raised chicks. Most species will also have clutch mates in the nest with them. This also helps them learn social skills. Once they fledge, there still is a great deal of parental care and teaching. Many parrots also have a crèche system, a flock of juvenile birds that move about together and learn about the environment, feeding, predators, and other important life skills. One of these skills is how to interact with other birds without conflict.

Hand raised birds have somewhat displaced imprinting. While filial imprinting on humans may have some advantages, sexual imprinting on humans has severe behavioral effects. Birds that are sexually imprinted on humans often pair bond with humans and will be aggressive to others, sometimes be aggressive to their perceived mate, or will defend a nest territory. Other birds may become frustrated and engage in self-damaging behaviors such as feather chewing or plucking. In addition, birds that are completely imprinted on humans may exhibit poor socialization with other birds. Many bird owners have multiple birds and one with poor social skills may either attack or be attacked by other birds.

Some studies have shown that parent raised birds that are handled by people while developing have pet qualities exceeding those of hand reared birds.[13,14] This can be slightly challenging to accomplish, but should be considered. Minimally, having birds raised with clutch mates or other babies will at least allow them to imprint on both humans and birds.

Assuming that a bird has well-developed social skills, the next concern is providing them with social opportunities. While bird owners like to believe they are providing this for their pets, in most cases, the social contact they provide is inadequate. Socializing a parrot is more than a full time job. The ideal solution is that birds be kept in a more normal social grouping. Companion birds can provide 24 hour socialization for each other. As discussed, they must be compatible and they must have some social skills with other birds. They also should be size and temperament matched. Although it can occasionally work, large birds with small birds can be a dangerous situation. Larger birds can easily tear the beak or a toe off of a smaller one. Likewise, a loud bird may be stressful for a nervous or quiet bird. Introductions are easiest with very young birds, and most birds that grow up together remain compatible provided that they don't conflict over a mate or resources. If breeding is not desired, the birds should be the same sex.[15]

The obvious benefit of having birds in pairs or groups is that it corrects for some of the limitations of human companionship. Companion birds can be together all of the time, reducing the complete reliance on the owner for companionship and protection. Although this may seem like a negative to some, severe behavioral problems such as nuisance vocalization, stereotypies, or feather damaging behavior may occur when birds rely too much on the owners' presence and attention.

Although there are exceptions, usually, breeding will have some negative effects on the pet characteristics of a parrot during the breeding season. Most breeding pairs will defend the nest territory from all others. This may result in aggressive behavior. If breeding is desired, the pair should be allowed to defend their territory. They should be conditioned to allow nest cavity inspection on a daily basis and should be trained to allow normal care procedures, but should otherwise be left alone.

In most cases, breeding and sexual behaviors are not desired. However many owners inadvertently encourage them constantly. For most wild birds, breeding occurs during a defined, short season. This may be signaled by changing photoperiod, rainy season, availability of certain food items, or simply food surplus. The presence of a mate is also a key stimulus. Pet birds typically are overfed on high calorie diets. This is one stimulus. Many are offered nest boxes, "snuggle" tube perches, or are allowed access to cabinets or other hollow structures that the bird may perceive as a potential nest site. Many owners interact with their bird in ways that are exclusively reserved for the pair bond in parrots. Reproductive instinct is strong in most species, and adding in these potent stimuli is enough to get the process started.

In order to discourage reproductive behaviors several strategies can be employed. Caloric intake should be carefully controlled. This is especially effective for suppressing egg laying, since most hens will increase food consumption in preparation for egg laying. Foraging can give the bird the impression that food is limited, even if higher amounts of food are present. Physical contact should be somewhat restricted. Alternative activities between bird and the owner should be encouraged. Training, playing, and other active games should replace much of the "cuddle" time. The petting that is done should be brief, used to reinforce good behaviors, and limited to non-sexual locations of the body. For those species that respond to photoperiod, this should also be regulated. A relatively short and consistent day length is desired.

Foraging Opportunities

The importance of foraging has become clear over the past decade. The difference in the time budget for this activity in wild birds vs. pet birds may be the greatest of any activity. This may be a significant reason for the frequency of displacement behaviors. Some research has shown foraging to reduce feather damaging behaviors in parrots. For these reasons, foraging should be considered the most important enrichment activity that can be provided. It is important that clients be guided on how to institute an effective foraging program. Simply telling owners to give foraging toys will not be effective. There are many types of foraging activities. These need to be matched to the feeding behavior of the species and the individual. For instance, many popular psittacids (e.g., budgerigars, cockatiels) are primarily ground feeding birds. They eat a lot of grass seeds and other items they find on the ground. These birds have not evolved extensive skills with holding food in the feet or opening complicated nuts. They do have keen eyes for spotting food particles on the ground. These birds will work best with a foraging tray, a flat pan filled with shredded paper or paper litter. Food can be sprinkled into this tray, allowing the bird to walk about and find the food particles. Tree feeding birds, by contrast may collect food from the ground but would bring it to a branch to eat it, or may find food items in the trees themselves. The will have greater dexterity with the foot and beak. They will have to open nuts, or sometimes solve other problems to retrieve their food. These birds are more adapted to commercial foraging toys. Foraging should also be introduced as a graduated program. A novice forager should not be given a complex foraging toy. Their inability to achieve rapid success will frustrate them and cause them to give up. The following is an excerpt from one of the author's client education handouts.

"Techniques for encouraging foraging

Wait until your bird is mastering one step before moving to a more challenging step.

Step 1: Reduce the total amount of food to just what your bird finishes in a day. Measure the food before putting it in the dish. Then in 24 hours measure what is left. If there is no food after 24 hours, give a little more the next day. If there is some food remaining reduce the food.

Step 2: Divide the food. Get 5-10 smaller dishes (preferably ones that fit in a ring holder). Initially put most in the original dish, and just a sprinkle in each of the others. Once your bird is eating from all of the dishes, start to divide the food more evenly between them.

Step 3: Mix in some non-edible materials as distracters. Some choices would be wooden beads, paper litter (i.e. CareFresh or Yesterday's News) or paper scraps. Initially just put a few in each dish. Gradually add more so your bird has to carefully sort through to separate food from debris.

Step 4: Cover the dishes. Take a paper towel and put it over the dish and tuck into the ring holder. If there is no ring, a rubber band can be used. Now your bird has to chew open the dish, then sort through the distracters to find the food.

Step 5: Wrap individual food items. Take a few food items and put in a small piece of paper and twist the ends. Place these in the dish with the other food items and the distracters.

Other ideas:

- Try hiding a few of your bird's favorite foods on or in other toys. This can encourage play and a more inquisitive nature.
- Put a food nugget in a paper cup and then roll the top of the cup around a cage bar.
- For a big challenge, hang a food item at the end of a string or rope so the bird has to pull it up (or snip the string to let it drop)
- There are lots of "puzzle toys" available for sale. After your bird masters the above steps, try a few of them to see what works best.
- Try a foraging tree. Set up a stand or tree with lots of little spots to hide a food nugget. Put food in some but not all of the spots. This keeps the bird guessing, and looking."

Once a bird gets proficient with foraging, the original food dish can be eliminated and all of the food can be offered via foraging opportunities. There are unlimited ways to present the food. Sometimes the food can be hidden in other toys, in holes drilled in perches, in the bedding, and other places that the bird can explore. Some spots will not have any food, others will. Foraging should always continue to challenge the bird. Ideally, finding and consuming the food should take up a large part of the day. Although it would be difficult to achieve the six hours of foraging wild birds may get, even 2-3 hours would be a huge benefit.

Positive Reinforcement Training

Traditionally, parrots have seemed difficult to train. This is perplexing given their intelligence. Part of the reason is likely to be the typical feeding practices for pet birds. Although verbal, physical, social, or even drama based reinforcements may be effective at times, it can be difficult to predict exactly what will motivate a parrot at times. As a consequence, training parrots has long involved less desirable methods that involve coercion and force rather than positive reinforcement. Mostly these techniques were used in good faith, but there is substantial evidence of their negative effects. Parrots become infinitely more trainable when food becomes a valuable resource and they become food motivated. Certainly other reinforcers are useful as well, but a food treat is generally very dependable, especially when working with a hungry (not starving) bird.

Positive reinforcement training provides a means for owners to communicate effectively with their birds and to allow the bird a degree of choice in what behavior they will exhibit. Undesirable behaviors can be replaced with desirable behaviors that are subsequently reinforced. Although the process of training is relatively simple, the devil is in the details. The ability to read the body language of the bird, developing timing with bridges and rewards, and even presenting cues that do not appear threatening requires practice. The following is another excerpt from a client education handout at the author's practice.

"Creating an eager student

- Start a foraging program (see separate handout on this topic).
- Limit the favorite food to a small quantity, and use it only for training.
- Break up the food to be used as a reward into small morsels (1/4 inch or less).
- Take away all food for about 1-2 hours before training sessions.
- Stop sessions before the bird loses interest in the treats.

Definitions

Cue: a signal given to the subject (your bird) that tells what behavior is desired
Verbal (e.g., step up)
Hand signals

Target: an object that the subject is taught to touch to get a reward.
Often the trainers hand or a colored stick or wand

Shaping: working through progressive approximations of the desired behavior

Bridge: a signal that the desired behavior was performed and that a reward is forthcoming
  Clickers
  Verbal (e.g., good)

Reward: a tangible reward for the performance of a behavior
  Food treats
  Praise (this is less tangible and therefore much less effective)

Jackpot: an especially large or desirable reward for completion of a complicated behavior, or one that the bird was initially very reluctant to perform.

Process

- Give the cue for the behavior to be taught.
- Use the target to guide the subject to do the behavior, or at least an approximation of it. Alternately the behavior can be "captured" when it naturally occurs.
- Once the behavior or the approximation is performed use the bridge to mark it.
- Give a reward.
- Repeat. As the pupil progresses, require a better performance before bridging and rewarding.
- Use a jackpot to reward a particularly big jump in progress."

Additional training for the owner is strongly recommended. There are many great books, videos, and online resources that can be used to improve training skills. The behaviors that are trained are far less important than the training itself. This is a great activity for clients to engage their birds without stimulating reproductive behaviors. Training some husbandry behaviors can greatly reduce stresses that will occasionally occur for any pet bird. If a bird allows the owner to file the nails daily with a fingernail file, then trimming with a hobby drill will rarely, if ever be required. Training a bird to defecate on cue will make it easier to integrate them into normal household routines. Training a bird to allow gentle restraint in a towel will facilitate veterinary exams or other procedures. Training to drink juice or almond milk from a syringe or other device will facilitate medicating a bird if an illness occurs. For flighted birds, recall is an important behavior to teach.

Physical Exercise

The benefits of physical exercise are well documented.[16-18] Likewise, the negative effects of inactivity are well known. As the number of aging birds increases, the frequency of certain degenerative diseases has grown by an even greater degree. While not all of these conditions are preventable, many are much less prevalent in animals that receive appropriate diet and exercise. Cardiovascular diseases in particular, such as atherosclerosis and congestive heart failure, are responsive to the effects of physical activity; they are less frequent in active animals. Obesity, hepatic lipidosis, and arthritis are other conditions that may result from a severe lack of physical activity. The musculature will be stronger with regular exercise. The joints require activity to nourish the articular cartilage. Bone density improves with physical activity as well. Physical activity has a great deal of psychological benefit as well. In humans, stress reduction is considered to be a major benefit of exercise. In animals, it can help reduce displacement behaviors. The expenditure of energy on healthy physical activity can only help reduce the use of energy on pathologic behaviors.

As in any animals or person, an examination should be performed to determine the safety of a new exercise program. Certain conditions may require restriction of exercise, even if it originated because of inactivity. Such conditions may include some orthopedic conditions, atherosclerosis, and heart failure. Some exercise may be beneficial in these cases, but they should be instituted very carefully. Forcing these birds to exercise may cause them to decompensate.

There are many types of exercise. For birds, the simplest and most obvious is flight. The largest mass of muscle in flying species is the pectoral musculature and this is the exercise they are best adapted to. There are complications to having a flighted bird as a pet. These include injuries, escapes, or difficulties with training. Most of these complications can be managed with appropriate training and flight skills. Recall (flight to the handler), flight to a perch, and other such games are easy ways to promote flight activity. An enclosure that is large enough to allow flight will passively encourage this type of exercise. In some cases, though, an owner may wish to keep the wings trimmed to prevent some of these complications. In these cases, wing exercise is more challenging. Wing flapping can be encouraged, but should be trained rather than forced. Dropping a hand while a bird is perched upon it will force a bird to flap. However, this can promote biting, sometimes just to hold on. Most birds will flap when the surface they are perched upon moves. Using swings, flexible branches, ropes, or other movable perching surfaces is a good way to encourage some passive wing flapping.

More so than any other avian species, parrots are also adapted to climbing. The hooked beak provides a strong hold on objects to allow for this activity. Parrots can hang from just the beak or from one toe. They can flip back on top of a branch with a few flaps of the wings. Even fully flighted parrots will often use their climbing abilities to negotiate a maze of branches rather than fly over them. Tarzan himself would be jealous of the gymnastic capabilities of a fit parrot! This can be readily encouraged by providing branches, ropes, ladders, and nearly any other type of structure a parrot can grip.

Beak activity is a relatively low energy form of exercise. It is fairly constant however. Parrot owners are fully aware of the destructive capability of psittacine beaks. Some birds may initially require encouragement to chew. Soft items like paper, cardboard or soft woods can be given to encourage this. As beak strength improves, harder woods can be given. Some may require a small nick or other starting spot to allow the beak to get purchase on the object at first. Also unique to parrots is the ability to manipulate items dexterously with the feet. This activity of often paired with chewing activity as items are often held up for the beak to engage. At least some of the chewing items should be loose to allow for this manipulation.

Self-Maintenance Behaviors

Although it is undesirable for these to take on obsessive roles in a parrot's life, preening and other self-maintenance behaviors are critical for health and welfare. Generally, no encouragement is required for birds to preen themselves. As long as there is sufficient space and the bird is healthy, preening will proceed normally. Bathing will encourage preening and for many birds is an excellent and vigorous activity in its own right. Birds vary in the method of bathing and several ways should be tried until the preferred method is found.[19]

Toenails are maintained by natural wear, but normal wear produces very sharp tips. This is true with the "grooming perches" and sandpaper perch covers as well as other surfaces. Although this is normal for the bird, it is highly uncomfortable for the handler. Unless the owner wants to wear falconry gauntlets, the nails will need to be kept trimmed. As previously discussed, birds can be trained to do this cooperatively.

Beaks also can be maintained by normal activity. This requires that some abrasive surfaces are made available. Using a bowl of stones as a foraging site, providing pumice stones, lava stones, and other items may be all that is required. Some birds may even take a stone (small enough to hold, too big to swallow) in the beak and scrape the stone for several minutes.

Immunologic Enrichment

In the early days of aviculture, "sterility" was considered to be essential and gram negative bacteria were considered highly pathogenic. While there certainly are pathogenic bacteria, many of the "infections" seen in the early days of avian medicine were secondary to poor nutrition, stress, underlying viral disease, and other factors. One of these factors may have to do with the very practice of overzealous hygiene. Microbial exposure is critical to the development of proper immune function.[20,21] Ideally, a balance should be stuck between sanitation and microbial exposure. Generally, the environment should be clean, not sterile. Disinfectants rarely need to be used, unless there has been a disease state. Removal of gross debris, feces, and discarded food should be undertaken frequently. Particulate beddings can be scooped in large enclosures without a complete change most of the time. Only a disease state should necessitate a total scrub down of the enclosure. Branches from the yard or forests can be a great resource for enrichment items. These are often avoided because of the fear of infectious agents. There are certainly some risks with this, but they can be minimized. Brush piles should not be harvested as they are sometimes used as latrines by raccoons or opossums. This presents a risk of *Baylisascaris or Sarcocystis*. Smaller branches can be baked in the oven to kill contaminants, but this may be unnecessary. The surfaces can be wire brushed to remove organic debris. If desired, a steam sprayer can be used to kill surface contaminants. The branch can then be sun-dried. This should be effective at removing the majority of pathogens. Certainly food and water should always be clean, but likewise need not be sterile.

Conclusions

Enrichment is a multidimensional process that can improve the lives of captive animals. Although there are challenges in providing enrichment for parrots, it is both critical and rewarding for the bird and the owner. The benefits can be seen in health, behavior, and overall quality of life. As advocates of avian welfare, it is critical for avian veterinarians to provide strategies for effective enrichment for their patients.

References

1. Kalmar ID, Janssens GPJ, Moons CPH. Guidelines and ethical considerations for housing and management of psittacine birds used in research. *ILAR J.* 2010;51:409-423.

2. de Andrade AA, de Azevedo CS. Effects of environmental enrichment in the diminution of abnormal behaviours exhibited by captive bluefronted Amazon parrots (*Amazona aestiva*, Psittacidae). *Rev Brasil Ornitologia* 2011;19:56-62.

3. Mason G, Clubb R, Latham N, et al. Why and how should we use environmental enrichment to tackle stereotypic behaviour? *Appl Anim Behav Sci.* 2007;102:163-188.

4. Coulton E, Waran NK, Young RJ. Effects of foraging enrichment on the behaviour of parrots. *Anim Welf.* 1997;6:357-363.

5. Meehan CL, Mench JA. Environmental enrichment affects the fear and exploratory responses to novelty of young Amazon parrots. *Appl Anim Behav Sci.* 2002;79:75.

6. Boyes RS, Perrin MR. Patterns of daily activity of Meyers parrot (*Poicephalus meyeri*) in the Okavango Delta, Botswana. *Emu* 2010;110:54-65.

7. Burger J, Gochfeld M. Parrot behavior at a Rio Manu (Peru) clay lick: temporal patterns, associations, and antipredator responses. *Acta Ethol.* 2003;6:23-34.

8. Barnard CJ. Flock feeding and time budgets in the house sparrow (*Passer domesticus* L.). *Anim Behav.* 1980;28:295-309.

9. Rozek JC, Danner LM, Stucky PA, et al. Over-sized pellets naturalize foraging time of captive orange-winged Amazon parrots (*Amazona amazonica*). *Appl Anim Behav Sci.* 2010;125:80-87.

10. van Zeeland YRA, Schoemaker NJ, Ravesteijn MM, et al. Efficacy of foraging enrichments to increase foraging time in grey parrots (*Psittacus erithacus erithacus*). *Appl Anim Behav Sci.* 2013;149:87-102.

11. Aydinonat D, Penn DJ, Smith S, et al. Social isolation shortens telomeres in African grey parrots (*Psittacus erithacus erithacus*). *PLoS One.* 2014;9:e93839.

12. Seibert LM. Social behavior of psittacine birds. In: Luescher AU, ed. *Manual of Parrot Behavior.* Ames, IA: Blackwell Publishing Professional;2008;43-48.

13. Aengus WL, Millam JR. Taming parent-reared orange-winged Amazon parrots by neonatal handling. *Zoo Biol.* 1999;18:177-187.

14. Schmid R, Doherr MG, Steiger A. The influence of the breeding method on the behaviour of adult African grey parrots (*Psittacus erithacus*). *Appl Anim Behav Sci.* 2006;98:293-307.

15. Meehan CL, Garner JP, Mench JA. Isosexual pair housing improves the welfare of young Amazon parrots. *Appl Anim Behav Sci.* 2003;81:73-88.

16. Alessio HM, Hagerman AE, Nagy S, et al. Exercise improves biomarkers of health and stress in animals fed ad libitum. *Physiol Behav.* 2005;84:65-72.

17. Brown MD. Exercise and coronary vascular remodelling in the healthy heart. *Exp Physiol* 2003;88:645-658.

18. Orimilikwe SO, Wong HYC, David SN, et al. Effect of exercise and guanethidine on plasma cholesterol and aortic atherosclerosis of atherogenic-fed cockerels. *Artery* 1983;12:60-73.

19. Murphy SM, Braun JV, Millam JR. Bathing behavior of captive orange-winged Amazon parrots (*Amazona amazonica*). *Appl Anim Behav Sci.* 2011;132:200-210.

20. Martin R, Nauta AJ, Ben Amor K, et al. Early life: gut microbiota and immune development in infancy. *Benef Microbes.* 2010;1:367-382.

21. Von Hertzen LC, Joensuu H, Haahtela T. Microbial deprivation, inflammation and cancer. *Cancer Metastasis Rev.* 2011;30:211-223.

The Endangered Species Act: Effects on Non-Native Species, Aviculture and Avian Veterinarians

Susan L. Clubb, DVM

Session #119

Affiliation: Rainforest Clinic for Birds and Exotics, 3319 E. Rd, Loxahatchee, FL 33470, USA.

In 2008 the US Fish and Wildlife Service (USFWS) was petitioned by "Friends of Animals" to add 19 species of psittacine birds to the Endangered Species List (ESA). Forty-two species of psittacines are currently listed on the ESA. For rare species not found in the United States, there is virtually no impact. However for species that are relatively abundant in the United States, this petition was a real game changer for aviculturists.

The USFWS was subsequently forced into evaluating the foreign (exotic) species named in the petition to see if they met the criteria for inclusion on the endangered or threatened species lists of the ESA. If a species is placed on the threatened list, USFWS is allowed to waive the ban on interstate commerce of that species [4(d) rule]. However, if listed as endangered, there is a prohibition on interstate sales without proper permits. Any interstate sale then requires a captive bred wildlife (CBW) permit for both the seller and the buyer or an interstate commerce permits for both covering the single transaction. In the past, these permits were more easily obtainable, but required annual reporting of all births, deaths and sales of listed species. Today it has become more difficult to acquire a permit, particularly if the goal is to obtain a pet or companion bird.

In recent years the blue-throated macaw (*Ara glaucogularis*), citron-crested cockatoo (*Cacatua sulphurea citrinocristata*), lesser Sulphur-crested cockatoo (*Cacatua suphurea sulphera*), red-vented cockatoo (*Cacatua hematuropygia*), Buffon's macaw (*Ara ambigua*) and military macaw (*Ara militaris*) have been listed as endangered with immediate restrictions on interstate commerce. The Moluccan cockatoo (*Cacatua moluccensis*), and umbrella cockatoo, (*Cacatua alba*), were listed as threatened with 4(d) rule, so will, at this point, not be restricted from intrastate sales. I have personally become aware of breeders who are selling producing pairs of these species in anticipation that they will be unable to sell the offspring.

Of great importance to avian veterinarians is that, in many states, (see list below) it is illegal to possess an ESA listed species as a pet. They are virtually illegal to own in states that follow the ESA federal guidelines to the letter of the law! This fact can serve to drive owners underground to the point that they would not feel comfortable seeking veterinary care for their birds. Even worse, they may have to give up long-term pet birds that are now illegal to own. Additionally, if strictly interpreted, listing a species as endangered could make the shipping or mailing of diagnostic samples (from animals of this species) to laboratories in another state as illegal as ownership!

At the time of writing we are in a comment period for listing of scarlet macaws (*Ara macao*). USFWS is proposing a split listing under which the South American subspecies would be listed as threatened with the 4(d) rule, and the Central American subspecies would be listed as endangered. This will create confusion as to the correct identification of these subspecies and restrictions applied. A decision is still pending on these proposals as well as on the listing of hyacinth macaws (*Anodohynchus hyacinthinus*).

In early 2016, the American Federation of Aviculture (AFA) convened an ad-hoc committee, of which I am a member, to explore issues of implementation of the ESA, Wild Bird Conservation Act (WBCA) and Convention on International Trade in Endangered Species of Wild Fauna and Flora (CITES), and how these laws and their accompanying regulations have and continue to impact aviculture in the United States. The AFA committee identified a group of issues and suggested some solutions that might be implemented to resolve them in an effort to enhance US captive-bred psittacine bird populations and further cooperation between the USFWS and aviculturists. Following is a synopsis of these issues and how they can affect captive populations of ESA listed species in the United States, as well as their impact on aviculturists and avian veterinarians.

American Federation of Aviculture

Discussion items - January 22, 2016 USFWS Meeting

AFA desires to collaborate with international, federal, and state authorities to ensure that regulators encourage not only the continuation of existing captive breeding, but also to promote future efforts to utilize captive breeding as a sustainable conservation tool. Regulatory mechanisms should also recognize the need for economic sustainability of captive breeding programs to ensure the longevity of genetically diverse populations.

ESA Issues

Issue 1: Captive breeding is an important conservation tool.

Conservation is not only important in the birds' native countries, but it is important in other countries that breed the species. Among adaptive species, captive breeding undertaken by aviculture is part of the survival matrix.

- Increases the world's total population of the species so that reintroduction to native habitats could be possible if needed. (Example: The Puerto Rican parrot - *Amazona vittata*).
- Safeguards against extinction. (Example: Spix's macaw- *Cyanopsitta spixxi*).
- Educates the population about the needs of endangered species so funds are contributed to conservation efforts in the birds' native countries.
- Reduces pressure from removal from the wild (poaching) if specimens can be supplied by US aviculture in a responsible captive-breeding program.

Recommendation: USFWS should recognize and appreciate captive breeding as a form of conservation and support the continued existence of healthy populations of captive bred animals in the United States, particularly the populations of parrots that are not receiving adequate protection in their range states.

Aviculture may in fact be the last refuge for some species.

Issue 2: ESA listing of non-native avian species and the subsequent interstate commerce regulations, permitting, and reporting requirements produce unintended consequences while providing no conservation benefit to the species in their native habitats.

ESA listings impede the exchange of specimens among aviculturists contributing to loss of genetic diversity and they restrict free movement of diagnostic and biologic samples between the states, and internationally.

Important unintended consequences of ESA listing non-native species include:

- Prevents breeders from augmenting their captive populations in order to maintain genetic diversity.
- Limits the number of US breeders willing to breed the ESA-listed birds.
- Creates internal US geographic boundaries for shipping, resulting in the 50 states being treated more like separate countries.
- Dictates who can breed what and where, rather than encouraging responsible, well managed breeding by aviculturists striving for robust, genetically diverse captive populations.
- Prevents interstate shipment of biologic samples from listed species to test for gender, disease, or to establish DNA relatedness. (This is an issue of importance to avian veterinarian who may inadvertently break the law by shipping diagnostic samples interstate.)
- Encourages hybridization and dilution of pure species genetics.
- Encourages illegal activities of moving birds, biologic samples and feather art across state line thereby damaging a multi-million dollar pet trade industry servicing 6.1 million birds in the United States. (Source: The 2014 - 2015 American Pet Products Association National Pet Owners Survey.)

Recommendation: USFWS should recognize that CITES and the 1992 WBCA are an adequate protection for non-native species. Regulation of activities with ESA listed species should be enforced only to the extent necessary to meet the minimum requirements of the law. AFA supports usage of 4(d) rules as the most appropriate mechanism to regulate ESA listed species of non-native birds.

AFA also recommended that USFWS convenes a working group, including stakeholders, to explore ways to improve the permitting processes for ESA, WBCA and CITES regulated activities. (Such a meeting has been proposed for the AFA annual meeting in August 2016)

Issue 3: ESA listings may trigger automatic prohibitions under state laws, thereby making ownership illegal.

Stricter state regulations limiting ownership of ESA listed species can cause legally obtained birds to become contraband. In the states listed in Table 1, stricter state requirements can render possession of ESA listed species illegal.

Table 1. Twenty-five states with regulations limiting ownership of ESA listed species.

Connecticut	Montana
Delaware	Nebraska
Hawaii	New Hampshire
Illinois	New Jersey
Indiana	New Mexico
Kansas	New York
Kentucky	North Carolina
Louisiana	Ohio
Maryland	Oklahoma
Massachusetts	Pennsylvania
Michigan	South Carolina
Mississippi	South Dakota
Missouri	

Note: Virginia recently successfully revised its regulations in response to concerns presented by aviculturists.

Wildlife authorities in these "unsafe" states often insist that the regulations only apply to native wildlife although this is not explicitly declared. Therefore, if challenged by humane groups, or neighbors with complaints, the regulations could become enforceable against ESA listed species of non-native (exotic) species.

This is vitally important for avian veterinarians because owners of ESA listed birds in "unsafe" states may be afraid to seek veterinary care due to fear of their possession of a contraband species being revealed.

Recommendation: USFWS should collaborate with the Association of Fish and Wildlife Agencies (AFWA) for better alignment of ESA, and CITES issues involving non-native species.

Issue 4: The listing of non-native species as "endangered" should not be applied to species being propagated in captivity in the United States.

Value must be assigned not only to the survival of a species in the wild, but also to the survival of that species in captivity. Maintaining diverse captive gene pools is critical to the survival of a species and should be promoted, enhanced and supported.

Recommendation: Captive-bred non-native species should not be listed as endangered when aviculture documents the existence of captive-breeding programs for such species. AFA recommends that if a listing is deemed appropriate such species should be listed as threatened and a 4(d) rule should be adopted at the time of listing. (The 4(d) rule may be applied to species listed as threatened, allowing unregulated interstate commerce to continue, but cannot be applied to species listed as endangered.)

Issue 5: Petitions to delist a species should be processed expeditiously.

Delisting species in response to a well-founded petition will reduce workload on FWS biologists.

Recommendation: Petitions to delist a species should take precedence over petitions to list species.

ESA and CITES Issue

Issue 6: Facilities registered by the CITES Secretariat to breed Appendix I species for commercial purposes are adversely affected by ESA listing.

CITES specifically recognizes captive breeding for commercial purposes, allowing trade in captive bred specimens.

Recommendation: Promulgate a regulatory exemption from export restrictions under ESA for registered facilities if the registered species becomes listed under the ESA. Alternatively, USFWS should provide a regulatory exemption under the ESA (or provide in the ESA regulations) that such species in a CITES registered facility would be provided a 4(d) rule exemption as a condition of the facility being registered.

ESA, CITES, and WBCA Issue

Issue 7: Current USFWS permitting procedures are burdensome, slow, complex, expensive, and intimidating.

Permitting challenges are a discouraging non-incentive result to captive- breeding of listed species, bringing about a decline in numbers of captive populations. (Permit processing time in the United States is significantly is slower than in most other countries.)

Recommendation: USFWS should convene a working group to review current permitting procedures with the goal of increasing efficiency, cost effectiveness, streamlining their copious paperwork, and eliminating unnecessary, burdensome, and costly requirements.

WBCA Issues

Issue 8: Original import documents are difficult, if not impossible, to obtain for founder stock imported generations or decades prior.

It is unreasonable that Appendix II founder stock must be documented back to original exports from range areas in order to import foreign captive-bred birds under an approved Cooperative Breeding Program (CBP) under the WBCA.

Recommendation: USFWS should implement means by which the burden of documenting legal founder stock of long standing populations can be reduced.

Issue 9: Regulations for importation of birds under the WBCA foreign captive-bred exemption have not been published 24 years after passage of the WBCA.

One of the major exemptions to the WBCA was for importation of captive-bred birds from approved foreign captive breeding facilities. American aviculturists need to have the ability to augment captive populations to increase genetic diversity.

Recommendation: USFWS should promulgate regulations to facilitate importation of birds under the WBCA from foreign breeding facilities.

So You Want to Start a Pet-Bird Rescue?

Terri Parrott, DVM

Session #042

Affiliation: St. Charles Veterinary Hospital, 39873 Hwy 27, Davenport, FL 33837, USA.

Abstract: The goal of any animal rescue is basic triage and care of the surrendered animal with the hope of fostering and re-homing the pet. Exams and testing must be focused on the species and overall condition of the pet bird. This paper breaks down the initial basic lab tests and exams needed to achieve an individual data base and care plan focused on the individual's needs.

Introduction

Pet bird rescue is very different from canine and feline rescue. For example, there are over 350 different species to deal with in just psittacine birds alone.[1] Pet bird households and dietary backgrounds vary significantly. The longevity of these species also vary greatly, as well as the lists of infectious diseases. Besides a basic exam, fecal and gram stain, many of these displaced birds will need species specific nutritional counseling, added blood testing for viral and fungal infections and ectoparasite exams. The foster program should also be set up with educated individuals that are aware of the species-specific behaviors and requirements of the birds they are caring for. Lastly, the goal should be re-homing and a better quality of life. If there are continuing pathological issues that cannot be alleviated or improved upon, the decision should be made to humanely euthanize these animals.

Intake Procedures

When first received all birds should have a weight (in grams) and photo recorded. The species, age and sex (if it can be verified) should be catalogued. The history from the previous owner is very important, as this will help in assessment of the nutritional status of the animal. The bird should be placed in a quiet and isolated area until examined by the veterinarian and daily weights and behaviors noted. If the animal has an obvious deformity or injury, it should be addressed immediately. Nail trims and beak trims also should be done if there is extreme length in appendages or difficulty seen ambulating, perching or handling food. These birds should be quarantined for 30 days at the foster facility before attempting to adopt.

Veterinary Exam

The bird should be observed in the carrier before attempting to remove it for the exam. Be aware of how the bird is perching and ambulating around the carrier while getting a history from the foster care intermediate. Ask about recent wing and nail trims before handling the bird.

All birds, regardless of species, should have a weight, heart rate, resting respiratory rate, capillary refill time, body scale index and pain assessment done by the veterinary team.[2] Initial lab tests would include a fecal and a gram stain. A *Chlamydophila* test is recommended on all fostered birds before adoption. The *Chlamydia* PCR test is easily obtained with a low associated cost involved for the rescue. When the bird is being examined the eyes, ears and oral cavity should be assessed, as well as the skin and feather condition. It is important to examine and palpate the coracoid, keel, uropygial gland and abdominal cavity, noting any abnormalities. While the bird is restrained, the wings should be extended and examined for pathology and feather condition, and the legs and feet should also be evaluated.

Species-Specific Testing

Cockatiels, budgerigars and lovebirds

If there are feather abnormalities, an ectoparasite exam, *Giardia* test (PCR) and a trichrome stain on the feces are highly recommended. *Circovirus* testing and complete blood count (CBC) with a serum chemistry also should be part of this initial work up (if possible). Some of these tests are expensive and may be declined until after the 30-day quarantine period and which allows time to determine if the problem with plumage is improving with better environmental and nutritional factors.

Conures, quaker parrots and other medium-sized birds

These birds should have a basic CBC and serum chemistry panel. If the bird has feather abnormalities the above tests should be discussed and initiated.

African grey parrots

These birds must have a CBC and serum chemistry panel. African grey parrots may exhibit behavioral issues that are a result of the previous environment. It is also documented that African grey parrots have metabolic problems with D3 and calcium regulation and absorption from the gastrointestinal tract. Many conditions can cause seizures in this species besides hypocalcemia.[3] Heavy metal toxicity, renal disease, hepatic disease and bacterial and viral infections are just a few differentials to be added to the plan and assessment if there is depression and/or neurologic abnormalities observed. Some birds may need extensive testing for *Circovirus*, ectoparasites and *Giardia* to rule out issues when there is feather pathology.

Amazon parrots

Many adult Amazon parrots will come into the rescue obese and with vitamin A deficiency as a result of "all-seed" diets. Identifying the problems with a good physical exam and complete chemistry with a CBC is very important. Gram stains will show squamous metaplasia in the choana with blunted papilla evident upon physical examination. The goal should be correction of the diet during the 30-day quarantine and a recheck before final adoption to a new home.

Macaws

Both larger and smaller species of macaws are susceptible to malnutrition and vitamin A deficiency. CBCs, chemistries, gram stains and cultures should be included in the basic exam. *Bornavirus* testing is also highly recommended for underweight birds. Many times blue and gold macaws (*Ara ararauna*) are over represented with feather abnormalities and self mutilatory behavior. *Giardia* testing, ectoparasite testing and *Circovirus* testing should be in the plan for the birds that have feather shearing or plucking.

Eclectus parrots

Basic exams and lab tests as described on other species should also be included in the eclectus parrot work up. This species also has a high requirement for vitamin A, and is also very susceptible to diseases like *Circovirus* and *Polyomavirus*. Many owners talk about "toe-tapping" self mutilatory behavior that is seen in caged eclectus parrots. This is a symptom and not a disease. Many issues can cause this mutilatory behavior.[4] Tests for ectoparasites, *Giardia* and *Polyomavirus* PCR testing should be in the plan for these affected individuals. Non-steroidal-anti-inflammatory medications as well as gabapentin have helped with self-mutilation.

Cockatoos

After the basic exam and tests mentioned above for the other species, *Circovirus* and *Bornavirus* should be performed on any of these rescued animals with poor condition noted either by their low body score or their plumage.

Lab Testing

In our practice we have specific labs that give us standardized results and can document their laboratory methods. This is by no means an endorsement. Our hospital runs over 10,000 outside avian tests yearly and has used the following laboratories for consistent and rapid results:

- For electrophoresis and aspergillus testing: University of Miami (Dr. Carolyn Cray).
- For ALL viral testing: University of Georgia (Dr. Branston Ritchie).
- For avian CBC and comprehensive chemistries: Antech and Idexx.
- For ectoparasite exams: University of Florida.

Summary

Working with avian rescue organizations can be both rewarding and frustrating. The communication between the foster care individuals with the veterinarians, as well as their board and administrators, on each and every patient is so important. Costs should be factored in every rescue depending on the species and condition of the bird. If the proper funding is not in place for this type of operation, not only do the birds suffer but there is the potential for adoption of a sick or infectious bird into a home. Pathology when rescued is a negative factor, both for cost and time involved with often a non-adoptable outcome for the bird. Leaving the bird to slowly deteriorate and die in foster care is not humane. Work the birds up, communicate the plan, monitor progress, re-evaluate with the medical staff and treat or euthanize accordingly.

References

1. Animal Diversity Web. Psittacidae, parrots, cockatoos and relatives. Available at: http:///www.Animal diversity.org/accounts/psittacidae/classification. Accessed April 27, 2016.

2. Parrott T. Standards of care and written protocols in a multi-species practice. *Proc Annu Conf Assoc Avian Vet*. 2015;323-328.

3. Antinoff N. A clinical approach to avian neurology and seizure disorders. *Proc Annu Conf Assoc Avian Vet*. 2007;201-216.

4. Echols S. Evaluating the avian integument: a physical and behavioral approach. *Proc Annu Conf Assoc Avian Vet*. 2015;97-106.

Long-Term Use of an Air Sac Cannula in a Griffon Vulture *(Gyps fulvus)*

M. Huynh, DVM, Dipl ECZM (Avian), and
C. Pignon, DVM, Dipl ECZM (Small Mammal)

Session #287

Affiliation: From Centre Hospitalier Vétérinaire Frégis, 43 Avenue Aristide Briand 94110 Arcueil, France (Huynh), Centre Hospitalier Universitaire Vétérinaire d'Alfort, 7 avenue du Général de Gaulle, Maisons Alfort, France (Pignon).

A 4-month-old griffon vulture (*Gyps fulvus*) was presented for an acute episode of respiratory distress. Radiography was performed and radiographs were unremarkable. Computed tomography scan highlighted a lesion in the right bronchus consistent with an upper airway obstruction (fungal or bacterial).

An air sac cannula was placed in the thoracic caudal air sac using a sterile 5.0 cuffed tracheal tube. Voriconazole 15 mg/kg and marbofloxacine 5 mg/kg were initiated. The vulture was hospitalized for supportive care; however, there was a poor tolerance of handling with repeated regurgitation. It was decided to release the bird in his aviary with the tube in place to decrease stress and minimize interactions. Voriconazole was switched to itraconazole and marbofloxacin were continued in the food. The air sac cannula was left in place for 14 days. The bird was re-examined, the cannula was retrieved and a control endoscopy was performed to evaluate sign of sacculitis. Antibiotic treatment was continued after the procedure for one more week and antifungal treatments were continued for 6 weeks. The bird recovered fully.

Air sac cannulation is a well known anesthetic technique as well as a therapeutic management of upper airway obstruction.[1,2] Presumptive diagnosis of aspergillosis was based on species sensitivity and response to treatment.[3,4] Voriconazole is a triazole antifungal medication which is considered the drug of choice for treating aspergillosis.[5] Location of the lesion in the bronchus preclude sampling for further diagnostics. Long term use of air sac cannula was a valid alternative to surgical treatment in the presented case.

References

1. Bennett RA, Harrison GJ. Soft tissue surgery In: Harrison BW, Harrison GJ,Harrison LR, eds. *Avian Medicine: Principles and Application.* Wingers Publishing;1994.

2. Brown C, Pilny AA. Air sac cannula placement in birds. *Lab Anim.* 2006;35:23-24.

3. Chege S, Howlett J, Al Qassimi M, et al. Opportunistic infection of Aspergillus and bacteria in captive Cape vultures (Gyps coprotheres). *Asian Pac J Trop Biomed.* 2013;3:401-406.

4. Jung K, Kim Y, Lee H, et al. *Aspergillus fumigatus* infection in two wild Eurasian black vultures (*Aegypius monachus* Linnaeus) with carbofuran insecticide poisoning: a case report. *Vet J.* 2009;179:307-312.

5. Di Somma A, Bailey T, Silvanose C, et al. The use of voriconazole for the treatment of aspergillosis in falcons (*Falco species*). *J Avian Med Surg.* 2007;21:307-316.

Prior Investment in Training Significantly Reduces Stress for Treatment of Unexpected Illness

Barbara Heidenreich

Session #060

Affiliation: Barbara's Force Free Animal Training, PO Box 150604, Austin, TX 78715, USA.

Abstract: A 44-year-old lesser sulphur-crested cockatoo (*Cacatua sulphurea*) presented with an obstructive aspergillosis granuloma at the base of the trachea. Prior to this the bird had been trained for behaviors such as accepting oral medications, accepting intra-muscular injections, allowing towel restraint, and standing on a scale. This paper will demonstrate how an investment in training significantly reduced stress and allowed voluntary participation in various aspects of treatment for an unexpected serious illness.

Introduction

In 2010 a then 38-year-old male lesser sulphur-crested cockatoo (*Cacatua sulphurea*) was integrated into a training program that was being established at a zoological park. Prior to that time the bird's primary routine consisted of stepping onto a dowel in the morning for transport from an off display holding area to perch in a public area. This was repeated in the evening for return travel to the off display enclosure. Prior to residing in the zoological park the bird had lived as a companion animal in a home. Methodologies utilized for handling in the home were unknown and the bird arrived with various behavior problems including aggressive behavior and lack of compliance or cooperation for stepping up, entering transport containers or voluntarily participating in any preventative healthcare procedures. Due to the behavior problems exhibited, handling at the zoo was implemented using a dowel and the bird was not permitted to step onto hands of staff members.

The Training Program

The behavior and training program established at the zoo in 2010 was based in positive reinforcement and other force-free methodologies. The cockatoo and other animals were incorporated into the program and began learning behaviors to facilitate day to day handling and husbandry procedures.

This included target training. Target training involves teaching an animal to orient a body part towards something. This can then be used to direct an animal where to go.[1] The cockatoo was trained to touch its foot against the finger of a staff member resting on the dowel. This was used to train the cockatoo to voluntarily step onto and move up and down the dowel rod. The cockatoo also learned to orient its beak towards a closed fist as a target. This allowed staff members to direct the cockatoo's head upright or direct the animal where to walk.

Targeting was then used to train simple behaviors such as moving from point A to B, turning around, and holding position. While the bird held position staff members could work on teaching the bird to allow wings to be pulled out from the body and also palpate pectoral muscles and legs.

In one session the bird was also trained to accept water from a syringe. This was done by slowly introducing a syringe using systematic desensitization and pairing it with desired food items. Once the cockatoo was comfortable accepting water from the syringe, juice was offered from the syringe. This was then generalized to many other types of fluids including those that were not palatable. This behavior was trained in anticipation of administering oral medications if ever needed.[2] This behavior was easily maintained for years after it was initially trained by offering various fluids from the syringe on occasion followed by desired food items.

Because of the excellent success keepers were having training this once non-compliant cockatoo they were inspired to attempt the goal of training for an intramuscular injection. At the time this facility was preparing to vaccinate for the West Nile virus and the goal was to train the bird to accept this vaccination without restraint. Because the team had not trained this behavior before, they were uncertain if the bird might attempt to bite when an injection was administered. Therefore, they included training the bird to allow a clear plastic cup to be placed on its head as a barrier between the beak and injection site at the pectoral muscle. However, this proved unnecessary for the procedure as the bird never showed an inclination to respond with aggressive behavior due to the training process being performed in such a manner as to never cause the bird to be pushed beyond its level of comfort. However, the cup training later proved useful for nebulization treatments.

Injection training included applying pressure to the chest using a dulled needle, snapping the chest with a rubber band to simulate stinging medications or higher volume of fluids, and piercing the skin with small gauge needles. Each approximation was always followed with the delivery of high value reinforcers and only attempted if the bird showed body language indicative of comfort. Once these steps were mastered injection of a small amount of saline was attempted. Upon successful completion of this step, the vaccination was successfully administered. The behavior was then maintained with occasional injections of saline or lesser approximations such as touching the chest with the dulled needle.[2]

Additional behavior goals included training for restraint in a towel and scale training. Both of these behaviors were trained in the veterinary hospital of the zoo. Training for restraint made it possible for the veterinary staff members to successfully collect blood samples with little to no stress. Scale training proved to be somewhat challenging due to the location of the immoveable scale. The arrangement of the scale created a small cavity like area that was attractive as a nest site. However, this was overcome by strategic thinking of the bird's trainer. Essentially the nest-like area was covered during training and the area was slowly uncovered overtime as the bird mastered the behavior.

The following list summarizes the behaviors the cockatoo was trained to present:

- Targeting;
- Step onto a dowel rod for transport;
- Lift each foot;
- Turn around;
- Allow tactile on chest, legs, head, feet, under wings, on back;
- Allow wings to be pulled out;
- Allow cup to be placed on head;
- Accept various fluids from a syringe;
- Allow intramuscular injection in pectoral muscle;
- Allow restraint in a towel; and
- Stand on a scale.

Illness Detected and Treatment

The scale behavior is what lead to the discovery of the cockatoo's symptoms. In November 2015 a staff member had retrieved the bird to bring to the veterinary hospital to demonstrate the scale training when an audible change in the bird's breathing was noticed. After collecting the bird's weight, the veterinary team and keeping staff members that had trained the bird worked together to immediately begin diagnostics and treatment. Veterinary staff members were able to successfully bring a stethoscope to its chest and back without restraining the bird. When restraint was required for blood collection, the bird sat calmly while a towel was draped over its body. Although restraint did require pressure to be applied to the bird, due to prior training the bird coped well with restraint and when released was relatively calm, responsive and was willing to accept food.

Later in that same day the bird accepted four injections without restraint as well as subcutaneous fluids without restraint. The bird also sat comfortably with its head in a cone shaped mask for nebulization. The cone was quite similar to the cup that had been used in injection training which may have contributed to the bird's comfort with this procedure.

It took surgery to diagnose an aspergillosis granuloma on the syrinx. A cannula was placed in an air sac to assist with breathing and the bird continued to voluntarily participate in accepting oral and inhaled medications with little to no stress, according to the veterinary team and staff members who treated the bird over the course of several weeks after initial observation of symptoms. Unfortunately, the bird did not survive the obstructive fungal lesion. It returned and the bird passed under anesthesia during a procedure to recheck the trachea.

Conclusion

Although ultimately the bird did not survive, its final weeks of receiving treatment were relatively stress free thanks to an investment in training prior to onset of illness. Over the course of six years the cockatoo only needed to learn approximately seven behaviors to make medical care relatively stress free, some of which were trained in one twenty-minute session. Once the behaviors were trained, maintaining them required a small investment of time. The result was a significant reduction in stress for the treatment of a serious medical condition.

Animal training is often considered a luxury or extra activity to include in animal's life as a means to add enrichment and entertainment. However, more and more examples demonstrate that training significantly contributes to high standards in animal welfare. Training and training programs should be as essential as daily feeding, cleaning and veterinary care. When professionals begin to embrace training as a part of well-rounded approach to providing best practices in animal care we can set a precedent for the future and reduce or eliminate stress for many animals in regard to medical care.

References

1. Heidenreich B. An introduction to the application of science based training technology. *Vet Clin North Am Exot Anim Pract*. 2012:371-385.

2. Heidenreich B. Training animals to accept oral medications and injections without restraint. *Proc ExoticsCon*. 2015;619-624.

Wound Management with Medical Maggots in a Peacock *(Pavo cristatus)*

Rhoda Stevenson, DVM, Dipl ABVP (Avian)

Session #032

Affiliation: Exotic Bird Hospital, Inc., 8820 Old Kings Rd., S. Jacksonville, FL 32257-5303, USA

Abstract: A peacock (*Pavo cristatus*) was presented for severe soft tissue injuries to the right wing from dog bite wounds. After 9 days of antibiotics and debridement, the infection was still out of control due to the devitalization of the tissues involved. Medical maggots were introduced into the wound for 3 days and then removed. There was significant improvement after the maggots were removed and with continued debridement and antibiotics the wound healed completely.

Clinical Report

A 4-year-old, 3.8-kg male peacock (*Pavo cristatus*) was presented with wounds sustained by a wild dog pack attack 2 hours earlier. He was able to stand but reluctant to walk. The worst wound was on the right wing where the propatagialis complex muscle, deltoideus major and minor, triceps brachii, coracobrachialis, superficial pectoral and humerus were exposed but there was no visible fracture. There were multiple bite wounds in the caudal area and all of the rectrices were missing. The cloaca was intact but there was a 14-cm laceration on the right side of the cloaca opening that did not penetrate the coelom although there was small amount of gross blood in the feces. The bird sat on the exam table while the wounds were flushed with sterile saline. Piperacillin and tazobactam (200 mg/kg IM q12h; Piperacillin and Tazobactam, Mylan Institutional LLC, Rockford, IL, USA), butorphanol tartrate (1 mg/kg IM q12h; Dolorex, Intervet Inc., d/b/a Merck Animal Health, Summit, NJ, USA), and meloxicam (0.5 mg/kg IM q12h, Meloxicam solution for injection, Putney Inc., Portland, ME, USA) were started. The bird was given saline (20 cc/kg SC q6h; 0.9% Sodium Chloride Injection USP, Hospira Inc., Lake Forest, IL, USA) and placed in an oxygen cage (1 L/min). The following day, the bird was tube fed hand-feeding formula (10 cc/kg PO q12h, Pretty Bird 19/8, Pretty Bird International Inc., Stacy, MN, USA) using an 18-french red rubber catheter because there was no substance to the droppings and it was not eating on its own. The wounds were flushed again with sterile saline.

On the third day the bird was anesthetized with sevoflurane (2 L oxygen flow, 4% sevoflurane, Sevoflurane, Patterson Veterinary USA) delivered with a face mask and the wounds in the caudal area were flushed with sterile saline, debrided and closed with 3-0 polysorb (Polysorb Lactomer 9-1, coated braided synthetic absorbable suture, Covidien LLC, Mansfield, MA, USA) in a simple interrupted pattern. The wound on the right wing had a large amount of necrotic tissue that was not yet able to be debrided and still had evidence of infection so it was flushed with sterile saline but not closed. The bird had started eating the following day, so the tube feeding was discontinued, the subcutaneous fluids were dropped to q8h and the injectable meloxicam was changed to oral meloxicam (0.5 mg/kg, Metacam, Boehringer Ingelheim Vetmedica, Inc., St. Josephs, MO, USA).

On the fifth day the right wing necrotic area was measured at 10 cm × 6 cm × 3 cm. The following day amikacin (15 mg/kg IM q12h; Amikacin compounded 50 mg/ml, Wedgewood Pharmacy, Swedesboro, NJ, USA) was added to try to control the infection after a CBC revealed a severe regenerative anemia and the uric acid was normal. The bird was losing weight so the tube feeding was started again. On day 7, enrofloxacin (10 mg/kg PO q12h;

Enrovet 10%, CIPLA MEDRO [PTY] LTD, Rosen Park, Bellville, RSA) and clindamycin (75 mg/kg PO q12h; Cleocin HCl 300 mg, Pfizer, Kalamazoo, MI, USA) were added to the medication regimen as the infection in the wing wound was still out of control. Radiographs did not show any evidence of a fracture or osteomyelitis.

On the ninth day, the bird was anesthetized with masked sevoflurane (2-liter oxygen flow, 4% sevoflurane) and after the right wing wound was copiously flushed with sterile saline, medical maggots (5-8/cm^2; Medical Maggots, Monarch Labs, Irvine, CA, USA) were placed in the wound which was covered with pantyhose material and taped in place. The wound was then covered with loose fitting gauze to help absorb the discharge. The gauze was changed twice daily for the next 3 days. The maggots visibly grew during that time. On the 13th day, the maggots were removed one by one with hemostats and placed in alcohol before disposal. This procedure took about an hour and the bird did not mind at all although there were attempts by the bird to eat the maggots as they were being removed. The wound was flushed with sterile saline and a non-adherent pad (Kendall Telfa, Tyco Healthcare Group, LP, Mansfield, MA, USA) was placed on it for several hours. The wound was re-explored to be sure there were no remaining maggots, flushed again with sterile saline and silver sulfadine (thin film; Silver Sulfadine, Ascend Laboratories, LLC, Montvale, NJ, USA) was applied. Healthy granulation tissue was visible and the exposed bone appeared healthy, although there was still necrotic tissue present in the wound. The piperacillin/tazobactam and amikacin were discontinued at this time.

The wound was debrided and flushed twice daily for the next 2 days as residual necrotic tissue was removed. On day 16, the bird was anesthetized with masked sevoflurane (2-L oxygen flow, 4% sevoflurane) and the remaining necrotic tissue was removed. A culture of the bone and muscle was taken and the wound was flushed with sterile saline before it was closed with 3-0 polysorb in a horizontal mattress suture pattern. Telfa pad, non-adherent bandaging tape (vetrap, 3M Deutschland Gonbh- Consumer Business, Neuss, Germany) and elastic tape (Elastikon, Johnson & Johnson Consumer Companies, Inc., Skillman, NJ, USA) were placed over the wound site after it was closed. This bandage was changed the following day and the areas that were not completely closed with the suture were flushed with sterile saline. The bandage was left off on day 18. The culture revealed a small amount of *Enterobacter* sp. and yeast from the wound site. The bacteria were sensitive to enrofloxacin, so that was continued, the clindamycin was discontinued, and fluconazole (10 mg/kg PO q12h; Fluconazole, Pack Pharmaceuticals, Buffalo Grove IL, USA) was started. There was some devitalization of the sutured wounds in the caudal area near the cloaca that had some visible fungal growth. Since this was relatively superficial at this stage, a thin film silver sulfadine was applied to those areas twice daily.

The wounds continued to heal and on day 27 the bird was sent home on enrofloxacin, fluconazole and silver sulfadine. Rechecks were done over the next month until the wounds had totally healed. The bird was able to flap the wing and had tail feathers coming back in.

Discussion

Medical maggots are not usually considered in mainstream medicine since antibiotics and debridement often work for most wound management. In human medicine, the beneficial effects of using larvae in the wounds were first noticed in 1557.[1] The first clinical application of maggot therapy was performed during the American Civil War. Sterile maggots were later used to prevent maggot-induced wound infections and the therapy was widely used for treatment of chronic or infected wounds across North America and Europe during the 1930s. With the introduction and widespread use of antibiotics in the 1940s, MDT (maggot debridement therapy) was largely forgotten until the rising incidence of antibiotic resistance in the late 1990s.[1,2]

MDT involves the application of sterile larvae of the blow fly (*Phaenicia [Lucilia] sericata*) to a wound site. The flies are bred in a controlled sterile moist environment and the newborn larvae should be used within 8 hours or stored in the refrigerator to slow the development.[1-3] It is important to ensure an optimal body temperature, adequate oxygen supply and moisture, although too much moisture may kill the larvae, to maximize the debridement. The maggots have two salivary glands which produce digestive enzymes and a powerful pump-like pharynx that enables them to suck liquefied food into their alimentary canal.[3] This saliva contains collagenases, trypsin-like and chymotrypsin-like enzymes that break down extracellular matrix components, including laminin and fibronectrin. The larvae use adduction of the mouth-hooks of the cephalopharyngeal skeleton to scrape and abrade the necrotic tissue and this may allow the salivary enzymes better penetration into the tissues. Wound debridement may be further assisted by larvae crawling around in the wound, dislodging necrotic material. Feeding larvae have a process of auto-disinfection in their alimentary canal that results in the destruction of bacteria including multiple-resistant *Staphylococcus aureus*, *Streptococcus* A and B, *Escherichia coli* and *Pseudomonas species*.[1-3] The larvae also secrete allantoin, ammonium bicarbonate and urea that alkalinize the environment and along with the physical disruption of the wound environment make it unsuitable for bacterial habitation. Additionally, the action of the maggots and the response of the host tissue leads to the production of large quantities of fluid that help to irrigate bacteria from the wound.

The clinical indications for use in human medicine include debridement of non-healing necrotic skin and soft-tissue wounds such as pressure ulcers, neuropathic foot ulcers, chronic leg ulcers, or non-healing traumatic or post-operative wounds.[4] During the use of medical maggots the wound should not be allowed to close over and dead maggots should not remain in the wound as they may trigger an allergic response or may become the nidus for further infection.[5] Larvae also may try to escape from the dressing, especially when they have debrided all necrotic tissue or when they are satiated. Escapes are more common after 48 hours when they are ready to leave and pupate. Contraindications for the use of medical maggots include use in patients that are allergic to fly larvae, brewer's yeast or soy protein. Patients with acute, rapidly advancing, life or limb threatening infections are not candidates for MDT. Maggots should not be used in wounds that are near large necrotic blood vessels or in patients that have a coagulopathy as some bleeding can occur during the debridement and rupture of the blood vessel could be life-threatening. The maggots may not survive if the patient receives insecticides or insect-toxic drugs.

The use of MDT in veterinary medicine is also indicated for difficult to debride areas like the hoof of a horse with severe laminitis where the minimally invasive treatment does not disturb the normal architecture.[3] A few dogs, cats and rabbits with chronic wounds have been treated with some success.[6] Sheep with recalcitrant foot rot and foot scald have been treated successfully with one application of larvae.[3]

MDT was chosen for this case because the location of the wound site made it difficult to effectively debride due to the multiple muscles, tendons, ligaments, fascia and bone that were involved. The position did not allow for good drainage of the wound since it was on the dorsal aspect of the proximal wing. A more aggressive mechanical debridement could have resulted in damage to structures necessary for the use of the wing. The wing could have been amputated but there was still concern for infection at the site of the amputation since it was so close to the body. The concern for sepsis was also present, even with the use of multiple antibiotics, since there was not good penetration into the wound due to the compromise in the blood supply. The potential for antibiotic resistance in the face of ineffective infection control at the wound site was also an issue in this case. The maggots were well tolerated by the patient and seemed to make a dramatic difference in the healing process.

References

1. Chan DCW, Fong DHF, Leung JYY, et al. Maggot debridement therapy in chronic wound care. *Hong Kong Med J.* 2007;13:382-386.

2. Davydov L, Pharm BS. Maggot therapy in wound management in modern era and a review of published literature. *J Pharmacol Pract.* 2011;24(1):89-93.

3. Jones G, Wall R. Maggot-therapy in veterinary medicine. *Res Vet Sci.* 2008;85(2):394-8.

4. Medical Maggots™. Monarch Labs, Irvine, CA 92614, USA.

5. Lau H, Lee F. Maggots in surgery. *Ann College Surgeons Hong Kong.* 2000;4:50-53.

6. Sherman RA, Stevens H, Ng D, et al. Treating wounds in small animals with maggot debridement therapy: a survey of practitioners. *Vet J.* 2007;173(1):138-143.

Proventriculotomy for Foreign Body Removal in Three Chickens *(Gallus gallus domesticus)*

Stephanie Lamb, DVM, Dipl ABVP (Avian)

Session #111

Affiliation: From Arizona Exotic Animal Hospital, 744 N. Center St. Suite #101, Mesa, AZ 85201, USA.

Abstract: Three domestic chickens presented for various signs and symptoms to the hospital for evaluation. Each was found to have metallic foreign bodies within the ventriculus. A proventriculotomy was performed to extract these foreign objects. Chelation therapy for suspected or confirmed lead toxicity was also initiated. All birds did well with the surgery in the immediate and long term post-operative time period. Chelation therapy was discontinued once blood lead levels were out of the toxic range.

Introduction

Raising backyard poultry has become popular in the last several years and, as a result, chickens are being seen more frequently at private veterinary practices for care. Often allowed to forage in yards, chickens can sometimes ingest foreign objects that owners are unaware were even present. Foreign objects can come in various shapes and sizes and be located at the crop, proventriculus, ventriculus or intestines. Based on the foreign objects shape, material and location, various methods and routes for removal are available.

Proventriculotomies have been described in birds and are indicated for the removal of foreign objects and toxic materials, to relieve gastric outflow obstruction, or for the complete or partial removal of gastric neoplasia.[1,2] Several texts have described various approaches and include a left lateral, transverse coeliotomy, and ventral midline with lateral flap incision to make an inverted L shape.[1-3] Endoscopy is another route that is a viable option to retrieve foreign objects. Evaluation of the proventriculus and ventriculus through an incision into the ingluvies has been described and prior reports have demonstrated the successful removal of foreign objects through this route.[4,5] Medical management has also been illustrated to allow for passage of foreign objects with the use of emollient cathartics, grit, and psyllium.[6,7]

The route chosen for removal of the foreign object is dependent upon numerous factors. Objects such as screws, nails and wires have their own inherent risks associated with removal through an endoscopy procedure. Therefore, a proventriculotomy to extract these types of objects may be a more appropriate route to take in order to avoid causing unintended perforation of the proventriculus. However, surgeon experience, the patient's status and size need to be considered on an individual case basis to determine which route is most appropriate.

Another consideration for gastrointestinal foreign objects that needs to be addressed is whether or not the foreign object is toxic for the patient. Metallic objects can contain lead which is well known to be toxic to birds.[6] Blood levels can be evaluated to assess for toxicity. Treatment for lead toxicity involves starting chelation therapy, providing supportive care for associated aliments and removal of the foreign lead object in some cases.[1,6] There are various chelating agents available for lead toxicity including calcium disodium ethylene diamine tetraacetate (CaEDTA), D-penicillamine, and dimercapto succinic acid (DMSA).[6]

The purpose of the study reported here was to evaluate the use of a proventriculotomy to remove foreign metallic objects from the ventriculus of backyard chickens. These metallic objects also contained lead and the toxicity from them required management as well. Proventriculotomies should be considered for chickens that have ingesta foreign objects that are clinically significant and cannot be safely removed by other means.

Clinical Report

Case 1

A 7-month-old Rhode Island Red female chicken presented for lethargy and laying an unshelled egg. The owners reported that she had been anorexic for approximately 12 hours. Upon physical examination she was tense on caudal coelomic palpation and had a soft egg shell and membrane fragments adhered to the vent mucosa.

Radiographs were taken and revealed a full appearing caudal coelom with a soft tissue to fluid opacity in the location of the caudal thoracic and abdominal air sacs. Although some gas was present in the intestines there was a loss of detail between organs. Within the gizzard there were mineral densities and two metal density objects that appeared to be screws. A third metallic density object was identified dorsal and caudal to the other metal and mineral densities. Blood was collected for a complete blood count, biochemistry profile and heavy metal toxin testing. A mild leukocytosis and mild anemia were identified. The calcium was within the reference limits but was considered low for a laying hen. The lead level came back as high and was greater than 65 μg/dl. The zinc level was within the reference ranges at 1.7 ppm.

The patient was admitted to the hospital and started on supportive care consisting of subcutaneous fluids (25 ml/kg SC q12h × 5 days; Lactated Ringer's injection, USP; Hospira Inc, Lake Forest, IL USA), calcium glubionate (25 mg/kg PO q24h × 21 days; Calcionate syrup, Rugby, Duluth, GA, USA), CaEDTA(35 mg/kg SC q12h × 5 days; CaEDTA, Diamondback Drugs, Scottsdale, AZ, USA), meloxicam (0.3 mg/kg PO q12h × 7 days; Metacam, Boehringer Ingelheim, Saint Joseph, MO, USA) and sucralfate (20 mg/kg PO q8h × 10 days; Sucralfate, Teva Pharmaceuticals, Sellersville, PA, USA). She was eating well in the hospital once supportive care was started so assist feeding was not necessary.

The patient was doing well with supportive care but given that the metallic objects within the ventriculus were screws, there was concern about possible perforation of the gastrointestinal tract. Options for removing these objects were discussed with the owner and included endoscopy or proventriculotomy. Since there were risks associated with endoscopy and possible failure to withdrawal the objects successfully, paired with the fact that the patient possibly had reproductive tract pathology, it was deemed more appropriate to perform a proventriculotomy to remove the metallic objects.

After 3 days of supportive care, the patient was taken to surgery to explore the caudal coelom and also perform a proventriculotomy. A ventral midline incision was made with a lateral flap incision at the cranial border extending towards the left to make an inverted L pattern. Once in the coelomic cavity the proventriculus and ventriculus were identified. Stay sutures were placed with 3-0 nylon (Weblon, Patterson Veterinary, Devens, MA, USA) in the tendinous portion of the ventriculus to help retract the organ closer to the surgery site. Gauze, which had been lightly moistened with saline, was packed around the proventriculus and ventriculus. An incision was made at an avascular region of the isthmus with a scalpel blade and extended slightly cranially and caudally into the proventriculus and ventriculus. The contents of the proventriculus and ventriculus were removed carefully. Three screws and several pieces of glass were also removed. Normal plant ingesta was removed as well. The surgery site was gently flushed with sterile saline and wiped clean of gross debris. The incision site was closed in a two layer pattern with 5-0 PDS (PDS II, Ethicon, Guaynabo, Puerto Rico, USA). The first layer was a simple continuous appositional pattern. The second layer was a cushing pattern to allow for slight inversion. The caudal coelom was explored and although the reproductive tract was large there was no evidence of pathology. The abdominal musculature was closed with 3-0 PDS (PDS II, Ethicon, Guaynabo, Puerto Rico, USA) in a simple continuous pattern. The skin was closed with 3-0 vicryl (Webcryl, Patterson Veterinary, Devens, MA, USA) in a simple continuous pattern. Recovery was slow but uneventful.

Post-operatively the patient continued on the previous supportive care with the addition of amoxicillin with clavulanic acid (125 mg/kg PO q12h × 10 days; Clavamox, Zoetis Inc., Kalamazoo, MI, USA). She was discharged from the hospital 2 days after the surgery. Two days after stopping chelation therapy, her lead level was rechecked and the values were within the reference range. It was recommended that although she was a pet bird that she not be used for meat consumption ever given her history.

This bird did come from a flock of 3 other individuals. It was recommended that the others be evaluated for heavy metal toxicity as well and that consumption of the eggs be withheld until heavy metal status was known for all individuals. Two of the other members of the flock tested high for lead and chelation therapy was started for them as well.

Case 2

A 7-month-old Buff Orpington hen presented for lethargy and disoriented behaviors of 1 day duration. On physical examination, she was ataxic and leaning toward the left. Radiographs were taken and revealed 5 large metallic objects within the ventriculus along with several smaller metallic fragments. The large objects appeared to be nails and screws. A complete blood count and biochemistry profile were evaluated. The CBC revealed a mild anemia and a moderate leukocytosis. The calcium level was within the reference ranges but was considered low for a bird actively laying eggs. Heavy metal toxin testing was not performed on entry to the hospital.

The patient was admitted to the hospital for supportive care similar to the chicken in the first report. Chelation therapy with CaEDTA was initiated as well for suspected heavy metal toxicity. Options of endoscopy or a proventriculotomy were discussed with the owner to have the foreign objects removed. A proventriculotomy was again elected given the concerns about possible damage to the gut while removing the sharp objects. The patient was admitted to the hospital and the surgery was performed that same day.

The proventriculotomy was performed as described for the bird in the first case report with the exception that at the end of the procedure a ¼ tablet of a 1 g sucralfate capsule was placed into the proventriculus before suturing the organ closed. Metallic objects and fragment of glass had been removed. A watch battery was also removed. The patients recovery went well and she was placed on subcutaneous fluids (25 ml/kg SC q12h × 5 days; Lactated Ringer's injection, USP; Hospira Inc, Lake Forest, IL USA), calcium glubionate (25 mg/kg PO q24h × 21 days; Calcionate syrup, Rugby, Duluth, GA, USA), CaEDTA (35 mg/kg SC q12h × 5 days; CaEDTA, Diamondback Drugs, Scottsdale, AZ, USA), pipercillin (100 mg/kg IM q8h × 10; Pipercillin and Tazobactam for Injection USP, Sandoz Inc., Princeton, NJ, USA), sucralfate (20 mg/kg PO q8h × 10 days; Sucralfate, Teva Pharmaceuticals, Sellersville, PA, USA) and torbugesic (1 mg/kg IM q8h × 5 days; Torbugesic, Pfizer Inc., New York, NY, USA).

After 5 days of chelation therapy a two day break was performed. At the end of this break a lead and zinc level were checked. The zinc level was within the reference ranges but the lead level was mildly elevated at 27 µg/dl. Chelation therapy was restarted for another 2 rounds of 5 days on treatment with 2 days off. A repeat lead level was taken at the end of this and it was found that the lead level had returned to the reference ranges. Chelation therapy was discontinued. The owners were advised to not consume the animal's meat of egg for at least 30 days after cessation of all medications.

Case 3

A 6-month-old Dutch Bantam hen presented for evaluation because another bird in the flock (case 2) had heavy metal toxicity and metal objects within the ventriculus. The patient was bright, alert and responsive on entry to the hospital and the physical examination appeared normal. She was an actively laying hen at the time.

Radiographs were taken and 3 linear metallic density objects were located in the ventriculus. There was also mineral density material as well. Due to financial constraints blood work was not evaluated on this patient.

The patient was taken to surgery and had a proventriculotomy performed similar to the second case above. Metallic objects were removed successfully. Recovery in this bird was uneventful and she was started on subcutaneous fluids (25 ml/kg SC q12h × 5 days; Lactated Ringer's injection, USP; Hospira Inc, Lake Forest, IL USA), CaEDTA (35 mg/kg SC q12h × 5 days; CaEDTA, Diamondback Drugs, Scottsdale, AZ, USA), sucralfate (20 mg/kg PO q8h × 10 days; Sucralfate, Teva Pharmaceuticals, Sellersville, PA, USA), pipercillin (100 mg/kg IM q8h × 10; Pipercillin and Tazobactam for Injection USP, Sandoz Inc., Princeton, NJ, USA) and torbugesic (1 mg/kg IM q8h × 5 days; Torbugesic, Pfizer Inc., New York, NY, USA).

After a 2-day break from a 5-day treatment course with CaEDTA blood was analyzed for lead and zinc. The zinc level came back in the normal reference ranges but the lead came back high at 84 µg/dl. CaEDTA chelation therapy was continued for 2 more rounds of therapy consisting of 5 days on CaEDTA and a 2-day break in between. After this treatment regimen lead levels were checked again and found to be in the normal ranges, therefore chelation therapy was discontinued. The owners were advised to not consume the animal's meat or egg for at least 30 days after cessation of all medications.

References

1. Rupley AE. *Manual of Avian Practice*. Philadelphia, PA: WB Saunders Co;1997.
2. Rubin JA, Runge JJ, Mison M, et al. Surgery. In: Speer BL, ed. *Current Therapy in Avian Medicine and Surgery*. St. Louis, MO: Elsevier;2016:631-667.
3. Echols MS. Soft Tissue Surgery. In: Greenacre CB, Morishita TY. *Backyard Poultry Medicine and Surgery: A Guide for the Veterinary Practitioner*. Ames, IA: John Wiley & Sons; 2015:220-259.
4. Lierz M. Diagnostic Value of Endoscopy and Biopsy. In: Harrison GJ, Lightfoot TL, eds. *Clinical Avian Medicine*. Vol 2. Palm Beach, FL: Spix Publishing;2006:631-652.
5. Hernandez SJ, Blasier M, Wilson H, et al. Endoscopic removal of (pro)ventricular foreign bodies in parrots. *Proc Annu Conf Assoc Avian Vet*. 2006;359-361.
6. Dumonceaux G, Harrison GJ. Toxins. In: Ritchie BW, Harrison GJ, Harrison LR, eds. *Avian Medicine: Principles and Applications*. Lake Worth, FL: Wingers Publishing;1994:1030-1052.
7. Lupu C, Robins S. Comparison of treatment protocols for removing metallic foreign objects from the ventriculus of budgerigars (*Melopsittacus undulatus*). *J Avian Med Surg*. 2009;23(3):186-193.

Section 14

AAV Posters

Clinical Exotic Animal Applications for F10 Antiseptic Solution: A Novel Agent in the USA - A Recent Addition to the Food and Drug Administration's Minor Use and Minor Species Index 900-007

Neil A Forbes, BVetMed, Dipl ECZM (Avian), FRCVS

Session #064

Affiliation: From Great Western Exotic Vets, Vets Now Referrals, 10-16 Berkshire House, County Park Estate, Shrivenham Rd, Swindon, SN12NR, UK.

Abstract: This paper describes the clinical applications of an alkyl dimethyl benzyl ammonium chloride (benzalkonium chloride) and poly (hexamethylene biguanide) hydrochloride topical solution in exotic animal practice. The product (F10 Antiseptic Solution, Health and Hygiene (Pty) Ltd) has been used in South Africa, Australia, the EU (including UK) and other parts of the world in exotic practice, over the past 15 years, for prevention of and control of lower respiratory infections in (excluding all patients bred for human consumption) avian patients, by fogging and nebulisation and of the upper tract by flushing and the aerosol environmental disinfection of premises (in particular where there is a risk of airborne infection). In these environmental applications the "sister" product is F10SC Veterinary Disinfectant which is approved and registered in the USA by the EPA.

Introduction

This product was first licensed for use as a hard surface disinfectant in 1994. In view of its great patient, user and tissue safety became used for topical patient treatment in the late 90's. The author has used this product with apparent efficacy and absolute safety, in many clinical exotic animal situations over the past fifteen years. Subsequently he acted as joint chair of the Minor Use and Minor Species (MUMS) Expert Scientific Committee, which evaluated the product for registration in USA by the Food and Drug Administration, under the MUMS scheme (Index 900-007; http://www.fda.gov/AnimalVeterinary/DevelopmentApprovalProcess/MinorUseMinorSpecies/ucm125452.htm). Working in the latter committee required a full assessment of all product and manufacturing safety, efficacy, clinical application data and publications relating to the product. The author believes that this product is a valuable addition to the therapeutic armament of the exotic animal clinician and wishes to share experiences and knowledge with colleagues who are not yet aware of the product.

Chemical Make-Up

Benzalkonium chloride (a quaternary ammonium compound) and polyhexanide (biguanide) are combined in this product. These two components are cell poisons and act on the cell membrane rupturing it with loss of essential components: destroy enzymes within the cell: disrupt DNA or RNA sequence: interact with other essential elements within the cell. They interact with SH bonds and rupture them. At a high concentration the antiseptic will in addition fix the proteinaceous components of the cell. Other minor constituents aid the penetration of the cell or spore wall or spore wall, acting as enzymatic poisons, chelation agents for certain essential metals, as well as fixation of certain proteins. As with most disinfectants/antiseptics F10 antiseptic solution (F10AS) is affected by the presence of gross organic matter being absorbed into the organic material, this effect is reduced by prior cleaning, or repeated flushing in a clinical situation. Published experiments have shown that different disinfectant active ingredients are better suited to efficient function in different environments (temperature and physical make up), moreover that the spectrum of activity is typically increased, by the combination of two active ingredients, without the necessity to increase the concentration of either active ingredient. The enhanced spectrum of activity, in terms of pathogens controlled and environments in which they are efficacious, is typically greater than either single active agent alone. Moreover as pathogen control can be achieved with a regular concentration, any potential risks to the environment and any animals which become exposed to the agents, is minimised. Whilst the debate over mono-therapy against poly-therapy rumbles on, it has been shown that poly-therapy does reduce the risk of pathogen drug resistance developing. Studies have shown single active products are more likely to lead to increased resistance, such as MRSA, whereas the overlapping nature of components in the formulation prevent this from happening to F10 disinfectants/antiseptics. This product, (F10 disinfectant) has been the only disinfectant in use, on a daily basis, for the past 15 years for the disinfection of hard surfaces in the Veterinary Academic Hospital, Onderstepoort, the Veterinary Faculty of the University of Pretoria, South Africa, where regular microbiologic surveys and challenge tests demonstrate its effectiveness against "super bugs."

Safety Data

Considering the active ingredients, against their published properties, one would anticipate a good safety profile, both to the target animal and also consequently within the environment. Standard quality controlled, government approved, independent GLP laboratory studies test certificates have been accessed, in respect of: dermal toxicity, skin sensitization, acute dermal irritation, acute oral toxicity, inhalation toxicity, acute eye irritation, all tests have been completed with no adverse reactions reported. No adverse reaction reports have been submitted in UK, Australia or South Africa, in respect of the use of F10 products, in the period from the introduction or products until the time of writing this paper. Having made due enquiries the author is also unaware of any adverse reaction reports from any other geographical region. In respect of lack of eye irritation, the author has used the product at a 1:1000 dilution, as a treatment for 'post cataract *Candida* conjunctivitis in an owl (*Bubo bubo*), when no proprietary topical therapeutic agent was available. The product was administered directly to the eye twice daily for 5 days. Clinical signs were completely resolved within 48 hours, and no adverse effect occurred at any point. Whilst not applicable to use in USA, tissue residue tests, have been conducted to international standards following oral administration in growing poultry, in the absence of any with drawl period, no inhibitory substances were detected in muscle tissues at post mortem. The product is manufactured in a regularly checked, audited and licensed GMP licensed facility, compliant with South African, Australian, EU and USA Standards. Each batch manufactured is sampled, concentration tested prior to release and stored.

Efficacy

Reference efficacy test certificates, from independent government approved laboratories, compliant with international standards, have been conducted. Bactericidal tests have been carried out in accordance with SANS 636-2001 and (South African) standard having a performance pass criteria of $>log^3$ reduction in microbial counts, the AFNOR (French) standard which is a $>log^4$ reduction in microbial counts, the European Union EN standard which is a $>log^5$ reduction in microbial counts, and the AOAC (USA) standard, a $>log^6$ reduction in microbial counts. MIC in-vitro tests have shown significant depth of performance as indicated. Generally testing is carried out at ambient temperatures with the exception of EN Standard tests where some were carried out at low temperatures 10°C and some DEFRA (APHA) tests at 4°C. The EN Standard for low temperatures is 10°C and at these temperatures bactericidal performance was equal to that at ambient temperatures; however, there was a significant fall off in performance at 4°C. Other tests carried out at elevated temperatures of 45°C showed a significant increase in performance. These tests have confirmed efficacy against: *Mycobacterium* spp., *Pseudomonas aeroginosa, E coli, Staph aureus, Enterococcus faecium, Penecillium verrucosum, Cladosporuium cladosporoides, Absidia corymbifera, Candida albicans*, enteroviruses, *Orthopoxvirus, Adenovirus*, HIV, canine *Parvovirus*, hepatitis B, *Leptospira* spp., *Campylobacter* spp., *Legionella* spp., Newcastle disease virus, feline *Herpesvirus*, *Acinetobacter anitratus*, IBD virus, MSRA, *Ornithobacterium rhinotracheale*, *Bacillus subtilus* spores, *Aspergillus niger*, *Trichophyton mentogrophytes*, *Klebsiella* spp., *Proteus vulgaris*, *Enterococcus hirae*, *Salmonella typhi*, vancomycin-resistant *Enterocioccus faecalis*, *Streptococcus pyogenes*, *Vibria cholera*, *Listeria monocytogenes*, *Staphylococcus epidermides*, *Pasturella cepacia*, *Micrococcus luteus*, *Salmonella abony*, *Klebsiella pneumonia*, *Corynebacterium xerosis*, *Citrobacter freundii*, *Enterobacter sakazakii*, *Enterobacter cloacae*, rabies virus, *Aspergillus fumigatusi Pasturella multocida*, pseudo-cowpox, *Clostridium pyogenes*, *Salmonella typhimurium*, avian influenza (H5N2), *Microsporum canis*, *Salmonella cholerasuis*, and *Clostridium difficile*.

Clinical Applications

Publications and the author's own experiences include the use of this product in the following patient groups (raptor, waterfowl, bustards, passerine and psittacine), for the prevention and treatment of diseases affecting the sinuses, nasal chamber, trachea, lung and air sacs, associated with the bacteria (*Bacillius* spp, *E.coli*, *Klebsiella* spp, *Pseudomonas* spp., *Entrococcus* spp, *Staphyloccocus* spp., *Streptococcus* spp.), *Mycoplasma* (*M. gallisepticum*), and fungi (*Candida* spp. and *Aspergillus* spp.), by nasal flushing to remove accumulated mucous and inflammatory material.

Using nebulisation therapy (or fogging of a whole room), to prevent bacterial and fungal infections of the lower respiratory tract (glottis, trachea, lungs and air sacs) and associated with the micro-organisms as listed above, where due to concurrent stress, (e.g., relocation), such diseases would otherwise be anticipated due to exposure or immune suppression. The use of nebulisation therapy as an adjunct to systemic therapeutic agents to assist in treating bacterial and fungal infections of the lower respiratory tract.

Treatments have included:

a. Fogging rooms (particle size <10u), 3x a day for 30 minutes, with 1:250 dilution of F10SC concentrate, containing rare and expensive falcons as a prevention for aspergillosis, by reducing environmental spore levels. This has proven invaluable, to reduce inhaled pathogen levels and prevent otherwise serious and frequently fatal respiratory infections.

b. Enclosing clinical bird respiratory cases suffering fungal, bacterial or viral infections in an air-tight environment and nebulising (particle size <2u), for 30 minutes 3x a day of 1:250 dilution from two to 8 weeks.

c. Flushing sinuses manually twice a day, 10 ml/kg, via each nares of 1:250 in saline dilution, by syringe into the sinuses via each nares, (whilst the bird is manually restrained, with body up, head down and mouth open), daily for 14 days.

d. Instilling 1 ml/kg, 1:250 dilution of F10SCAS concentrate, direct into the trachea, or direct into the air sac on a daily basis, for topical treatment of *Aspergillus* spp. lesions for 14 days

e. Daily, 1:250 saline dilution of F10SCAS concentrate, flushing of articular joints, tendon sheaths, general soft tissues and cutaneous wounds, which have been affected by bacterial infection, for a 10-day period.

f. Daily spraying, 1:250 dilution of F10SCAS concentrate, of birds to treat skin or plumage conditions with a bacterial (including MRSA), yeast or fungal etiology.

g. The flushing of contaminated surgical sites, prior to closure, or placement of drains.

In no cases have adverse effects, attributable to the treatment been experienced or observed, whilst infections, on occasion caused by highly resistant organisms (e.g., voriconazole-resistant *Aspergillus fumigatus* and multiple antibiotic resistant *Pseudomonas* spp. infection), have been controlled, achieving a successful outcome. Several publications also have reported use of the product for the treatment of respiratory disease in small companion and zoological mammals (e.g., rodents and lagomorphs). A 2-year study on the application of F10SC/F10AS in the prevention and treatment of chytrid in amphibians, in all three life stages (showing it to be efficacious and safe) has been completed and publication an appropriate peer reviewed international journal is pending.

The same disinfectant chemical preparation also is licensed under the MUMS scheme, (see www.fda.gov/downloads/AnimalVeterinary/.../UCM440051.pdf) as a long-acting germicidal barrier ointment, for use in raptors, captive passerine birds, psittacine birds, Felidae, captive non-food-producing minor species of hoofstock, nonhuman primates, captive Ursidae (bears), elephants, hippopotamuses, hyena and reptiles including captive tortoises, snakes and lizards. For use as a topical antiseptic ointment for control of, or prevention of, reinfection to surface wounds susceptible to bacterial, fungal or viral infections (e.g., *Staphyloccocus aureus*, *E.coli*, *Pseudomonas* spp., *Aspergillus* spp.) in particular where access to a patient for daily application is challenging, such that weekly applications are more practicable.

Conclusions

This product has been widely used in exotic veterinary practice for more than 15 years, by a significant number of well-respected clinicians on several continents. No adverse reactions have been reported. Clinical applications and efficacy have been published in many peer reviewed publications, as well as in other professional and special interest publications.[1-10] The author commends the application of this product in maintenance of veterinary facility infection control and biosecurity, in particular by fogging for the control of airborne infectious agents (e.g., *Circovirus* and *Chlamydia psittaci*), the agent being safe of staff, patients and is non corrosive to metals (i.e., working parts of automated spray/fogging equipment). The author further recommends the use of this product in the prevention, control and treatment of respiratory infections in a range of (non-food consumption), minor species in zoo and exotic animal practice, in line with recommendations above.

References

1. Bailey TA, Sullivan T. Aerosol therapy in birds using a novel disinfectant. *Exotic DVM*. 2001; 3(4):17.

2. Chitty J. Respiratory disease in exotics and small mammals. *Vet Times*. 2005;35(38):1-4.

3. Drake G, Koeppel K, Barrows M. F10SC nebulization in the treatment of "red leg" syndrome in amphibians. *Vet Rec*. 2010;166:593-594.

4. Forbes NA, Lloyd E, Temperley JP. Macro-Broth Minimum inhibitory concentration (MIC) and agar disc-diffusion zone of inhibition determination on F10SC disinfectant. *Proc BVZS Autumn Conf*. 2005;98-101.

5. Lampen F, Bailey TA, Combreau O. Medical rehabilitation and quarantine of illegally traded Macqueen's bustard (*Chlamydotis macqueenii*) in the UAE. *J Avian Med Surg.* 2005;19(1):35-45.

6. Samour JH, Naldo JL, Wernery U, Beer M. Highly pathogenic avian influenza H5N1 phenotype infection in a saker falcon (*Falco cherrug*). *Falco.* 2007;30:14-16.

7. Samour J. Management of raptors. In: Harrison GJ, Lightfoot TL eds. *Clinical Avian Medicine.* Palm Beach, FL: Spix Publishing;2006:915-934.

8. Smith S, Forbes NA. Treatment of pyotraumatic dermatitis infected with methicillin-resistant *Staphylococcus aureus* in three pet psittacines. *Proc Annu Conf Eur Assoc Avian Vet.* 2006;1-5.

9. Stanford M. Recombinant omega interferon in combination with F10 nebulisation for the treatment and prevention of *Circovirus* infection in African grey parrots. *Vet Rec.* 2003;154:435-436.

10. Verwoerd DJ. F10: clinical uses in an avian model with individual aspergillosis in gyrfalcons and fungal and bacterial air sacculitis in ostriches examples/case studies. *Proc Annu Conf Br Vet Zool Soc.* 2001.

Clinical Exotic Animal Applications for F10 Germicidal Wound Spray with Insecticide: A Novel Agent in the USA - A Recent Addition to the Food and Drug Administration's Minor Use and Minor Species Index 900-007

Neil A Forbes, BVetMed, Dipl ECZM (Avian), FRCVS

Session #065

Affiliation: From Great Western Exotic Vets, Vets Now Referrals, 10-16 Berkshire House, County Park Estate, Shrivenham Rd, Swindon, SN12NR, UK.

Abstract: This paper describes the clinical applications of an alkyl dimethyl benzyl ammonium chloride (benzalkonium chloride) and poly (hexamethylene biguanide) hydrochloride topical antiseptic solution, combined with cypermethrin, a synthetic pyrethroid insecticide (used for control of agricultural insect pests, as spot treatment to control insect pests in houses, and as an insect repellent for animals), in exotic animal practice. The product (F10 Germicidal Wound Spray with Insecticide, Health and Hygiene (Pty) Ltd) has been used in South Africa, Australia, the EU (including UK) and other parts of the world in exotic practice, over the past 15 years, for the prevention and control of bacterial, fungal or viral superficial infections, in particular where myiasis (flystrike) is present or presents a risk.

Introduction

The antiseptic base product contained, was first licensed for use as a hard surface disinfectant in 1994. In view of its great patient, user and tissue safety, it became used as a topical patient antiseptic treatment in the late 90's, specifically in situations where bacterial, fungal or viral infections were present, or where there was a risk of such infections occurring. Cypermethrin was added to this product for use in the control and prevention of myiasis (flystrike) of skin wounds. The author has used this product with apparent efficacy and absolute safety, in various clinical exotic animal situations over the past fifteen years. Subsequently he acted as joint Chair of the MUMS Expert Scientific Committee, which evaluated the product for registration by the FDA, under the MUMS scheme (Index 900-007; www.fda.gov/downloads/AnimalVeterinary/.../UCM440055.pdf). Working in the latter committee necessitated a full assessment of all product and manufacturing safety, efficacy, clinical application data and publications relating to the product. The author believes that this product is a valuable addition to the therapeutic armament of the zoo, wildlife and exotic animal clinician and wishes to share experiences and knowledge with colleagues who are not yet aware of the product.

Chemical Make-Up

Benzalkonium chloride (a nitrogenous cationic surface-acting agent belonging to the quaternary ammonium compound group) and polyhexanide (a cationic biocide also known as PHMB and hexamethylene biguanide) are combined in this product. These two components are cell poisons and act on the cell membrane rupturing it with loss of essential components: destroy enzymes within the cell: disrupt DNA or RNA sequence: interact with other essential elements within the cell. They interact with SH bonds and rupture them. At a high concentration the antiseptic will in addition fix the proteinaceous components of the cell. Other minor constituents aid the penetration of the cell or spore wall or spore wall, acting as enzymatic poisons, chelation agents for certain essential metals, as well as fixation of certain proteins. As with most disinfectants / antiseptics F10 antiseptic solution (F10AS) is affected by the presence of gross organic matter being absorbed into the organic material, this effect is reduced by prior cleaning, or repeated flushing in a clinical situation. Published experiments have shown that different disinfectant active ingredients are better suited to efficient function in different environments (temperature and physical make up), moreover that the spectrum of activity is typically increased, by the combination of two active ingredients, without the necessity to increase the concentration of either active ingredient. The enhanced spectrum of activity, in terms of pathogens controlled and environments in which they are efficacious, is typically greater than either single active agent alone. Moreover as pathogen control can be achieved with a regular concentration, any potential risks to the environment and any animals which become exposed to the agents, is minimised. Whilst the debate over mono-therapy against poly-therapy rumbles on, it has been shown that poly-therapy does reduce the risk of pathogen drug resistance developing. Studies have shown single active products are more likely to lead to increased resistance, such as MRSA, whereas the overlapping nature of components in the formulation prevent this from happening to F10 disinfectants/antiseptics. This product, (F10 disinfectant) has been the only disinfectant in use, on a daily basis, for the past 15 years for the disinfection of hard surfaces in the Veterinary Academic Hospital, Onderstepoort, the Veterinary Faculty of the University of Pretoria, South Africa, where regular microbiologic surveys and challenge tests demonstrate its effectiveness against "super bugs". Cypermethrin is a synthetic pyrethroid insecticide used for the control of agricultural insect pests in houses, as spot treatment to control insect pests in houses, and as an insect repellent for animals, is added to this product in order to treat and/or prevent myiasis. The F10 antiseptic consituents in the formulation together with cypermethrin (0.25%) and piperonyl butoxide combine synergistically to achieve an efficacious anti-myiasitis effect, at a much lower cypermethrin dose rate compared with that commonly used, thereby achieving a very wide safety margin.

Safety Data

Considering the active ingredients, against their published properties, one would anticipate a good safety profile, both to the target animal and also consequently within the environment. Standard quality controlled, government approved, independent GLP laboratory studies test certificates have been accessed, in respect of: dermal toxicity (rats), skin sensitization (guinea-pigs), acute dermal irritation (guinea-pigs), acute oral toxicity (rats), inhalation toxicity (rats), acute eye irritation (rabbits), all tests have been completed with no adverse reactions reported. The LD50 for oral dosing of rats was between 2000-5000 mg/kg, a dose of 4-10 L/kg bodyweight of ready to use product. Further details of safety tests are detailed in the FOI link as provided above. No adverse reaction reports have been submitted in UK, Australia or South Africa, in respect of the use of F10 products, in the period from the introduction or products until the time of writing this paper. The product is manufactured in a regularly checked, audited and licensed GMP licensed facility, compliant with South African, Australian, EU and USA Standards. Each batch manufactured is sampled, concentration tested prior to release and stored.

Efficacy

Reference efficacy test certificates, from independent government approved laboratories, compliant with international standards have been conducted. Bactericidal tests have been carried out in accordance with SANS 636-2001 and (South African) standard having a performance pass criteria of $>\log^3$ reduction in microbial counts, the AFNOR (French) standard which is a $>\log^4$ reduction in microbial counts, the European Union EN standard which is a >log^5 reduction in microbial counts, and the AOAC (USA) standard, a >log^6 reduction in microbial counts. MIC in-vitro tests have shown significant depth of performance as indicated. Generally testing is carried out at ambient temperatures with the exception of EN Standard tests where some were carried out at low temperatures 10ºC and some DEFRA (APHA) tests at 4ºC. The EN Standard for low temperatures is 10ºC and at these temperatures bactericidal performance was equal to that at ambient temperatures, however there was a significant fall off in performance at 4ºC. Other tests carried out at elevated temperatures of 45ºC showed a significant increase in performance. These tests have confirmed efficacy against: *Mycobacterium* spp., *Pseudomonas aeroginosa, E. coli, Staphylococcus aureus, Enterococcus faecium, Penecillium verrucosum, Cladosporuium cladosporoides, Absidia corymbifera, Candida albicans*, enteroviruses, *Orthopoxvirus, Adenovirus*, HIV, canine *Parvovirus,* hepatitis B, *Leptospira* spp., *Campylobacter* spp., *Legionella* spp., Newcastle disease virus, feline *Herpesvirus, Acinetobacter anitratus*, IBD virus, MSRA, *Ornithobacterium rhinotracheale, Bacillus subtilus* spores, *Aspergillus niger, Trichophyton mentogrophytes, Klebsiella* spp., *Proteus vulgaris, Enterococcus hirae, Salmonella typhi*, vancomycin-resistant *Enterocioccus faecalis, Streptococcus pyogenes, Vibria cholera, Listeria monocytogenes, Staphylococcus epidermides, Pasturella cepacia, Micrococcus luteus, Salmonella abony, Klebsiella pneumonia, Corynebacterium xerosis, Citrobacter freundii, Enterobacter sakazakii, Enterobacter cloacae*, rabies virus, *Aspergillus fumigatus, Pasturella multocida*, pseudo-cowpox, *Clostridium pyogenes, Salmonella typhimurium*, avian influenza (H5N2), *Microsporum canis, Salmonella cholerasuis*, and *Clostridium difficile.*

The in vitro insecticidal studies reviewed were a South African Bureau of Standards (SABS) knockdown study with adult *L. cuprina* (Australian sheep blowfly). Two additional studies (Dr MG Snyman 'The F10 Germicidal Wound Spray with insecticide field trial summary' and Prof. VD Linde 'F10 Wound Spray with insecticide and germicidal barrier ointment with insecticide final efficacy report to determine the effectiveness of F10 brand antiseptic wound spray with insecticide against the larvae and ova of *L. cuprina*). In the knockdown study, 100% mortality was achieved against adult flies within 24 hours after exposure to the drug. In the study using *L. cuprina* larvae, 89.2% mortality was achieved, with the remaining 10.8% of larvae considered moribund, within 24 hours after exposure to the drug. There was 100% mortality in the treated larvae 120 hours after exposure to the drug. In the study using *L. cuprina* ova, no treated eggs hatched after exposure to the drug.

Clinical Applications

This product is recommended for treatment in the following situations.

For use in zoo, and (non-food-producing animals) wildlife rehabilitation animals (i.e., captive wildlife housed in permanent enclosures for treatment or educational purposes and for the breeding of exotic and endangered species). The product is a broad-spectrum antiseptic and long-acting barrier to prevent re-infection with bacterial, fungal or viral pathogens, whilst also controlling or preventing myiasis.

It may be usefully applied to wound sites closed by primary and or secondary intention during the healing period which could be at risk from gram-positive or gram-negative bacterial or fungal infections, in particular where myiasis is a risk. The wound sites of (non- food-producing) hoof stock and other large animals in captive wildlife enclosures that have been anaesthetized with remote injection techniques such as darting which can result in infection at the dart site due to drawing in hair and skin into the dart wound. For the treatment of bite or claw wounds, horn injuries, blunt trauma, lacerations, abrasions, puncture wounds especially in species that are not easily handled or observed. For use along the absorbable sutures line to reduce the incidence of infection. The product incorporates a broad spectrum long acting insecticide to repel insects which might otherwise cause irritation and re-infection of the wounds. It is particularly useful for the prevention for 'fly strike' and treatment of 'fly strike wounds' and the elimination of maggot infestations.

The product is documented to have been used safely and efficaciously in mammals: zoo and captive wildlife (non-food producing) minor species (e.g. lion, leopard, cheetah, tiger, elephant, rhinoceros, giraffe, camel, polar bear, gorilla, primate, hippopotomous, eland, oryx, addax, gazelle, wild sheep, Barbary sheep, mouflon, sitatunga, hyena). The product is documented to have been used safely and efficaciously in birds (raptors, waterfowl, bustards, passerine birds, psittacine birds, and non-food-producing ostrich) as well as in reptiles (tortoise).

The product has been used by daily topical application to clean, contaminated or infected wounds where access is easily achieved. Also by topical application, to a range of infected, open wounds or abrasions (secondary to surgery, trauma (including horn gauges and bite wounds), tranquiliser dart injections, etc.), affecting patients representative of species and general groups listed above, at a frequency appropriate to the clinical and physical situation. The product has proven safe and efficacious in the control of myiasis by treatments at 7- to10-day intervals, especially useful in wildlife, exotic and zoo species, where tranquilisation is necessitated for handling and application. In no cases have adverse effects, attributable to the treatment been experienced or observed. Direct application of the product to maggot infested tissue effects maggot depopulation, without any need for additional operator intervention and in the absence of deleterious tissue effects. The product is not licensed for use in cats and dogs and is contraindicated for use in domestic cats.

The same disinfectant and insecticide chemical preparation is also licensed under the MUMS scheme, (see www.fda.gov/downloads/AnimalVeterinary/.../UCM440060.pdf) as a long-acting germicidal barrier ointment with insecticide. Whilst the insecticide has no longer duration of action in this product, the antibacterial and antifungal activity is extended to one week, being particularly useful where ongoing risk of bacterial or fungal infection exists, where only infrequent patient handling is practicable.

Conclusion

This product has been widely used in zoo, wildlife and exotic veterinary practice for more than fifteen years, by a significant number of well-respected clinicians on several continents. No adverse reactions have been reported. Clinical applications and efficacy have been published in many peer reviewed publications, as well as in other professional and special interest publications.[1-9] The author commends the application of this product in the control and prevention of superficial bacterial, fungal and viral infections, in particular where myiasis is a risk and where daily patient handling is contraindicated or impractical.

References

1. Barrows M. F10, a novel product range most suited to zoological medicine, *The Facts*. 2007;6. Available at: http://www.healthandhygiene.co.za. Accessed July 3, 2016.

2. Forbes NA, Lloyd E, Temperley JP. Macro-broth minimum inhibitory concentration (MIC) and agar disc-diffusion zone of inhibition determination on F10SC disinfectant. *Proc Annu Conf Br Vet Zool Soc*. 2005;98-101.

3. Gardner B, Le Rochais C. F10 used to treat abscesses in two African elephants. *The Facts*. 2009;13. Available at: www.healthandhygiene.co.za. Accessed July 3, 2016.

4. Lampen F, Bailey TA, Combreau O. Medical rehabilitation and quarantine of illegally traded Macqueen's bustard (*Chlamydotis macqueenii*) in the UAE. *J Avian Med Surg*. 2005;19(1):35-45.

5. Le Roux FE. An alternative localised therapy with extended clinical application. Proc Annu Conf Br Sm Anim Vet Assoc. 2004.

6. Samour J, Naldo J. The use of F10 in falcon medicine: practical applications. *Falco* 2010;35:21.

7. Slabber M. Some experiences and success with F10 products in equine practice. *The Facts.* 2008;10. Available at: http://www.healthandhygiene.co.za. Accessed July 3, 2016.

8. Stanford M. Use and safety of F10 in exotics. Proc Annu Conf Br Vet Zool Soc. 2006.

9. Zsivanovits P, Forbes NA, Saunders R, Higston S. Suggestions for optimising recovery and minimising disease risks in raptor rehabilitation facilities. *J Wildl Dis*. 2004;44:427-433.

Mycobacteriosis in Three Pet Psittacine Bird Species: A Case Series

Anna E. McRee, DVM,
Christine T. Higbie, DVM,
Javier G. Nevarez, DVM, PhD, Dipl ECZM (Herpetology), Dipl ACZM,
Thomas N. Tully, DVM, MS, Dipl ABVP (Avian), Dipl ECZM (Avian)

Session #120

Affiliation: From the Department of Veterinary Clinical Sciences, School of Veterinary Medicine, Louisiana State University, Baton Rouge, LA 70803, USA.

In 2015, 3 psittacine birds were presented within 30 days, each with differing clinical signs and patient histories. A 12-year-old male eclectus parrot (*Eclectus roratus*) was presented for weakness, depression, and anorexia of 2 days duration. On presentation it was determined to have a heart murmur, severely elevated white blood cell count (93.9 K/µl) with a left shift (2.8 K/ µl bands), and anemia (30%). Severe hepatomegaly was noted on radiographs, ultrasound, and CT scan. Fine needle aspirate of the liver revealed intracellular acid-fast bacilli identified as *Mycobacterium avium*. A 20-year-old female double yellow-headed Amazon parrot (*Amazona aestiva*) was presented for a 1-month history of lethargy and weight loss despite a good appetite. The parrot's total white blood cell count was (16.8 K/µl) and PCV (35%). Necropsy revealed extensive granulomatosis of the small intestines, lungs, liver, spleen, and medullary cavities of the long bones, with intracellular acid-fast bacilli identified as *Mycobacterium genavense*. Lastly, a 1.5-year-old female black-headed caique (*Pionites melanocephala*) was presented with a 1-day history of lethargy, depression, and reclusiveness. On presentation, it had a heart murmur, distended coelom, palpable thickening of the coelomic organs, and increased lung sounds bilaterally. Necropsy revealed mycobacteriosis of the liver, spleen, small intestines, pericardial fat, and bone marrow identified as *Mycobacterium genavense*. The importance of advances in *Mycobacterium* spp. identification, continued presence of this organism in companion avian populations, difficulty in obtaining a definitive ante mortem diagnosis, and conflicting recommendations regarding treatment are thought provoking medical areas of focus in this recent case series.

Eustrongylides Induced Esophagitis, Ventricular Perforation and Celomitis in an Australasian Darter *(Anhinga novaehollandiae)*

Michelle Sutherland, BVM&S, MANZCVS (Avian Health), Cert. AVP (ZooMed),
Patricia J. Macwhirter, BVSc, PhD, FANZCVS (Avian),
David Phalen, DVM, PhD, Dipl ABVP (Avian)

Session #175

Affiliation: From the Burwood Bird and Animal Hospital, 128 Highbury Rd, Burwood, Vic 3125, Australia (Sutherland, Macwhirter), and The Avian, Reptile and Exotic Pet Hospital, University of Sydney, 410 Werombi Rd, Camden, NSW 2570, Australia (Phalen).

Nematodes in the genus *Eustrongylides* have their adult lifecycle phase in piscivorous birds.[1] Waterbirds such as herons,[2-5] egrets,[2-6] and mergansers[7] have been described as the definitive hosts for several Eustrongylides species. Fatal cases of verminous peritonitis caused by *E. ignotus* and *E. tubifex* have been responsible for historical mass morbidity and mortality events in herons and egrets in the United States,[2-5] with *E. ignotus* reported to perforate the stomach of its host.[4] Parasitism of Australasian waterbirds in natural environments has not been the subject of extensive research, and the lifecycles and distribution of *Eustrongylides* spp. in Australia are poorly understood.[8] Within Australian avifauna *E. excisus* has been reported only in cormorants.[1,9]

A wild Australasian darter (*Anhinga novaehollandiae*) presented weak, dehydrated and unable to fly. Survey radiographs and a fecal examination were performed. Radiographs showed a soft tissue opacity in the mid-left cranial celom and homogeneous soft tissue opacity throughout the celomic cavity. Fecal wet smear and floatation revealed *Capillaria* and *Eustrongylides* spp. oocysts. The bird died despite supportive treatment. On gross necropsy *Eustrongylides* worms had perforated the ventriculus causing celomitis and were present throughout the thorax. Histologically, severe granulomatous inflammation was associated with living and dead adult *Eustrongylides* in the muscularis of the ingluves, proventriculus and ventriculus. A diagnosis of severe diffuse verminous ventriculitis, esophagitis with ulceration, and perforation of the ventriculus was made. The *Eustrongylides* and *Capillaria* spp. described represent a new host record for the Australasian darter.

References

1. Spalding M, Forrester DJ. Eustrongylidosis. In: Atkinson CT, Thomas NJ, Hunter B, eds. *Parasitic Diseases of Wild Birds*. Ames, IA: Wiley Blackwell;2008:289-315.

2. Wiese JH, Davidson WR, Nettles VF. Large scale mortality of nestling ardeids caused by nematode infection. *J Wildl Dis*. 1977;13:376-382.

3. Locke LN. Heron and egret losses due to verminous peritonitis. *Avian Dis*. 1961;5:135-138.

4. Winterfield RW, Kazacos KR. Morbidity and mortality of great blue herons in Indiana caused by *Eustrongylides ignotus*. *Avian Dis*. 1977;21:448-451.

5. Spalding M, Bancroft GT, Forrester DJ. The epizootiology of eustrongyloidosis in wading birds (Ciconiiformes) in Florida. *J Wildl Dis*. 1993;29:237-249.

6. Roffe TJ. Eustrongylides epizootic in common egrets (*Casmerodius albus*). *Avian Dis*. 1988;32:143-147.

7. Locke LN, DeWitt JB, Menzie CM, Kerwin JA. A merganser die-off associated with larval Eustrongylides. *Avian Dis*. 1964;8:420-427.

8. Chapman A, Hobbs RP, Morgan DL, Gill HS. Helminth parasitism of *Galaxias maculates* (Jenyns, 1842) in southwestern Australia. *Ecol Freshwat Fish*. 2006;15:559-564.

9. Pollard DA. The biology of a landlocked form of the normally catadromous salmoniform fish *Galaxias maculates* (Jenyns) VI: effects of cestode and nematode parasites. *Aust J Freshwat Res*. 1974;25;105-120.

Co-Occurrence of Mycoplasmas and Pigeon Herpesvirus-1 Infection in Racing Pigeons *(Columba livia)*

Tom Hellebuyck, DVM, PhD, Dipl ECZM (Herpetology),
Stephan Göbel, DVM,
Frank Pasmans, DVM, PhD, Dipl ECZM (Herpetology),
Connie Adriaensen, PhD,
An Martel, DVM, PhD, Dipl ECZM (Wildlife Population Health)

Session #179

Affiliation: From the Department of Pathology, Bacteriology and Avian Diseases, Faculty of Veterinary Medicine, Ghent University, Salisburylaan 133, B-9820 Merelbeke, Belgium.

Oropharyngeal swabs were collected from 438 live racing pigeons (*Columba livia*) housed at 220 lofts in the western part of the Netherlands and investigated using PCR for the presence of *Mycoplasma* species and pigeon herpesvirus-1 (PHV-1). In 30.9% of the sampled pigeon lofts signs of respiratory disease were observed during a 6-month period prior to the sampling and in 8.6% of the pigeon lofts signs of respiratory disease were present at the moment of sample collection. A total of 30.6% of the tested pigeons belonging to 48.6% of the pigeon lofts tested positive for PHV-1 and 39.8% of the pigeons representing 54.5% of the pigeon lofts tested positive for the presence of *Mycoplasma* spp. In 15.8% of the pigeons belonging to 34.0% of the pigeon lofts, the co-occurrence of *Mycoplasma* and PHV-1 infection could be demonstrated. The number of pigeon lofts with co-occurrence of *Mycoplasma* spp. and PHV-1 infection was remarkably higher than the number of lofts where the presence of only one of both infections could be demonstrated. Neither the presence of *Mycoplasma* species, PHV-1 or the co-occurrence of mycoplasmas and PHV-1; however, seemed to be associated with respiratory disease.

Parrots with Traumatic Spinal Fractures: Three Cases and Their Outcomes

Leila Marcucci, DVM, Dipl ABVP (Avian)

Session #215

Affiliation: From Bay Area Bird Hospital, 2145 Taraval Street, San Francisco, CA 94116 USA.

This poster presents three cases of spinal fractures in psittacine birds and their outcome. The first, a 2-year-old sun conure (*Aratinga solstitialis*) presented for sudden onset of bilateral leg paresis. The owner was unaware of any trauma although his roommate later admitted to flinging the bird against the wall. This synsacral fracture was non-displaced and incomplete. Of the three cases, this patient's recovery was the most prolonged with the least return to normal function. The second and third cases involve white-fronted or spectacled Amazons (*Amazona albifrons*) in their early thirties. In both cases the fractures resulted from falls. Each of these birds presented with palpable spinal displacement and readily visible fractures on radiograph. Both birds had less severe neurologic deficits that the sun conure and made a more complete return to normal function. Each of these birds had been on a seed-based diet for most of their lives. Supportive care was the only treatment provided but each case required different treatments for differing lengths of time.

Gap in pagination due to formatting issues.

Pages 407-862

Section 32

Saturday Hybrid Master Classes

Troubleshooting Poor Anesthetic Recovery: Persistence Pays

Angela M. Lennox, DVM, Dipl ABVP (Avian), Dipl ABVP (Exotic Companion Mammal), Dipl ECZM (Small Mammal)

Session #257

Affiliation: From Avian & Exotic Animal Clinic, 9330 Waldemar Road, Indianapolis, IN 46268, USA.

Exotic companion mammals have a reputation for increased surgical risk. A 2006 study comparing anesthetic death rates of various pet species reported a rate of 0.17% for dogs, 0.24% for cats, 1.39% for rabbits, and 3.8% for guinea pigs.[1] Explanations for increased surgical death rate abound, and include greater intolerance to pain and stress, fewer numbers of well, elective procedures in this species, and the ability to hide underlying illness.

Surveys of traditional domestic pets indicates the majority of anesthetic deaths occur during the post-operative period. Careful monitoring and attention to detail can help rapidly identify the patient experiencing a less than ideal recovery; quick identification of underlying etiologies can lead to improved survival rates. This presentation emphasizes the role of the veterinary technician as part of a team-based approach to reducing loss during the recovery period.

Incidence of poor anesthetic recovery can be dramatically reduced by paying careful attention to pre- and peri-surgical details, which include the following:

1. Adequate patient preparation: Few surgeries in exotic companion mammals are urgent, with the exception of conditions such as urinary or gastrointestinal obstruction. In most cases there is time to perform a complete pre-anesthetic evaluation, including an examination just prior to surgery, as condition can rapidly change, and pre-anesthetic blood work.

2. Vascular access: Vascular access is considered a standard of care for human and traditional pet species undergoing any anesthetic procedure. With recent increased utilization of techniques such as placement of intraosseous catheters, vascular access should be attempted in every exotic surgical patient as well. A general recommended surgical fluid rate is 10 ml/kg/hr; however, ideal rates for each species have not been determined.

3. Careful selection of pre-anesthetic protocols: There is no legitimate indication for single agent inhalant anesthesia in exotic companion mammals. Pre-anesthetic agents have been shown to reduce the concentration of inhalant general anesthetics, which is linked with greater patient safety. Pre-anesthetics contain analgesics, which provide pre-emptive analgesia.

4. Careful selection of induction and maintenance agents.

5. Intubation of patients whenever possible.

6. Proficient and continuous anesthetic monitoring.

7. Adequate analgesia: As pain is associated with poor anesthetic recoveries in many species; multi-model analgesia should be considered, including pre-emptive analgesia, local blocks, constant rate infusion (CRI) of analgesics, and effective post-surgical administration.

Recovery

Patients should be monitored carefully post surgery, until they are up and about, walking, and taking food. Ideally, a designated recovery technician monitors recovering patients continuously at first, and then intermittently as recovery progresses. When recovery is delayed, every effort must be made to attempt to find a cause and provide a remedy. In the author's opinion, many patients are lost in the recovery period due to inadequate analgesia and hypovolemia.

Causes of Poor Recoveries

There are a number of factors associated with poor anesthetic recovery. These factors are described below.

Use of pre-anesthetic agents

Use of pre-anesthetic drugs usually results in a delayed recovery, when compared to no or fewer agents. This is often concerning in practices just beginning to add new protocols and drug combinations. Ideally, the recovery period is slow, steady and calm, as opposed to a rapid recovery with struggling. This must be distinguished from a poor recovery that could benefit from reversal of one or more pre-anesthetic agents; experience and familiarity with the protocol in a particular species are often the best guides. Agents commonly used in ECM that are reversible include dexmedetomidine and midazolam. The reversal of opioid drugs is a controversial; complete reversal of analgesic agents is never recommended; however partial reversal may be considered.

Hypovolemia

The hypovolemic patient demonstrates a low or unobtainable blood pressure. In patients where acquisition of blood pressure is impractical, other clinical markers should be considered, including decreased mentation, pale mucus membrane color and delayed capillary refill time. The hypovolemic patient is weak and responds poorly to stimulus. Optimal fluid therapy is critical for treatment of hypovolemia associated with anesthesia or blood loss. While little information exists on specific guidelines for treatment of hypovolemia in various exotic species, information can be extrapolated from with other traditional pet species. The endpoint of fluid resuscitation is restoration of normal blood pressure and normalization of other parameters. The following strategy outlines treatment of hypovolemia in exotic companion mammals

Reverse pre-anesthetic agents, if applicable (see above). Administer rapid intravenous infusion of warmed isotonic crystalloids at 10-15 ml/kg, followed by colloids (Hetastarch, 6%, B Braun Medical Inc., Irvine, CA) at 5 ml/kg over 5-10 minutes. Boluses of isotonic crystalloids (10-15 mg/kg) and colloids (5 ml/kg) are continued until systolic Doppler blood pressure reads above 60-70 mmHg.

An alternative to initial treatment with crystalloids is the use of 7.2-7.5% hypertonic saline at 3 ml/kg as a slow bolus. Hypertonic saline quickly draws fluids from other compartments into the intravascular space.[2] The effect is maintained with follow up administration of crystalloids as discussed above.

If blood pressure cannot be obtained, look for improvement in other clinical markers.

Anemia due to hemorrhage

The patient with ongoing blood loss post surgery presents similarly to the patient with hypovolemia; features include, pale mucus membranes, poor mentation and delayed CRT. The surgeon should have some indication that hemorrhage was possible based on the nature of the surgery or the recovery technician may observe ongoing blood loss from a surgery site.

Determination of PCV is usually not helpful in detecting acute blood loss. General fluid support may be adequate for treatment of mild to moderate anemia (see above).

More severe hemorrhage may require blood transfusion or the use of colloids with oxygen carrying ability such as Oxyglobin, which is currently unavailable in the United States. With the exception of the ferret, exotic companion mammals are reported to have distinct blood types; however, the likelihood of transfusion reaction after a single transfusion is not high. The risk of reaction must be weighed against the risk of withholding transfusion. Sources of blood donors include the ill pet's housemates, rescue organizations, or pet stores. The author keeps a list of owners willing to provide blood donors in exchange for clinic credit. Blood is collected from healthy donors under sedation if necessary with 1 ml acid citrate dextrose (ACD) per 10 ml blood, maximum 10% of blood volume based on calculated body weight. Whole blood is administered via IV or IO catheter.

An example of blood transfusion following hemorrhage is outlined below:

- Place an IV catheter or IO catheter of the recipient if not already established.
- Sedate or anesthetize the donor.
- Collect blood from the donor (7-10% of body weight) via the largest available and safely accessed vein into one or several 1-3 ml syringes prepared ACD 1cc/10 ml blood.
- Administer whole blood via the IO catheter manually over 3-5 minutes, or with a small precision infusion pump.

Hypoglycemia

True hypoglycemia is not commonly seen in rabbits or rodents, but may be seen post operatively in the ferrets with insulinoma. Even ferrets that demonstrate normoglycemia pre-surgically may experience marked hypoglycemia due to fasting prior to surgery or decompensation due to stress, anesthesia and surgery. It should be kept in mind ferrets undergoing pancreatic nodulectomy or partial pancreatectomy may also demonstrate marked hyperglycemia, which often takes 24-72 hours for normalize. Other causes of hypoglycemia include septicemia, which can occur in any patient, including rabbits and rodents.

Hypothermia

Loss of heat prior to and during surgery may continue well into recovery; therefore temperature should be monitored until the patient is upright and moving. Hypovolemic mammals below a certain temperature threshold are not able to obtain normalvolemia, and fluid resuscitation will not be successful. Rewarming methods include external warming and infusion of warmed IV fluids.

Some practitioners have noted hyperthermia in ferrets that may occur up to an hour or more post surgery. This may be similar to hyperthermia as a result of opioid administration in cats.

Pain and discomfort

Pain has long been associated with poor surgical outcomes in traditional pet species, and likely applies to exotic species as well. The painful patient differs from the hypovolemic patient in that the animal is generally not weak, but demonstrating tense, painful body posture. It should be kept in mind that the duration of analgesia of analgesics are unknown for all exotic species under all conditions; therefore, if analgesia is administered as part of pre-anesthesia, and the preparation/surgery period is more than a few hours, the patient may be recovering without benefit of analgesia. Multi-model protocols are extremely important (constant rate infusion (CRI) of analgesics, local analgesia). For patients with poor recoveries due to suspected pain, administer a dosage of an opioid IV, if possible, and monitor carefully for the next 15-30 minutes.

Cardiac disease

Animals with pre-existing occult cardiac disease may decompensate as a results of anesthesia and surgery. Patients in heart failure often exhibit respiratory distress. Careful handling and additional sedation may help with obtaining diagnostic imaging in order to confirm diagnosis.

Trouble Shooting

Treatment often is required for surgical patients experiencing slow recovery from anesthesia. A trouble-shooting guide for slow recoveries as a result of the factors described above is presented in Table 1.

Table 1. Trouble shooting guide for slow recoveries in surgical patients.

Potential cause	Diagnosis	Treatment
Administration of pre-anesthetics	Rule out other causes of slow recovery	None required; monitor carefully Reverse if prolonged
Pain	Rule out other causes of slow recovery; note painful posture with rapid cardiac and/or respiratory rate. Consider especially if time of last dose of opioid is more than 3-4 hours prior.	Administer low dose opioid IV or IO and observe response over next 10 minutes
Hypovolemia	Take indirect blood pressure if possible, observe peripheral vein turgor, evaluate mm/CRT	Administer crystalloids and colloids at shock rates IV or IO as indicated above
Hypoglycemia	Measure blood glucose with hand-held glucometer	Administer glucose PO if patient is swallowing; otherwise IV, IO
Hemorrhage	Look for evidence of hemorrhage, see section on hypovolemia above, note mucus membrane color	Administer colloids and/or whole blood

References

1. Brodbelt DC, Blissitt KJ, Hammond RA, Neath PJ, Young LE, Pfeiffer DU, Wood JL. The risk of death: confidential enquiry into perioperative small animal fatalities. *Vet Anaesth Analg*. 2008;35(5):365-373.

2. Bowles H, Lichtenberger M, Lennox A. Emergency and critical care. *Vet Clin North Am Exotic Anim Pract*. 2007;10(2):345-394.

Drug Dosing in Exotic Animals: Are We Using Evidence-Based Medicine?

Marcy J. Souza, DVM, MPH, Dipl ABVP (Avian), Dipl ACVPM

Session #017

Affiliation: From the Department of Biomedical and Diagnostic Sciences, College of Veterinary Medicine, University of Tennessee, Knoxville, TN 37996, USA.

Abstract: Evidence-based medicine can be challenging when limited research has been performed to determine appropriate dosing of medications in many exotic species that are kept in captivity. This talk will discuss some inappropriate recommendations that are routinely made in clinical practice and steps to reduce the likelihood of making medical decisions based on hearsay alone.

Introduction

Ideally, dosing recommendations for all animals, including exotic animals, would be based on controlled clinical trials. Unfortunately, these types of studies are rare for exotic species. Often, we decide our dosing based on clinical experience and personal communications with other veterinarians; this strategy may or may not provide accurate information. A personal communication of success with a particular drug regimen can become ingrained in the literature despite any real evidence of success.[1] Therefore, practitioners must carefully evaluate the references for reported dosing regimens in the literature and gauge how much evidence is truly behind the recommendations.

Treatment Examples

Mycobacteriosis in pet birds

The treatment of mycobacteriosis in pet birds is controversial: some practitioners feel it appropriate to treat these animals, while others feel that euthanasia is necessary due to the potential for zoonotic infection. Regardless of whether these animals should be treated due to the zoonotic potential of *Mycobacterium* species, do practitioners have enough information to appropriately treat and cure this infection without creating drug-resistant, and potentially zoonotic, strains? And what role does owner compliance play in the treatment of these animals?

In humans, therapy for mycobacteriosis typically consists of 6 to 9 months of treatment with approximately 4 drugs given at various intervals, some daily and others 1 to 3 times per week.[2] If appropriate administration of drugs is not performed, patients quickly relapse with clinical disease and drug resistance often occurs.

In an exotic animal formulary that describes protocols used to treat mycobacteriosis in birds, 7 references are listed for drug regimens; however, none are from pharmacokinetic and pharmacodynamic studies or clinical trials.[3] Most are articles, book chapters, or proceedings, which are not necessarily peer reviewed. Additionally, in this resource, a footnote states that a clinical trial evaluating the treatment of mycobacteriosis in ring-necked doves with a multi-drug protocol was unsuccessful.[3] The lack of research behind these dosing recommendations in combination with the challenge of owner compliance certainly should factor into practitioner decisions as to whether treatment is truly best for the patient and owner.

Tramadol use in rabbits

Analgesia can be difficult to assess in animals, particularly those that are considered prey species. A reasonable amount of research has been performed examining tramadol use in dogs and cats, and recommended doses are typically 4 to 5 mg/kg PO 2 to 3 times per day. Anecdotally, a dose of 10 mg/kg PO 1-2 times per day has been recommended for rabbits. Because of this recommended dose, the author (MS) performed a research study with others to examine the pharmacokinetics of oral dosing (11 mg/kg) of tramadol in rabbits. Plasma concentrations were below those associated with analgesia in humans.[4] Another study in Hispaniolan Amazon parrots found that plasma concentrations associated with analgesia were similar to those reported in humans,[5] but we do not know the therapeutic concentrations in rabbits. Administration of 4.4 mg/kg IV led to a statistically significant reduction in the amount of isoflurane needed to maintain a surgical plane of anesthesia, but the amount was not clinically significant.[4] In a separate unpublished study, rabbits that had been spayed and treated post-operatively with tramadol (11 mg/kg PO) were noticeably in more pain than rabbits treated with tramadol (11 mg/kg PO) and meloxicam. Despite evidence that 11 mg/kg PO is unlikely to reach therapeutic concentrations, a recent review recommended a dose of 10 mg/kg PO 1 to 2 times per day based on personal experience and anecdotal reports.[6] A pain evaluation scale is currently being developed by Flecknell, and this tool could aid in the evaluation of efficacy of analgesics in rabbits. For more information about recognizing pain in laboratory animals, please visit: http://www.ncbi.nlm.nih.gov/books/NBK32656/.

Conclusion

It is up to the practitioner to be well informed on how to make decisions for patients. Critical evaluation of the literature is of utmost importance when making decisions. Although it is often easiest to perform a Web search for dosing recommendations, practitioners should make sure they know what those recommendations are based on. Are they based on a personal communication after treatment of a few patients? Or are they based on peer-reviewed research? Also, whenever given the opportunity, practitioners should contribute to clinical research. Whether it be by submitting samples, providing records review, or participating in a clinical trial, YOUR participation in the process can improve and increase the knowledge needed for the treatment of exotic pets.

References

1. Kukanich B. Clinical interpretation of pharmacokinetic and pharmacodynamic data in zoologic companion animal species. *Vet Clin North Am Exot Anim Pract.* 2011;14:1-20.

2. Centers for Disease Control and Prevention. Treatment of tuberculosis: American Thoracic Society, CDC, and Infectious Diseases Society of America. *MMWR Morb Mortal Wkly Rep.* 2003;52(RR-11). Available at: http://www.cdc.gov/mmwr/PDF/rr/rr5211.pdf. Accessed April 11, 2016.

3. Carpenter JW, Marion CJ, eds. *Carpenter's Exotic Animal Formulary.* 4th ed. St. Louis, MO: Elsevier Saunders;2013.

4. Souza MJ, Greenacre CB, Cox SK. Pharmacokinetics of orally administered tramadol in domestic rabbits (*Oryctolagus cuniculus*). *Am J Vet Res.* 2008;69:979-982.

5. Souza MJ, Sanchez-Migallon Guzman D, Paul-Murphy JR, Cox SK. Pharmacokinetics after oral and intravenous administration of a single dose of tramadol hydrochloride to Hispaniolan Amazon parrots (*Amazona ventralis*). *Am J Vet Res*. 2012;73:1142-1147.

6. Barter LS. Rabbit analgesia. *Vet Clin North Am Exot Anim Pract*. 2011:14:93-104.

How, What, and Why of Tracking the Acute-Phase Response

Carolyn Cray, PhD

Session #144

Affiliation: From the University of Miami Miller School of Medicine, Department of Pathology and Laboratory Medicine, Division of Comparative Pathology, P.O. Box 016960 (R-46), Miami, FL 33101, USA.

Abstract: Tracking the acute phase response has demonstrated clinical value in avian, reptile, and exotic mammal species and the use of acute-phase protein biomarkers is now well established in companion animals, large animals, and non-domesticated mammals. Protein electrophoresis has been a mainstay and is still a high value tool in avian species but newer methodologies allow for the opportunity to view changes in specific major acute-phase proteins. The how, what, and why of acute-phase protein testing will be discussed as well as other possible biomarkers of interest. These measures, including electrophoresis, will be contrasted to traditional blood tests.

Introduction

The acute-phase response (APR) is an integral part of the early innate immune response which is also inclusive of physical barriers, phagocytes, complement, and toll-like receptors.[1-3] The APR is a complex systematic inflammatory process that begins with a local stimulus such as a response to injury or infection and then the production of numerous cytokines which are produced in part by the first responding macrophages and monocytes. Some of these cytokines signal the liver to initiate or upregulate the expression of acute-phase proteins (APP). It has been estimated that more than 200 APP may be produced during the APR.[1] Each APP has a unique biological function from acting as an opsonin to activate complement and enhance phagocytosis (C-reactive protein, CRP), enhancing chemotaxis and aiding in tissue repair (serum amyloid A, SAA), and binding hemoglobin to minimize oxidative activity (haptoglobin, HP).[2] The goal of this response is to promote healing and restore homeostasis.[2] These proteins have been found to highly conserved with observations of these or similar proteins from mollusks to turtles to mammals.

The expression of APP can increase or decrease with an ongoing APR.[2-4] Major APP can increase 10-1000-fold and are often present in negligible levels in normal animals. The increase is rapid and often occurs within 24 hours. When a suitable response is made in the inflammatory stimuli, the decrease is often equally rapid due to a good negative feedback system and a short half-life. In many companion and large animals, SAA and CRP act as major APP. Mild and moderate APP are often present in detectable levels in normal animals. Increases may be 2-10 fold for moderate APP and less than 2-fold for minor APP. Expression is often delayed to 4-6 days after insult and elevated levels will persist after resolution of the inflammatory process. HP is often a minor to moderate APP in most species.

Albumin is classified as a negative APP as it decreases in ongoing APR. In some mammals, transferrin has also been classified as a negative APP.

APP biomarkers contrast the sensitivity of monitoring other more traditional means of measuring inflammation including the total white blood cell count, neutrophil count, and the A/G ratio. When gold standard agents of inflammation are injected into laboratory animals (e.g., turpentine oil, Freund's adjuvant, lipopolysaccharide), there can be a mild transient increase in WBC but a true leukocytosis was not observed until 4-7 days later. Serum albumin will decrease but not reaching minimum concentrations until day 5. In contrast, CRP and SAA increase several hundred fold by 48 hours. Furthermore, the features of APP – including duration of response, magnitude of response, and short half-life – make these markers very viable prognostic indicators. In addition, their application in wellness exams has also been demonstrated.

As the origin of the APP stimulus may be infection, inflammation, trauma, stress or neoplasia, APP are rarely solely diagnostic for any one disease or etiology.[2] While it may not necessarily be fair to compare these independent measures given their different timelines of expression, many studies have been conducted in dogs and horses. In a study from dogs with various inflammatory processes showed a correlation of $r = 0.44$ between WBC counts and CRP and, in a study of horses, the correlation between WBC counts and SAA was weak ($r = 0.11$).[2] In the latter study, SAA had the highest diagnostic accuracy for inflammation vs. the other measures at 75%.[2]

An ongoing APR can be reflected in fraction changes in protein electrophoresis.[1] Newer automated and semi-automated methodologies including enzyme-linked immunosorbent assay, immunoturbidity, and colorimetry has been employed to quantitate specific APP. In the antibody based protocols, methods and reagents must be individually validated for each species.[6] This must be done in a stepwise fashion with the assay of samples from known clinically normal and abnormal animals. If reactive reagents are identified and fully vetted at the bench level, the study can now include a large number of samples to determine reference intervals and assess the clinical impact of using these biomarkers with patients. Of possible future interest is the validation of lateral flow devices and other point-of-care testing for these biomarkers.

Birds

Studies in chickens have dominated the literature on APP expression in birds.[7,8] With injection of gold standard inflammatory agents, changes in SAA, transferrin, PIT-54 (analog of HP), and alpha-1 acid glycoprotein have all been documented. In our laboratory experience, finding reagents with wide avian species cross reactivity has been difficult although the colorimetric reagents used to quantitate HP appear to be valid. Increased SAA expression has been documented in falcons with aspergillosis.[9,10] It appears that this APP is a major APP in this species.

Our current impression is that protein electrophoresis provides very good sensitivity for ongoing APR. It provides a broad impression of different globulin changes and specifically quantitates albumin which is an important and sensitive negative APP in birds.[11] In a comparison of total WBC counts versus EPH abnormalities in African grey parrots (judged by a change in the A/G ratio), a marginal correlation was observed between significantly elevated WBC counts and significantly decreased A/G ratio (C. Cray, oral communication, May 2016). In 49% of the cases with a decreased A/G ratio, an elevated WBC count was found. Notably, 43% of the cases showed a normal WBC count when a significant decrease in the A/G ratio was present. Similarly, the majority of the cases showed a normal WBC count when a mild to moderate decrease in the A/G ratio was present. This data indicates a differential sensitivity between these two methods and supports the use of electrophoresis in routine clinical pathology investigations of avian species.

Reptiles

In soft-shell turtles experimentally infected with gram negative bacteria, SAA expression (at the RNA level) was found to increase 1000-fold.[12] Attempts to validate commercially available SAA antibodies in most reptiles have failed to date. Hemoglobin binding activity, quantitated by the colorimetric assay for HP, has been demonstrated in box turtles, loggerhead turtles, inland bearded dragons, and rattlesnakes (C. Cray, oral communication, May 2016).[13,14] As additional reagents for major APP become available, studies should consider the relative sensitivity of protein electrophoresis vs. APP assays in these species.

Exotic Mammals

Rats and mice have interesting differences in APP expression and mice produce another major APP named serum amyloid P.[4] There has been little to no description of APP expression in guinea pigs, hamsters, and gerbils.

CRP expression has been described in rabbits with suspected *Encephalitozoon cuniculi* infection.[15] Elevated CRP levels acted as an adjunct diagnostic test to improve the specificity of serologic titers against this organism and also served as a prognostic indicator.[16] Recent studies indicate also the utility of newly described SAA-1 reagents (C. Cray, oral communication, May 2016). In a study inclusive of samples from various clinically abnormal rabbits (C. Cray, oral communication, May 2016), the sensitivity and specificity of CRP for clinical abnormality was 92% and 93%, respectively. The correlation between SAA-1 and CRP was significant, $r = 0.72$, $P < 0.0001$. While WBC counts were found to be specific (100%), the sensitivity was only 25%.

Ferret SAA was cloned and found to be similar to other mammalian SAA. In studies from our laboratory, SAA was found to be increased 2-5 fold in clinically abnormal ferrets.[17] A 10-fold increase in HP was observed in a ferret with myeloma.

Other Species

APP expression has been well documented at the RNA level in several species of fish. Attempts to document expression in a trauma model in koi were not successful and may have been related to lack of reagent cross reactivity and experimental design.[18] Reference intervals have been generated for the white-spotted bamboo shark and some commercial CRP reagents have been validated for use in the bonnethead shark (C. Cray, oral communication, May 2016).[19] A commercial ELISA for SAA quantitation has also been validated for use in salmon. A wide range of published and ongoing studies have validated the use of APP in many wildlife species and in some species, these tests have been implemented as part of the standard of care for captive animals, for prognostic value in rehabilitation setting, and in population studies of free-ranging animals.

Other Markers

The field of proteomics is rapidly growing so it is expected many more biomarkers will be identified to gauge the acute-phase response and assist in the diagnosis of specific diseases. In a spectroscopy study of samples from falcons with aspergillosis, 3-hydroxybutytrate and LDL are significantly altered versus normal falcons.[20] Lipoprotein fractions were also found to be significantly altered in clinically abnormal cownose rays.[21] These observations underlie the opportunities for further study in these areas in other species.

References

1. Kaneko JJ. Serum proteins and the dysproteinemias. In: Kaneko JJ, Harvey JW, Bruss ML, eds. *Clinical Biochemisty of Domestic Animals*. 5th ed. San Diego, CA: Academic Press. 1997;117-138.

2. Cray C. Acute phase proteins in animals. *Prog Mol Biol Transl Sci*. 2011;105:113-150.

3. Eckersall PD. Proteins, proteomics, and the dysproteinemias. In: Kaneko JJ, Harvey JW, Bruss ML, eds. *Clinical Biochemistry of Domestic Animals*. 6th ed. San Diego, CA: Academic Press;2008:117-155.

4. Cray C, Zaias J, Altman NH. Acute phase response in animals: a review. *Comp Med*. 2009;59:517-526.

5. Belgrave R, Dickey M, Arheart KL, et al. Assessment of serum amyloid A (SAA) testing and its clinical application in a specialized equine practice. *J Am Vet Med Assoc*. 2013;243:113-119.

6. Cray C. Biomarkers of inflammation in exotic pets. *J Exot Pet Med*. 2013;22:245-250.

7. O'Reilly EL, Eckersall PD. Acute phase proteins: a review of their function, behaviour, and measurements in chickens. *World Poult Sci J*. 2014;70:27-44.

8. Chamanza R, van Veen L, Tivapasi MT, et al. Acute phase proteins in the domestic fowl. *World Poult Sci J*. 1999;55:61-71.

9. Caliendo V, McKinney P, Bailey T, et al. Serum amyloid A as an indicator of health status in falcons. *J Avian Med Surg*. 2013;27:83-89.

10. Fischer D, Van Waeyenberghe L, Cray C, et al. Comparison of diagnostic tools for the detection of aspergillosis in blood samples of experimentally infected falcons. *Avian Dis*. 2014;58:587-598.

11. Cray C, Wack A, Arheart KL. Invalidty of albumin measurement of specimens from clinically ill birds by bromcresol green methodology. *J Avian Med Surg*. 2011;25:14-22.

12. Zhou X, Wang I, Feng H, et al. Acute phase response in Chinese soft-shelled turtle (*Trionyx sinensi*) with *Aeromonas hydrophila* infection. *Dev Comp Immunol*. 2011;35:441-451.

13. Flower JE, Byrd J, Cray C, Allender M. Plasma electrophoretic profiles and hemoglobin binding protein reference intervals in the Eastern box turtle (*Terrapene carolina carolina*) and influences of age, sex, season, and geography. *J Zoo Wildl Med*. 2014;45:836-842.

14. Dickey M, Cray C, Norton T, et al. Assessment of hemoglobin binding protein in loggerhead sea turtles (*Caretta caretta*) undergoing rehabilitation. *J Zoo Wildl Med*. 2014;45:700-703.

15. Cray C, Rodriguez M, Fernandez Y. Acute phase protein levels in rabbits with suspected *Encephalitozoon cuniculi* infection. *J Exot Pet Med*. 2013;22:280-286.

16. Cray C, McKenny S, Perritt E, Arheart KL. Utility of IgM titers with IgG and CRP quantitation in the diagnosis of suspected *Encephalitozoon cuniculi* infection in rabbits. *J Exot Pet Med*. 2015;24:356-360.

17. Ravich M, Johnson-Delaney C, Kelleher S, et al. Quantitation of acute phase proteins and protein electrophoresis in ferrets. *J Exot Pet Med.* 2015; http://dx.doi.org/10.1053/j.jepm.2015.04.010.

18. Christiansen EF, Cray C, Lewbart G, Harms C. Plasma protein electrophoresis and acute phase proteins in koi (*Cyprinus carpio*) following exploratory celiotomy. *J Exot Pet Med.* 2015;24:76-83.

19. Krol L, Allender M, Cray C. Plasma proteins and selected acute phase proteins in white-spotted bamboo sharks (*Chiloscyllium plagiosum*). *J Zoo Wildl Med.* 2014;45:782-786.

20. Pappalardo L, Hoijemberg PA, Pelczer I, et al. NMR-Metabolomics study on falcons affected by aspergillosis. *Curr Metabolomics.* 2016;2:155-161.

21. Cray C, Rodriguez M, Field C, et al. Protein and colesterol electrophoresis of plasma samples from captive cownose ray (*Rhinoptera bonasus*). *J Vet Diagn Invest.* 2015;27:688-695.

Section 33

Wednesday
Hybrid Track

Louisa Asseo, DVM and
James Brogan, DVM, Dipl ABVP, CertAqV
Moderators

Development of an Online Tumor Database for Zoological and Exotic Species: Roundtable Discussion

Ashley Zehnder, DVM, Dipl ABVP (Avian), PhD,
Tara M. Harrison, DVM, MPVM, Dipl AVZM, Dipl ACVPM,
Cassondra Bauer, DVM, MS, Dipl ACVPM,
Ryan Colburn, DVM,
Joanne Paul-Murphy DVM, Dipl ACZM, Dipl ACAW,
Michelle G. Hawkins, VMD, Dipl ABVP (Avian),
Catherine Pfent, DVM, PhD

Session #048

Affiliation: From Stanford University, School of Medicine, Stanford, CA 94305, USA (Zehnder), North Carolina State University, College of Veterinary Medicine, Raleigh, NC 27607, USA (Harrison), WIL Research, Ashland, OH 44805, USA (Bauer), John Ball Zoo, Grand Rapids, MI 49504, USA (Colburn), School of Veterinary Medicine, University of California, Davis, CA 95616, USA (Paul-Murphy, Hawkins), Lincoln Memorial University-College of Veterinary Medicine, Harrogate, TN 37752, USA (Pfent).

Diagnoses of cancer are increasing in frequency in exotic animal medicine and zoological institutions. Increasingly, there are parallels between human and animal cancer and therapeutics.[1,2] Further, improved preventive medicine and earlier diagnoses of cancer in zoological species has expanded the therapeutic options in zoo animals. Cancers have been diagnosed and treated in fish, reptiles, birds, and all forms of mammals.[3-6,7] Reported cancers include lymphoma, papillomas, carcinomas, sarcomas, and melanomas. There are likely to be many more cases that have gone unreported, whether as a case report or case series. Treatments in zoo animals have been facilitated by use of vascular access ports for delivery of chemotherapy, and for anesthesia for radiation treatments.[8-10] Through operant conditioning, and the development of newer, safer medications, treatment of zoological species is increasingly feasible. The Exotic Species Cancer Research Alliance, in collaboration with Stanford University, North Carolina State University, and U.C. Davis is developing a data base for clinicians in private practice and in zoological institutions to contribute cases, query for parameters related to tumor type, treatment types, adverse effects of treatment and outcomes. Institutions are encouraged to participate through case contributions to this database, which will include diagnoses, plus annotation of any therapeutics employed and outcomes observed. Through these collaborative efforts, we hope to discover trends in cancer incidence, better understand treatment and therapeutic options for exotic and zoological species, and provide a long-term archive of study material for future research.

References

1. Fukumoto, S, Hanazono K, Fu DR, Endo Y, Kadosawa T, Iwano H, Uchide T. A new treatment for human malignant melanoma targeting L-type amino acid transporter 1 (LAT1): a pilot study in a canine model. *Biochem Biophys Res Commun.* 2013;439:103-108.

2. Ranieri G, Gadaleta CD, Patruno R, Zizzo N, Daidone MG, Hansson MG, Paradiso A, Ribatti D. A model of study for human cancer: spontaneous occurring tumors in dogs. Biological features and translation for new anticancer therapies. *Crit Rev Oncol Hematol.* 2013;88:187-197.

3. Abou-Madi N, Kern TJ. Squamous cell carcinoma associated with a periorbital mass in a veiled chameleon (*Chamaeleo calyptratus*). *Vet Ophthalmol.* 2002;5:217-220.

4. Anderson ET, Kennedy-Stoskopf S, Sandy JR, Dorn B, Boyette T, Harms CA. Squamous cell carcinoma with vascular invasion in a diamondback rattlesnake (*Crotalus adamanteus*). *J Zoo Wildl Med.* 2010;41:745-748.

5. Bertelsen MF, Steele SL, Grondahl C, Baandrup U. Pheochromocytoma in a white rhinoceros (*Ceratotherium simum*). *J Zoo Wildl Med.* 2011;42:521-523.

6. Dillberger JE, Citino SB, Altman NH. Four cases of neoplasia in captive wild birds. *Avian Dis.* 1987;31:206-213.

7. Lombardini ED, Hard GC, Harshbarger JC. Neoplasms of the urinary tract in fish. *Vet Pathol.* 2014;51(5):1000-1012.

8. Kent MS. The use of chemotherapy in exotic animals. *Vet Clin North Am Exot Anim Pract.* 2004;7:807-820.

9. Rassnick KM, Gould WJ, Flanders JA. Use of a vascular access system for administration of chemotherapeutic agents to a ferret with lymphoma. *J Am Vet Med Assoc.* 1995;206:500-504.

10. Roberts WG, Klein MK, Loomis M, Weldy S, Berns MW. Photodynamic therapy of spontaneous cancers in felines, canines, and snakes with chloro-aluminum sulfonated phthalocyanine. *J Natl Cancer Inst.* 1991;3:18-23.

Characterization of Gout Versus Pseudogout Crystals in Exotic Animals

Kevin Keel, DVM, MS, PhD, Dipl ACVP, and
Ash Sundaram, MS

Session #075

Affiliation: From the UC Davis School of Veterinary Medicine, Davis, CA 95616, USA.

Abstract: Gout, pseudogout and hydroxyapatite deposition disease (HADD; syn tumoral calcinosis, calcinosis circumscripta) are characterized by the accumulation of crystals in soft tissues and may represent a diagnostic challenge in the evaluation of histopathology material. These lesions occur in a variety of species and can be difficult to differentiate in some instances. Gout is defined by the deposition of monosodium urate crystals whereas pseudogout and HADD are typically associated with accumulation of calcium pyrophosphate dihydrate ($Ca_2P2O_7.2H2O$) or hydroxyapatite ($Ca_{10}(PO_4)_6(OH)_2$), respectively. The appearance of pseudogout has been variably described as amorphous, globular to fibrillar material that is basophilic to amphophilic with giant cells or chondroid metaplasia. Descriptions of gout are just as varied and range from amorphous, eosinophilic material to feathery, amphophilic material. As with the deposition of calcium salts, urate deposition often incites a foreign body inflammatory response. We hypothesized that the microscopic appearance of gout, pseudogout and HADD on HE-stained sections is insufficient to differentiate these lesions with a significant level of specificity. To test this hypothesis we used Von Kossa stain and nonaqueous alcoholic eosin stain (NAES) to evaluate cases diagnosed with gout (n=14), pseudogout (n=7) or both (n=1). Thirteen birds and nine reptiles were represented. The NAES method prevented the dissolution of urates that often occurs in typical histologic processing. The Von Kossa stain allowed the confirmation of divalent cations, represented by calcium salts in pseudogout.

References

1. Shidham V, Chivukula M, Basir Z, Shidham G. Evaluation of crystals in formalin-fixed, paraffin-embedded tissue sections for the differential diagnosis of pseudogout, gout, and tumoral calcinosis. *Mod Pathol.* 2001;14:806-810.

2. Jones YL, Fitzgerald SD. Articular gout and suspected pseudogout in a basilisk lizard (*Basilicus plumifrons*). *J Zoo Wildl Med.* 2009;40(3):576-578.

Practical Application of Photobiomodulation (Laser Therapy) in Exotic Pet Practice

Robert D. Ness, DVM

Session #333

Affiliation: From Ness Exotic Wellness Center, 1007 Maple Ave, Lisle, IL 60532, USA.

Laser therapy is becoming a common modality in veterinary practice. Therapeutic lasers are being routinely used for the treatment of everything from simple surgical incisions to severe osteoarthritis. Potential applications continue to expand as practitioners utilize this revolutionary technology to treat virtually any condition that exhibits pain or inflammation. In addition, many practices routinely incorporate laser therapy in any wound management or arthritis protocol. It would not be surprising to find this technology to be considered the standard of care in veterinary practice in the not so distant future.

Photobiomodulation, also known as laser therapy, utilizes a specific wavelength of near-infrared light to penetrate tissues and elicit a biochemical and biological response. The laser light differs from ordinary visible light by being monochromatic (single wavelength), collimated (tight narrow beam) and coherent (uniform orderly beam). As a result the laser is an amplifier of light which sustains a stimulated emission of photons (energy packets of light). The resultant effects of photobiomodulation include reduced pain, reduced inflammation and accelerated healing. Effective treatment depends on delivering an appropriate dose of laser energy into the affected tissue. Patients with acute inflammatory or traumatic conditions typically respond with a single or few treatments sessions at relatively low target doses. Chronic or deep tissue conditions require a higher target dose and more frequent treatment sessions, with some necessitating ongoing maintenance therapy.

Protocol settings for exotic species were typically extrapolated from existing protocols for cats and dogs, but some laser companies have established protocols and incorporated the settings for bird and exotic pets into their therapeutic laser units. Consideration for the anatomic and physiologic differences in these exotic species required further adjustments of the protocols. For example, adjustments are made to further decrease the avian settings to take into account their thin, delicate skin and feather architecture, as well as their higher metabolic rates. Conversely, the dosage settings for reptiles are relatively increased to accommodate for their thicker skin and lower metabolic rates. The small mammals required fewer modifications except for body size and weight when utilizing the feline settings. As a precaution, certain Laser units for exotics have a pulsed frequency and generally lower power pre-sets than the comparable protocols for traditional companion species.

Success in treating conditions depends on delivering an appropriate dose of laser energy into the affected tissue. Insufficient dosing results in treatment failure. Overdosing wastes time and practice resources, however it is not detrimental to the patient. Target doses are expressed in J/cm^2. One joule (J) is the energy delivered with one Watt of power in one second. Established target doses are 3-4 J/cm^2 for superficial conditions (conditions in tissue that we can see) and 8-10 J/cm^2 in deep conditions (conditions beneath the skin). Power and time are inversely proportional in calculating the total energy (J) delivered. Treatment time can be reduced by increasing power. Likewise, treatment time increases if the power (Watts) is reduced. When treating superficial conditions with a lower target dose, power usually remains relatively low since achieving a target dose of 3 - 4 J/cm^2 does not take long. Many superficial conditions affect sensitive tissue and a lower power setting is more comfortable for the patient. When treating deep tissue conditions, power is increased to reduce the time required to achieve a target dose of 8-10 J/cm^2. Some laser treatment protocols, such as wound or abdominal disorders, require an estimation of the treatment area when calculating the target dose of a laser treatment. A simple technique for estimating the patient's surface treatment area is by comparing the size of the area to be treated relative to a 3 × 5" index card, which has an approximate area of 100 cm^2.

Small Exotic Mammal Laser Therapy

Exotic small mammal protocols are modified versions of the established feline settings for many of the common conditions seen in both types of patients. Similar to other companion animals, therapeutic dosages established for trauma and wound settings are based on lesion size, typically starting at 1-2 J/cm^2. These settings are used for pododermatitis in rabbits and traumatic wounds in ferrets, for example. However, photobiomodulation target dosing protocols for conditions such as arthritis, pain, and swelling/edema are based on patient size and anatomic location of the body. Spondylosis in aged rabbits resulting in hind limb weakness and paresis responds well to arthritic dosage settings, at a therapeutic dose up to 10 J/cm^2. The quality of life in these older patients is greatly improved when utilizing the laser at the arthritis and chronic pain setting by reducing pain and inflammation, which results in improved mobility. This is also the setting used for the unfortunate condition of spinal or pelvic trauma experienced in rabbits, however these cases may require more frequent laser sessions for a positive outcome. Photobiomodulation is even beneficial when incorporated in the treatment protocol for gastrointestinal stasis in rabbits and guinea pigs. A therapeutic target dose of 7-10 J/cm^2 is recommended, with the area based on treating the entire abdominal region. Utilizing photobiomodulation in these cases has greatly improved the recovery of these patients in practice by reducing inflammation and providing abdominal pain relief. Similar results are observed in cases of cystic calculi in ferrets, rabbits and guinea pigs.

Avian Laser Therapy

Avian patients equally benefit from the use of photobiomodulation. More precautions should be taken when treating a bird then with most other species because of their thin dermis and feathering. As a result, the established settings in birds are relatively lower to comparable indications in mammal species. Wound and trauma settings are quite useful in treating pododermatitis in pet birds and raptors in addition to self-mutilated wounds commonly encountered in Quaker parrots and cockatoos. These superficial lesion cases require a target dosage of 2-4 J/cm^2 for treatment, with the number of sessions and frequency dependent on the severity and chronicity of the lesion. Of course, the underlying conditions must also be addressed, but the laser treatments will expedite the recovery. In fact, laser settings have been provided to facilitate recovery from feather picking by stimulating the quiescent feather follicles and reduce inflammation in those cases involving folliculitis. Feather follicle stimulation target dosages can range from 2-3 J/cm^2 to as high as 10 J/cm^2, depending on severity and chronicity. Protocols for fractures, arthritis and other painful conditions are based on the size of the bird as well as the anatomic location of the problem, but typically range between 4 – 8 J/cm^2 in most cases. In clinical experience, fractures which are treated with photobiomodulation tend to recover approximately 25-50% faster than those managed by traditional means alone.

Reptile Laser Therapy

Reptile protocols are approached differently at this time, due to the variety of species seen in practice and the wide range of size and shapes. Individual consideration must be implemented for variables such as the shell of a tortoise, thick scales of a blue-tongue skink or the thin dermis of a leopard gecko lizard. However, a wide variety of conditions can be readily treated in any of these reptile species. Cases that are generally classified as superficial lesions in terms of laser protocol settings include rat bite wounds, blister disease, thermal burns, and infectious stomatitis. These lesions can be treated with an initial target dose of 1-3 J/cm^2 in uncomplicated cases to as high as 8-10 J/cm^2 for rat bite lesions having deep wounds with significant muscular involvement. Shell fractures in turtles and tortoises heal more rapidly with use of photobiomodulation, starting with a target dose of 4-6 J/cm^2 for cracked shell to 10-12 J/cm^2 for deep shell wounds penetrating the coelomic cavity. Deep seated conditions, such as metabolic bone disease, fractures and arthritis, would be treated with a target dose ranging from 6-10 J/cm^2, or even higher if severe and chronic. As with birds and mammals, anything that causes pain or inflammation can be addressed with laser therapy.

Photobiomodulation is indicated in many conditions commonly encountered in avian and exotic practice. As with our dog and cat patients, the incorporation of a therapeutic laser in daily practice will greatly enhance the quality of life in so many patients by expediting the healing of lesions as well as reducing inflammation and managing pain.

Establishing a Fish Practice in an Existing Exotics Practice

Mike Corcoran, DVM, CertAqV, and
Katie Haman, DVM, MSc

Session #218

Affiliation: From VCA Wakefield Animal Hospital, 19 Main St., Wakefield, MA 01880, USA and Washington State Department of Fish and Wildlife, 1111 Washington Street NW, Olympia, WA 98501, USA.

Abstract: Adding aquatic animals to an existing exotic animal practice is relatively simple and can add a fun, challenging aspect to the practice. This article will share the authors' experiences and give some advice to practitioners wanting to explore this field.

Introduction

Exotic animal practitioners are experienced with taking what is known regarding more commonly treated species and applying the principles of good quality medicine to less frequently studied species. This often requires learning new anatomy, physiology and husbandry. Fish and other aquatic animals are no different. Continuing education is available from a number of different sources. Aside from a small investment in new equipment and some marketing, the addition of aquatic animal clients to an existing exotic animal practice is seamless.

Supplies Required to Add Aquatic Animal Patients

Very basic supplies required to start seeing fish as patients requires a minimal investment. Of course, as the practice is expanded more specialized supplies may be beneficial or required for additional procedures. In this section we will discuss materials in some depth and provide a table listing basic minimal supplies and additional supplies that are useful as you come into contact with more advanced cases and expand the practice. We will also discuss the role of veterinarians in aquaculture and fish population health. Expanding a practice to include not only pet fish, but also food fish is easily accomplished; the most difficult component will most likely not be the medical aspect, but instead mastering the elements of husbandry and fish culture that are frequently the primary cause of clinical disease, both in pet and food fish.

Minimum supply list

Many of the items on this list will already be available, but it's wise for some to be dedicated and it's also included on this list to highlight some important considerations. Gloves, cover slips and suture scissors will already be in a practice, but there are some special considerations. Gloves should always be used when handling aquatic animals (for the protection of the patient and the staff). Gloves used for handling fish should not contain powders or lotions. These can adversely affect the protective slime coat of the patient. Nitrile gloves are a great choice. If powdered gloves are the only glove available, then the powder must be rinsed before contacting the fish. Plastic cover slips are recommended for obtaining skin (slime coat) cytology samples due to the decreased risk of iatrogenic trauma during the procedure. Suture scissors or iris scissors are used to obtain gill biopsy samples as part of the basic examination. It is advisable to have a dedicated pair (or several) that can be decontaminated appropriately and will not be subject to use with adhesive tapes or other non-clinical use.

Other supplies required to get started include some type of aquarium or holding tank for sedation, air stones, nets, water testing kits and some specific medications. These will be discussed in further detail in the following sections.

Holding tanks and aquariums

There is no telling how an aquatic animal will be presented to you at the hospital. Fish have been brought for appointments in the normal bowl or aquarium that acts as their habitat and have also been brought in plastic bags, coolers, coffee mugs and anything else that will hold water. The practitioner should be prepared to receive animals in many different ways and be able to hold them in the hospital safely for many hours. In a beginning practice, many inexpensive options exist. Try to have varying sizes for different fish from bettas to koi. Small "Critter Keepers" can be used for smaller fish and any type of plastic storage container can be utilized for larger fish. Some type of cover should be available to prevent escape during the excitatory phase of sedation and the container should allow easy handing and good visualization of the patient.

As the practice develops, it may also be wise to invest in more semi-permanent solutions to allow for longer stays on surgery patients or even hospitalization of animals. Aquariums varying from 20 gallons to 40 gallons could be kept cycled at all times in a designated treatment room if the case load is large, or collapsible containers are available. Filtration will be discussed later, but should be considered when selecting larger and more permanent setups. For large fish, collapsible show tanks for koi are relatively inexpensive and take up little storage space even for several hundred gallons. Smaller store display tans or breeder tanks can be used as well providing they allow for disinfection between animals.

If dealing with high value koi, there are some specialized pieces of equipment that should be considered. Keepers of these fish use certain tools to ensure gentle handling of show quality fish, and your use of the same helps distinguish you as a professional in their eyes. Round holding tanks are standard for handling koi out of the pond. The use of sock nets for transportation between tanks allows decreased stress and decreased chance of trauma. These items can be obtained from many breeders or pond suppliers. Using the owner's own equipment should always be the first choice for sterility, but the practitioner working with these animal frequently should have their own supplies when needed.

Air stones and filtration

Maintaining good water quality is an essential part of aquatic animal health. Water quality testing will be discussed below, but some consideration must be given to how to maintain good quality. In a new practice that is seeing only occasional outpatients, a simple air stone system can provide aeration in the holding tank or in the owner's shipping container while the patient is in the hospital for the appointment. Good air stones are available for as cheap as $20 from any pet store. Battery operated models are available for transportation.

For longer duration stays, some type of mechanical, biological and/or chemical filtration system may be required. An understanding of filtration systems and the needs of the practice will dictate the type of system required and how to install the system.

Medications

Many medications utilized in the practice will be useful for aquatic animals. Formularies have good doses for Meloxicam, Ceftazidime and Enrofloxacin among many others. Referencing formularies is wise to utilize what is already in the inventory.

For food fish, FDA regulations on the use of drugs/chemicals in aquaculture must be considered. There are only a few drugs approved for use in aquaculture (http://www.fda.gov/AnimalVeterinary/DevelopmentApprovalProcess/Aquaculture/ucm132954.htm). The use of prescriptions and veterinary feed directives (VFDs) is commonplace in aquaculture, and with the impending changes to VFDs (2017), aquaculture facilities will need to rely more on veterinary services to obtain medications and treat sick fish populations.

In the private practice, there are some additional drugs that will need to be ordered specifically for fish. Most contact with fish will require sedation for a thorough examination and any required treatment. MS-222 (http://www.wchemical.com/tricaine-s-ms-222.html) is a very safe anesthetic agent that can be used in the water to sedate most fish. It has a relatively short established withdrawal time (21 days) for food fish and a wide margin of safety in most animals. It is inexpensive and has a good shelf life. With most freshwater sources it must be buffered during use to counteract the acidity. Sodium bicarbonate in the form of ordinary baking soda should be purchased at the same time. If used in saltwater, MS222 need not be buffered. Always test the pH of the MS222 solution before sedating or anesthetizing fish.

Another anesthetic that is commonly used in aquaculture is Aqui-S 20E (http://www.aqui-s.com/index.php/aqui-s-20e). Currently available under an INDA, Aqui-S 20E can be used in food fish in the field for immediate release, or in an aquaculture setting with 72-hour withdrawal period.

Some type of water conditioner should also be on hand at all times. Read the bottle to make sure that it will remove chloramines in addition to chlorine. Always encourage owners to bring enough extra water from home to do any testing and use the same source water for anesthesia, but be prepared to utilize dechlorinated water from the hospital when this doesn't happen.

External parasites are a common finding in fish medicine. There are a number of water based medications that can be used for the parasites. Western Chemical (http://www.wchemical.com) and National Fish Pharmaceuticals (http://www.nationalfishpharm.com/) are good sources. Again, the specific caseload and regional diseases will dictate what products are most useful.

Water testing supplies

Of course husbandry is important to address with any patients and improper husbandry is at least a contributing factor to a great deal of exotic animal disease. Good water quality is vital to the health of aquatic animals. A good quality water testing kit should be on hand. At a minimum the aquatic animal veterinarian should be able to evaluate temperature, pH, hardness, alkalinity and the biological filter (ammonia, nitrite and nitrate) using an accurate kit. As the practice expands, other parameters may become important. Later tests could include dissolved oxygen, various minerals and electrolytes, conductivity and oxidation/reduction potential.

Kits come in many forms. As a professional, the accuracy is important. An inexpensive but fairly accurate option would be a kit like the API Freshwater Master Kit (pH, ammonia, nitrite and nitrate) and individual tests for alkalinity and total hardness (Available through API, Red Sea and other companies). As soon as possible it's wise to upgrade from these kits to get a more professional kit. HACH (http://www.hach.com) makes a middle range professional kit that costs just over $300, the FF-1A that is recommended by the author (MC).

Husbandry and water quality is perhaps one of the most important aspects of fish health, especially in a captive rearing environment like a hatchery or aquaculture facility. In aquaculture, water quality doesn't just include above mentioned variables that are easily measured, but also factors such as flow, turbidity, pounds of fish per gallons/minute, and density index (or number of fish relative to rearing space and flow).

Most often, diseases seen in a population of fish are due to suboptimal environmental conditions that lead to increased stress and decreased immune function, making the population more susceptible to pathogens that otherwise do not cause disease. There are numerous examples of this in aquaculture, and changes to husbandry and/or fish culture can drastically improve fish health and minimize the need to medically treat.

Including Aquaculture in Your Practice

To include an aquaculture or hatchery facility as a client in your practice you will need to visit the farm. Making a mobile fish practice is simple – a kit including slides, coverslips, scalpel blades, an anesthetic, gloves, stain (Diff Quick), and a microscope should be sufficient for most visits to an aquaculture facility. This enables the veterinarian to collect the basic samples (skin scrapes, gill clips) and conduct necropsies to look for pathogens in internal organs. Further diagnostics, if warranted, can be done back in the clinic or by sending samples to appropriate laboratories.

Working as a veterinarian in aquaculture requires more than medical knowledge and diagnostic capabilities. One must also appreciate the complexities of fish culture, husbandry, feeding, and the life history of the species being reared. Many diseases are preventable, and if conditions are addressed and changed accordingly, will not manifest as population-level disease and will not require medical treatment. In this capacity, it is the role of the veterinarian to work closely with fish culturists to ensure best practices are employed to maximize fish health.

The use of drugs in aquaculture and food fish is regulated by the FDA. As a veterinarian working in aquaculture, you must know what drugs are approved for what species and diseases and employ them accordingly. Withdrawal periods must always be met. There are few drugs approved for use in aquaculture, so options are limited. This further highlights the importance of preventing disease and maintaining good population health rather than relying on medications.

Expanding your practice to include aquaculture can be very rewarding and lucrative. Aquaculture is a rapidly growing market, and can sustainably provide food for a growing human population. Veterinarians can play a vital role in this, helping ensure healthy fish and sustainable practices as the field expands.

Marketing a Fish Practice

Start marketing your fish practice to the existing clients in your practice. Fish are the most common pet in the United Sates, there's a good chance that many of your existing clients have fish in addition to the pets you treat. You already have a relationship and trust established. Use your existing social media to feature any wet pets or share interesting articles regarding aquatic animal medicine.

Contact pet stores, pond stores and aquarium stores in the area, starting with any with which you have an existing relationship. Most fish owners try to utilize the pet stores first when dealing with an illness, not always because it is cheaper, but often because they are unaware of veterinarians as an option. Many store employees welcome the chance to refer to a veterinarian and report that they are uncomfortable recommending the OTC products in the store. In addition to a referral source, offering a good discount on services to the store can give you additional opportunities to learn while demonstrating to the staff of the stores what can be offered.

It's also a good idea to research hobbyist groups in the area. Koi clubs are very common as are aquarium hobbyist groups. Go to meetings when possible and offer to be a speaker. Not only will they feel more comfortable contacting you with any concerns with their own fish, but these groups are often contacted by outsiders for advice regarding fish disease. Send letters or other announcements to other veterinarians in the area for referrals and see about listing on local veterinary medical associations.

If interested in the ever-growing aquaculture field and population health of fish, reach out to local facilities. If you are unaware of aquaculture facilities or hatcheries in your area, contact the state department of fish and wildlife or natural resources for a list of licensed facilities. A local or regional US Fish and Wildlife Service office will also have information on federal facilities and potentially private aquaculture businesses in the area. With the pending changes to the FDAs rules on VFDs, there is going to be an increased need of veterinary services in aquaculture.

Finally, there are groups (listed below) that are good sources for support and education.[1-5] They also have listings of veterinarians under a find-a-vet section of their websites.

References

1. American Association of Fish Veterinarians. Available at: http://www.fishvets.org. Accessed July 6, 2016.
2. World Aquatic Veterinary Medical Association. Available at: http://www.wavma.org. Accessed July 6, 2016.
3. Roberts HE. *Fundamentals of Ornamental Fish Health*. Ames, IA: Wiley-Blackwell;2010.
4. Noga EJ. *Fish Disease: Diagnosis and Treatment*. Ames, IA: Wiley-Blackwell;2010.
5. Saint-Erne N. *Advanced Koi Care for Veterinarians and Professional Koi Keepers*. Glendale, AZ: Erne Enterprises;2010.

Section 34

Hybrid Posters

European and American Clinical Cooperatives: A Novel Veterinary Student Study Abroad Program

Steve Thompson, DVM, Dipl ABVP (Canine and Feline),
Zdenek Knotek, MVDr, PhD, Dipl ECZM (Herpetology),
Vladimir Jekl, MVDr, PhD, Dipl ECZM (Small Mammal)

Session #034

Affiliation: From Departments of Veterinary Clinical Sciences and Veterinary Teaching Hospital, College of Veterinary Medicine, Purdue University, 625 Harrison St, West Lafayette, IN 47907, USA (Thompson), and the Avian and Exotic Animal Clinic. Faculty of Veterinary Medicine, University of Veterinary and Pharmaceutical Sciences, Palackeho trida 1946/1, Brno, Czech Republic 612 42, CZ (Knotek, Jekl).

Purdue University's College of Veterinary Medicine (PVM) and the University of Veterinary and Pharmaceutical Sciences Brno (UVPS) in the Czech Republic entered into a sister school relationship with a memorandum of understanding in 2008. The focus was initially to develop clinical cooperatives with visitor exchanges hosting faculty and veterinary students. Initially, several fourth- and fifth-year UVPS Brno veterinary and food hygiene students visited Purdue through their Erasmus charter for 1 month each year and rotations were included through PVM and the College of Agriculture.

The PVM's inaugural study abroad course for credit expanded on this positive cooperation and became a 3-week faculty-led program with a focus on exotic pet medicine and surgery as well as exposure and participation in zoological and conservation medicine. Twenty-nine veterinary students who had completed their first or second year, have participated in this 3-week, 3-credit course. UVPS serves as the home base for many networked cooperatives that have expanded during visits in 2010, 2011, 2013 and 2015. The heart of the program remains 10-12 days of exotic pet clinics with reptiles, small mammals and birds at UVPS with an annual caseload exceeding 5000 visits. Students were responsible for patient care of outpatient and hospitalized pets and observed numerous surgical procedures. All cases were discussed in the clinic and in-depth after hours. In 2015, Prof Knotek and Dr. Jekl became adjunct professors within the Department of Veterinary Clinical Sciences in the PVM in recognition of their ongoing teaching cooperatives with PVM students. On-site experiences also have occurred with wildlife and care of the collection at Zoo Brno, a large metropolitan zoo with an adjacent wildlife rehabilitation facility and Vienna's Tiergarten Schonbrunn, the oldest zoo in the world. Behind-the-scenes tour opportunities have developed with Zoo Praha as well. Networking has expanded to include the equine and camelid services from UVPS in Brno. Prof Knotek's joint appointment with the University of Veterinary Medicine in Vienna (VetMedUni), the second oldest veterinary university in the world, allowed students to work with avian species, reptiles and aquaculture species in addition to an active spring wildlife rehabilitation service on site. Networking in Austria has since expanded to include the Messerli Research Institute and their behavior studies through the Clever Dog, Corvid and Kea labs. It is anticipated that the program will include VetMedUni rotations with exotic small mammals, equines and imaging in 2017.

UVPS has concurrent veterinary masters and doctoral degree study programs taught in English so communication with faculty has been very good considering the native language is not English and none of the PVM students or faculty have spoken Czech. The International Veterinary Student Association (IVSA) Vienna chapter has been helpful in expanding social interactions with veterinary students in Austria. PVM added a formal IVSA chapter after initial cooperation with IVSA Vienna. Cultural opportunities have included the Gregor Mendel museum in Brno, the birthplace of the study of genetics, as well as morning training of the Lipizzaner stallions at the Spanish Riding School in Vienna.

Acknowledgments: The authors would like to acknowledge the support of their rectors, deans, department heads, international program offices, colleagues and spouses. Purdue University Study Abroad and International Learning grants, Purdue University Study Abroad Initiative scholarships, Indiana VMA, SAVMA Study Abroad scholarships, PVM Spates Pet Bird Endowment and PVM Office of Lifelong Learning have provided funding and assistance that has offset some of the faculty and student costs for this study abroad course while also enhancing UVPS faculty visits to benefit PVM students onsite in West Lafayette. Brno faculty have been speakers related to exotic pets and diversity during Purdue CE conferences and national SAVMA and APVMA symposia.